Microorganisms for Sustainability

Volume 32

More information about this series at http://www.springer.com/series/14379

Dilfuza Egamberdieva • Nils-Kåre Birkeland •
Wen-Jun Li • Hovik Panosyan
Editors

Microbial Communities and their Interactions in the Extreme Environment

Springer

Editors
Dilfuza Egamberdieva
Faculty of Biology
National University of Uzbekistan
Tashkent, Uzbekistan

Nils-Kåre Birkeland
Department of Biological Sciences
University of Bergen
Bergen, Norway

Wen-Jun Li
School of Life Sciences
Sun Yat-Sen University
Guangzhou, China

Hovik Panosyan
Faculty of Biology
Yerevan State University
Yerevan, Armenia

ISSN 2512-1901 ISSN 2512-1898 (electronic)
Microorganisms for Sustainability
ISBN 978-981-16-3730-8 ISBN 978-981-16-3731-5 (eBook)
https://doi.org/10.1007/978-981-16-3731-5

This Springer imprint is published by the registered company Springer Nature Singapore Pte Ltd.
The registered company address is: 152 Beach Road, #21-01/04 Gateway East, Singapore 189721, Singapore

Foreword

Extremophilic microorganisms are fascinating by virtue of their ability to thrive under conditions that are hostile to most life. These conditions include physical stresses such as high and low temperature, chemical stresses such as high osmolarity or chaotropicity, high or low pH, and toxic heavy metals. These stresses limit ecosystem diversity and select for microbes that are phylogenetically and physiologically distinct from their less extremophilic counterparts. Their success in these environments depends on a myriad of biochemical mechanisms, many of which serve to stabilize macromolecules or to remove offending chemistry from the cytoplasm. These mechanisms are interesting in their own right, but also hold the potential to be exploited for the betterment of humankind through biotechnology, agriculture, or bioremediation.

This book brings together an outstanding team of scientists from Armenia, Georgia, Norway, Kazakhstan, Kyrgyzstan, Tajikistan, Uzbekistan, Turkey, China, India, and Pakistan with interest and expertise in different aspects of extremophiles—community ecology, organismal biology, biochemistry, and biotechnology—embodying diverse research that makes extremophiles so exciting. Within the book's 14 chapters are several studies focusing on understudied extreme environments throughout Central Asia, making substantial contributions to a growing body of knowledge of extremophiles in this region. But these chapters and others focused on organismal biology and biochemistry have global implications because the principles underlying extremophily know no boarders. Three chapters focus on halophiles, including microbial diversity in saline-alkaline soils and salt mines, and the potential roles of soil halophiles for sustainable agriculture. Five chapters focus on thermophiles, with three exploring microbial diversity in geothermal springs in Central Asia, one reviewing the biology and biotechnology of the genus *Thermus*, and another highlighting the role of thermophiles in next-generation biotechnology. Another chapter focuses on the other temperature extreme: the diversity, ecology, and biotechnology of psychrophiles. Four chapters spotlight research on heavy metals, including biochemical adaptations to resist heavy metal toxicity and biotechnological applications of extremophiles in metal bioleaching and bioremediation. Finally, one chapter describes the distribution of purple photosynthetic bacteria in Armenia and their biotechnological applications. I am sure you will enjoy the diverse and exciting research described in this book and share in my enthusiasm for continued international collaboration on extremophiles and their applications.

School of Life Sciences and Nevada
Institute of Personalized Medicine
University of Nevada, Las Vegas
Las Vegas, NV, USA

Brian P. Hedlund

Contents

About the Series Editor

Naveen Kumar Arora Fellow of International Society of Environmental Botanists (FISEB), PhD in Microbiology, is Professor and Head of the Department of Environmental Science at Babasaheb Bhimrao Ambedkar University (A Central University), Lucknow, Uttar Pradesh, India. He is a renowned researcher in the field of environmental microbiology and biotechnology. His specific area of research is plant-microbe interactions, particularly plant growth promoting rhizobacteria. He has more than 75 research articles published in premium international journals and several articles published in magazines and dailies. He is an editor of 25 books, published by Springer. He is a member of several national and international societies, Secretary General of the Society for Environmental Sustainability, in editorial boards of 4 journals, and reviewer of several international journals. He is also the editor in chief of the journal *Environmental Sustainability* published by Springer Nature. He has delivered lectures in conferences and seminars around the globe. He has a long-standing interest in teaching at the PG level and is involved in taking courses in bacteriology, microbial physiology, environmental microbiology, agriculture microbiology, and industrial microbiology. He has been advisor to 134 postgraduate and 11 doctoral students. He has been awarded for excellence in research by several societies and national and international bodies/organizations. Although an academician and researcher by profession, he has a huge obsession for the wildlife and its conservation and has authored a book, *Splendid Wilds*. He is the President of the Society for Conservation of Wildlife and has a dedicated website www.naveenarora.co.in for the cause of wildlife and environment conservation.

About the Editors

Dilfuza Egamberdieva is a Head of Joint Uzbek-China Key Laboratory "Ecosystem and Biomes" at the National University of Uzbekistan. She graduated in biology from the National University of Uzbekistan and received her PhD in Agricultural Sciences from the Humboldt University of Berlin, Germany. She conducted her postdoctoral studies at the Universities of Finland, Italy, UK, Netherlands, and Germany. She authored five books and coauthored over 150 publications related to soil and plant microbiomes. In 2013, she received The World Academy of Sciences Award in Agricultural Sciences. In 2018 she has been elected as a member of the committee at High Level Panel of Experts on Food Security and Nutrition (CFS).

Nils-Kåre Birkeland graduated with a Dr Philos. degree from the University of Oslo, Norway. He has work experience from four universities in Norway and as a visiting scientist in the USA and Japan. Currently, he is Professor of Microbiology at the University of Bergen. His main research areas are extremophiles and microbial biotechnology. He has isolated and described a number of novel taxa of extremophiles and analyzed the molecular mechanisms for high-temperature adaptations and microbial diversity in extreme environments. He is appointed as an editorial board member for five peer-reviewed international journals, acts as a reviewer for more than 10 journals, and published more than 110 research papers.

Wen-Jun Li received his PhD in microbiology from Shenyang Institute of Applied Ecology, Chinese Academy of Sciences. He is currently working as Distinguished Professor in School of Life Sciences, Sun Yat-Sen University, Guangzhou, China. His publications include three monographs, 15 authorized patents, and more than 830 research articles. His research is mainly focused on microbial diversity of those terrestrial extremophilic environments, by using culture-dependent and culture-independent methods, and on mechanisms of extremophilic actinobacteria to adapt those unusual environments. He was awarded the WFCC (The World Federation for Culture Collections) Skerman award for microbial taxonomy in 2007, and other six provincial and ministerial level awards for his outstanding research contributions on the field of microbial systematics.

Hovik Panosyan graduated with a degree in Biology from YSU, Armenia. He received his PhD in microbiology from the IB NAS of Armenia. Currently, he is Associate Professor and Faculty Member at YSU. His main research areas are Microbial Ecology and Biology of Extremophiles. He has received numerous research fellowships and awards provided by FEBS, FEMS, NFSAT, and DAAD. He is currently ISME ambassador of Armenia. He has work experience at the UB (Norway), LMU Munich (Germany), UNLV (USA), and IBC CNR (Italy). He has published more than 65 research papers in peer-reviewed journals, 4 books, and 25 chapters.

Chapter 1
Extremophiles in Saline Environment: Potential for Sustainable Agriculture

Dilfuza Egamberdieva, Jakhongir Alimov, Burak Alaylar, Mehmet Karadayi, and Naveen Kumar Arora

Abstract Soil salinity is a major issue world-wide degrading agriculture lands and disturbing soil biological process. Exerting adverse effects on seed germination, root system, nutrient acquisition plant physiology leads to drastic reduction in plant growth and soil productivity. The adverse effects of salt stress on soil microbial activity, diversity, and numbers have been studied extensively. The understanding of the adaptive properties of soil microbes makes it possible to use them in restoring abandoned salt affected lands. The salt-tolerant microorganisms are essential components of carbon, nitrogen and phosphorus cycling. Soil microbes are known to play important role in soil biochemical processes, nutrient cycling through their ability to fix atmospheric nitrogen, solubilize phosphate, or by enhancing decomposition of plant residues. Over the past decades, plant associated microorganisms have been utilized to enhance plant growth and resistance to versatile abiotic stresses such as drought, salinity and temperature maintaining agricultural productivity under abiotic stresses. These stress tolerant microbes have a great biotechnological potential to improve soil productivity and plant health of saline soils under arid conditions.

Keywords Salinity · Soil microbes · Diversity · Nutrient cycle · Stress tolerance · Plant nutrients

---

D. Egamberdieva (✉) · J. Alimov
Faculty of Biology, National University of Uzbekistan, Tashkent, Uzbekistan

B. Alaylar
Department of Molecular Biology and Genetics, Faculty of Arts and Sciences, Agri Ibrahim Cecen University, Agri, Turkey
e-mail: balaylar@agri.edu.tr

M. Karadayi
Department of Biology, Faculty of Science, Ataturk University, Erzurum, Turkey

N. K. Arora
Department of Environmental Science, BBA University, Lucknow, UP, India

D. Egamberdieva et al. (eds.), *Microbial Communities and their Interactions in the Extreme Environment*, Microorganisms for Sustainability 32,
https://doi.org/10.1007/978-981-16-3731-5_1

1.1 Introduction

Salinity is an abiotic stress of global scale having substantial effect on plant growth, and development which result in significant loss of crop productivity (Ahmad et al. 2018; Marasco et al. 2013). Soil salinity results from natural and anthropogenic factors that reduce deteriorate agricultural fields reducing production and result in degraded lands (Pitman and Lauchli 2002). Soil salting is a dramatically growing environmental problem caused by both natural and human activities. Natural activities of soil salinity can be considered as inter-mixing of fresh and sea water, low rainfall, weathering (Akram et al. 2020), and anthropogenic activities are inadequate drainage systems, improper irrigation methods, intensive farming practices and climate change (Machado and Serralheiro 2017; Mwando et al. 2020). About 33% of irrigated agricultural lands around the world are affected by salinity (Shrivastava and Kumar 2015; Zhang et al. 2019; Litalien and Zeeb 2020). Salinity exerts adverse effects on seed germination, root system, nutrient acquisition plant physiology and leads not only to drastic reduction in plant growth, but also severe impact on soil fertility (Kaymakanova 2009; Latef et al. 2016). For example, salinity has negative effect on plant physiological processes including photosynthesis, through reduction in the leaf turgor and leaf surface area (Chaves et al. 2009; Tanveer and Shah 2017). Moreover, salt stress induce ionic unbalance, water stress, nutritional imbalance, ion toxicity, decreased cell division and expansion and degradation of main metabolic process in plants (Shrivastava and Kumar 2015; Mwando et al. 2020; Wani et al. 2020; Arora et al. 2020). An increase of the reactive oxygen species (ROS) production and ion accumulation cause disturbance in the cellular ion balance under salt stress (Ahmad et al. 2018). Also, oxidative damage plays a pivotal role in production of reactive oxygen species and restricted carboxylation. Na^+ uptake enhancement plays negative role in leaf development and early leaf abscission due to ion toxicity. Plants' exposure to excessive salinity may result in nutrient imbalances and unavailability of minerals to the plants. For instance, N, Ca, K, P, Fe, and Zn are one of the most essential and required elements for plants and their deficiencies under salinity/ osmotic stress cause adverse effects on plants (Kamran et al. 2019; Mwando et al. 2020). Salinity induces plant responses toward survival rather than growth by activating the interacting pathways such as synthesis of stress proteins and antioxidants and accumulation of compatible solutes (Sandhya et al. 2010).

Soil microorganisms including eukaryotes and prokaryotes assist in nitrification, ammonification, oxidation, nitrogen fixation and solubilizing of different type of minerals (Egamberdiyeva et al. 2001; Salwan et al. 2019). Microorganisms particularly those inhabiting the extremes of pH, hydrostatic pressure, ionizing radiation, low oxygen levels, temperature extremes, vacuum pressure, microgravity, extremes of pH, existence of heavy metals, aridity, and salinity develop mechanisms to survive and grow in such conditions (Irwin 2010; Saralov 2019; Egamberdiyeva 2005; Merino et al. 2019; Cho et al. 2015; Jiang et al. 2019). Abiotic stresses also affect soil biological activity such as microbial biomass and enzymes and several other activities in the rhizosphere (Egamberdieva et al. 2010). Moreover, salinity has negative effect on the diversity and physiological properties of soil and plant microbiome, which play an important role in soil productivity and plant health (Etesami 2018; Arora et al. 2020). Plant and soil microbiome has been extensively used for improvement of plant growth, stress tolerance, and soil productivity under various abiotic stress conditions (Cho et al. 2015; Khan et al. 2019; Egamberdiyeva et al. 2004; Egamberdieva et al. 2019). The microbial composition in the rhizosphere is diverse and differs greatly from one plant species to another (Lugtenberg et al. 2001).

1.2 Microbial Diversity Under Saline and Drought Conditions

Soil microorganisms contribute in soil nutrient mineralization making them available for plant uptake, through maintaining biogeochemical cycle (Nelson and Mele 2007). These microbes have various adaptation strategies to drought and salt stress through diverse physiological acclimation mechanisms (Mendes et al. 2013). There are many reports on the diversity of salt-tolerant plant beneficial bacteria which include various species such as *Acetobacter*, *Azospirillum*, *Arthrobacter*, *Bacillus*, *Pseudomonas*, *Rhizobium*, *Serratia* and *Stenotrophomonas* (Choudhary et al. 2016; Latef et al. 2016).

Several dominant microbial groups were reported in salinated soil of China such as *Acidobacteria*, *Actinobacteria*, *Bacteroidetes*, *Chloroflexi*, *Gemmatimonadetes*, and *Proteobacteria* (Lijuan et al. 2017). In other study, *Chlorobi*, *Chloroflexi*, *Cyanobacteria*, *Deferribacteres*, *Firmicutes*, *Gemmatimonadates*, *Nitrospira*, *Planctomycetes*, *Proteobacteria*, *Spirochaetes*, *Tenericutes*, *Verrucomicrobia*, were observed in salt affected soil of Sicily (Canfora et al. 2014). Yang et al. (2016) observed highly dominant microbial groups such as *Loktanella* and *Kordiimonas* in extreme soil conditions with high salt concentrations. In saline soils of Armenia several salt-tolerant bacteria including *Halobacillus*, *Piscibacillus*, *Bacillus*, and *Virgibacillus* were reported (Panosyan et al. 2018). *Pseudomonas* species were found in salt affected soils of Uzbekistan with salt tolerance up to 5% NaCl) (Egamberdieva et al. 2009).

Plant rhizosphere is a nutrient rich ecosystem for microbes, where they colonize and use nutrients and proliferate (Lugtenberg et al. 2013). It is reported that plant growth promoting rhizobacteria (PGPR) are able to colonize the rhizosphere, the root surface, or even superficial intercellular spaces of the plants (Kamilova et al. 2005; Latef et al. 2016) and facilitate to develop a beneficial association with the plants. Over the past decades, plant associated microorganisms have been utilized for enhanced plant growth and resistance to versatile abiotic stresses such as drought, salinity and temperature maintaining agricultural productivity (El-Esawi et al. 2019; Khan et al. 2019; Kang et al. 2019; Berg et al. 2013). It is believed that PGPR effect on plant growth and physiology directly or indirectly through several mechanisms such as: nitrogen fixation, production of siderophores, phytohormones, osmolytes, ACC deaminase enzymes, and phosphate solubilization. PGPR also synthesize extracellular enzymes such as cellulase, protease, pectinase, which play a vital role in plant protection against phytopathogens (Berg et al. 2013). Others have indirect roles protecting the plant against soil-borne diseases, which are mainly caused by pathogenic microorganisms (Shurigin et al. 2019; Cho et al. 2015; Qessaoui et al. 2019), maintaining nutrients availability to plants (Etesami 2018) and upregulation of antioxidant enzymes which protect plants from oxidative stress (Islam et al. 2016). Moreover salt-tolerant PGPR modulate phytohormones such as auxin, gibberellin, and cytokinin levels in plant playing critical role in maintaining plant health under stress conditions (Egamberdieva et al. 2017; Khan et al. 2019). Liu et al. (2017) observed dominant microbes such as *Acidobacteria*, *Actinobacteria*, *Bacteroidetes*, *Firmicutes*, *Gemmatimonadetes*, and *Proteobacteria* in the plant root tissues grown in salinated soil. Among microorganisms, root associated bacteria have great potential for increase in plant performance under harsh conditions both directly and indirectly. The genera *Agrobacterium*, *Azospirillum*, *Bacillus*, *Burkholderia*, *Enterobacter*, *Klebsiella*, *Ochromobacter*, *Pseudomonas*, *Rhizobium*, *Serratia*, and *Streptomyces* are among the most beneficial and commonly used for improving the yield of agricultural crops under saline conditions (Ahmad et al. 2008; Egamberdieva et al. 2019). In a study, Ansari et al. (2019) utilized the salt-tolerant bacterial strains of *Hartmannibacter diazotrophicus* and *Pseudomonas* sp. on *Medicago sativa* L. plant under salinity conditions. *Planococcus rifietoensis*, an alkaliphilic bacterium is reported to enhance plant growth and yield of wheat crop under salinity stress (Rajput et al. 2013). Liu et al. (2019) studied endophytic bacteria associated with halophyte *Lycium ruthenicum* and characterized endophytic bacterial diversity and plant growth enhancement and protection. They observed, dominant diversity included *Burkholderiales*, *Corynebacteriales*, *Glycomycetales*, *Enterobacteriales*, *Micrococcales*, *Streptosporangiales*, *Pseudonocardiales*, *Propionibacteriales*, and *Rhizobiales*. *Sesbania cannabina* grown in salinated soil of China is colonized with bacteria belonging to the genera *Agrobacterium*, *Ensifer*, *Neorhizobium, and Rhizobium* (Li et al. 2016).

1.3 Plant Growth Promotion and Stress Tolerance

There are many reports and evidences about the improvement of plant growth, development, and stress tolerance under extreme environmental conditions (Table 1.1).

According Zhang et al. (2017) root associated microbes that colonize rhizosphere and plant tissue synthesize diverse signaling molecules which modulate plant physiological processes as well respond to signal molecules secreted by plant roots. Induced systemic tolerance (IST) is widely used in microbe mediated plant tolerance to abiotic stresses (Meena et al. 2017). Bacterial inoculation of cotton with *Pseudomonas* species resulted in improvement of plant growth, and nitrogen, phosphorus

Table 1.1 Plant associated microbes and their ability to stimulate plant growth and stress tolerance

Microorganisms	Plant	Beneficial effect	References
Trichoderma harzianum	*Suaeda salsa* L.	Plant growth, nutrient uptake, antioxidant enzymes	Li-Hua et al. (2016)
Arthrobacter woluwensis, Microbacterium oxydans, Bacillus megaterium, Bacillus aryabhattai	Soybean (*Glycine max* (L.) Merr)	Plant dry weight, root system, nutrient uptake	Khan et al. (2019)
Bacillus subtilis	Wheat (*Triticum aestivum* L.)	Plant growth, induction of systemic resistance	Lastochkina et al. (2017)
Bacillus subtilis	Fennel (*Foeniculum vulgare*)	Plant growth, nutrient uptake, seed yield, essential oil content	Mishra et al. (2016)
Enterobacter cloacae strain	Canola (*Brassica napus*)	Plant growth, nutrient uptake and chlorophyll content, and antioxidant enzyme activity	Li et al. (2017)
Bacillus megaterium, Pantoea agglomerans	Maize (*Zea mays*)	Plant growth, increased hydraulic conductance at root surface	Gond et al. (2015)
Achromobacter xylosoxidans	Maize (*Zea mays*)	Root and shoot biomass, nutrient uptake	Danish et al. (2020)
Bacillus amyloliquefaciens	Chinese cabbage, radish, tomato	Plant growth, nutrient uptake	Kim et al. (2017)
Serratia plymuthica Stenotrophomonas rhizophila, Pseudomonas fluorescens, P. extremorientalis, P. fluorescens P	Cucumber	Phytohormone and proline modulation, plant growth	Egamberdieva et al. (2011)
Azospirillum spp.	*Z. mays*	Plant growth, water uptake, stress tolerance	Garcia et al. (2017)
P. denitrificans PsD6, *M. bullata* MpB46, *A. tumescens*	Pea	Plant growth, nutrient uptake	Egamberdiyeva and Hōflich (2003a)
Mycobacterium phlei Mycoplana bullata	Wheat	Plant growth, nutrient uptake	Egamberdiyeva and Höflich (2003b)
Bacillus amyloliquefaciens SQR9	Maize	Enhanced solute accumulation, enhanced antioxidant enzyme activities, increased expression of salinity stress response genes	Chen et al. (2016)
Enterobacter ludwigii	*Festuca arundinacea*	Plant growth, stress tolerance, nutrient acquisition	Kapoor et al. (2017)
Pseudomonas putida, P. extremorientalis, P. chlororaphis, P. aureantiaca	Wheat	Plant growth, nutrient uptake	Egamberdieva and Kucharova (2009)

(continued)

Table 1.1 (continued)

Microorganisms	Plant	Beneficial effect	References
Bacillus	Pepper (*Capsicum annuum* L.)	Root, shoot biomass, stress tolerance, reduces ethylene in plant	Wang et al. (2018)
Curtobacterium flaccumfaciens	Barley	Plant biomass	Cardinale et al. (2015)
Thalassobacillus denorans, *Oceanobacillus kapialis*	Rice	Increased germination and growth of root and shoot, developed pigment system, reduced Na^+ ion accumulation	Shah et al. (2017)
Enterobacter sp.	Wheat	Plant growth, nutrient uptake	Sorty et al. (2016)
B. licheniformis	Maize (*Zea mays*)	Enhances plant water use efficiency	Akhtar et al. (2020)
Aspergillus fumigatus	Chickpea	Plant growth, stress tolerance, nutrient uptake	Khan et al. (2011)
Bacillus subtilis	Thal tree (*Acacia gerrardii*)	Plant growth, stress tolerance, nutrient uptake	Hashem et al. (2016)
Enterobacter sp.	Rice (*Oryza sativa*)	Plant growth promotion, salt stress tolerance	Sarkar et al. (2018)
Serratia marcescens	Rice (*Oryza sativa*)	Root, shoot biomass, stress tolerance, modulation of plant physiology	Singh and Jha (2016)

and potassium uptake under saline soil conditions (Egamberdieva and Kucharova 2009). *P. denitrificans* PsD6, *M. bullata* MpB46 and *A. tumescent* increased root growth, shoot biomass, as well as N, P, and K acquisition in wheat under saline soil conditions (Egamberdiyeva and Höflich 2003a). Khan et al. (2019) reported halotolerant rhizobacterial strains mitigate the adverse effects of NaCl stress in soybean seedlings. *Arthrobacter woluwensis* (AK1), *Microbacterium oxydans* (AK2), *Arthrobacter aurescens* (AK3), *Bacillus megaterium* (AK4), and *Bacillus aryabhattai* (AK5) were found to be highly tolerant to salt stress and showed some of the plant growth promoting approaches such as enhanced production of indole-3-acetic acid (IAA), gibberellin (GA), and siderophores and improved phosphate solubilization under saline conditions. Verma et al. (2020) reported that a halotolerant *Rhizobium radiobacter* (LB2) having zinc and phosphate solubilizing activity enhanced lettuce cultivation under salinity stress. In 2019, Khademian and coworkers studied PGPR strains to mitigate hazardous effects of salinity in sesame (*Sesamum indicum* L.). They found that PGPR application improved the absorption of essential nutrients in plants. The microbial inoculants also modify physiological processes in pants, e.g., biosynthesis of organic acids, soluble sugars, antioxidant enzymes (Hashem et al. 2016). For example, *Trichoderma harzianum* stimulate plant growth of *Suaeda salsa* L. through modulation of biological active compounds which are responsible for stress toölerance of plants (Li-Hua et al. 2016). The inoculation of plants with beneficial microbes increase root system, root hairs thereby improve water uptake under extreme conditions (Marasco et al. 2013). Gond et al. (2015) observed an increased hydraulic conductance at root surface of *Zea mays* after inoculation with *Bacillus megaterium* and *Pantoea agglomerans* under salinity stress conditions. In other study Mishra et al. (2016) demonstrated an increased essential oil synthesis in fennel (*Foeniculum vulgare* Mill.) and nutrient uptake by *Bacillus subtilis* under saline soil condition. An increased seed germination, plant biomass, nutrient uptake, concentration of proline, and antioxidant enzyme activity of canola was observed after inoculation of plants with *Enterobacter cloacae* strain under saline soil conditions (Li et al. 2017). In other study the plant biomass, and development of *Hordeum secalinum* were increased by inoculation of *Curtobacterium flaccumfaciens* under salt stress condition (Cardinale et al. 2015). The endophytic bacteria isolated from the root of chickpea grown in saline soil were identified as *Bacillus cereus* NUU1, *Achromobacter xylosoxidans* NUU2, *Bacillus thuringiensis* NUU3, and *Bacillus subtilis* NUU4 (Egamberdieva et al. 2017). These bacterial isolates were able to survive in the rhizosphere of chickpea grown in saline soil and improved plant biomass, symbiotic performance of plant with rhizobia, and N, P, K uptake in salt affected soil. Moreover, the endophytes showed antagonistic activity against phytopathogen *Fusarium solani* and thus were able to reduce the infection rate of root rot in chickpea caused. The inoculation of pepper with *Bacillus* sp. increased plant dry weight, shoot and root length, increased proline production and antioxidant enzyme activities under salt stress (Wang et al. 2018). Upadhyay et al. (2012) demonstrated an improved plant biomass of wheat by inoculation of *Bacillus subtilis* and *Arthrobacter* sp. Similar observations were reported by Hahm et al. (2017), as *Microbacterium oleivorans*, *Brevibacterium iodinum*, and *Rhizobium massiliae* increased plant height, fresh weight, dry weight, total chlorophyll and proline contents and the activity of several antioxidant enzymes of pepper under salinity stress. In another study, salt-tolerant *Planococcus rifietoensis* producing IAA and compatible solutes increased plant growth and stress tolerance of wheat under saline soil conditions (Rajput et al. 2013). Several root associated bacteria including *Arthrobacter*, *Bacillus*, *Burkholderia*, and *Pseudomonas* have been reported to improve plant growth through increasing proline synthesis under stress (Choudhary 2012). The salt-tolerant strain *Pseudomonas* sp. TSAU1 stimulated plant growth of *Galega officinalis L.* under salt stress (Egamberdieva et al. 2013, Fig. 1.1).

1.4 Mechanisms of Plant Growth Stimulation

Plant associated microbes which promote plant growth and stress tolerance use direct and indirect mechanisms (Cho et al. 2015). Increased nutrient uptake by plants inoculated with effective bacteria has been attributed to the production of plant growth regulators by the bacteria at the root interface, which stimulated root growth and facilitated greater absorption of water and nutrients from the subsoil (Haas and

Fig. 1.1 The effect of salt-tolerant PGPR strain *Pseudomonas* sp. TSAU1 on plant growth of *Galega officinalis* L. at 50 mM NaCl condition

Défago 2005). The traits include the production of phytohormones such as auxins and gibberellins (Egamberdieva 2012a, b; Rajput et al. 2013; Liu et al. 2019), production of low molecular weight organic acids and exopolysaccharides (Naseem et al. 2018), production of siderophores (Sarwar et al. 2020), synthesis of osmoprotectants, exopolysaccharides (Berg et al. 2013; Mishra et al. 2016), 1-aminocyclopropane-1-carboxylate (ACC) deaminase (Glick 2014; Egamberdieva et al. 2011), modulation of antioxidant enzymes (Hashem et al. 2016), antagonism against phytopathogens (Abdullah et al. 2017) and synthesizing cell wall degrading enzymes such as pectinase, β-1,3-glucanase, and chitinase (Shoda 2000; Cho et al. 2015).

Many root associated beneficial microbes, including salt-tolerant PGPR are able to synthesize low molecular weight siderophores under iron-deficient conditions. Sarwar et al. (2020) isolated root associated bacteria from the groundnut, among them 25% bacterial isolates were able to produce siderophores. The improvement of plant growth and nutrient uptake by siderophore producing bacteria was observed by Pahari et al. (2016).

Plants are more susceptible to plant fungal pathogens such as *Fusarium*, *Verticillum*, *Rhzoctonia* under abiotic stress condition. Root associated beneficial microbes with antifungal activity able to control plant fungal disease (Deketelaere et al. 2017; Abdullah et al. 2017; Egamberdieva et al. 2017). The root associated bacteria isolated from saline soil were screened for their antagonistic activity against *Fusarium solani* and bacterial isolates *Serratia plymuthica*, *Stenotrophomonas rhizophila*, *Pseudomonas fluorescens*, *P. extremorientalis*, and *P. fluorescens* were found to be effective to control cucumber root rot (Egamberdieva et al. 2011). In other study salt-tolerant *Bacillus velezensis* reduced fungal mycelium of *Verticillum dahlia* and decreased Verticillium wilt olive (Castro et al. 2020). Sugarcane associated microbes with antifungal activity against *C. falcatum* showed biological control ability of red rot disease (Backer et al. 2018). Amna et al. (2020) reported antagonistic activity of *Bacillus xiamenensis* isolated from the sugarcane rhizosphere against phytopathogenic fungi, such as *Colletotrichum falcatum*, *Fusarium oxysporum*, *Fusarium moniliforme*, *Rhizoctonia solani*, *Macrophomina phaseolina*, and *Pythium splendens*. The bacterial isolates were able to control sugarcane red rot and enhanced plant growth under salt stress condition. The synthesis of various extracellular antifungal metabolites and/or volatile molecules inhibits fungal mycelia (Haidar et al. 2016). Similarly, inoculation of *Pseudomonas* sp. induced systemic resistance in sunflower plants against *Macrophomina phaseolina* and enhanced plant growth under salinity stress (Tewari and Arora 2016, 2018).

Root associated microbes produce EPS which contain glucose, galactose, and mannose, helping them to survive in extreme soil conditions (Schmidt et al. 2015). Dodd and Perez-Alfocea (2012) studied how bacterial EPS protect bacterial cells, and found that microbial EPS improve biofilm formation in the rhizosphere which prevent entry of Na^+ into plant roots. In an earlier report, EPS producing bacterial strains of *Aeromonas hydrophila* and *Bacillus* sp. improved salt stress tolerance in wheat by reducing Na^+ accumulation (Ashraf et al. 2004). Similar observation were reported by Qurashi and Sabri (2012), where salt-tolerant EPS producing bacterial strains *Halomonas variabilis* (HT1) and *Planococcus rifietoensis* (RT4) form biofilms, improved plant growth development and stress tolerance in chickpea plants under saline soil condition. In another study, Fatima et al. (2020) reported an EPS producing *Alcaligenes* sp. involved salt tolerance and growth promotion of rice under saline conditions. Similarly, Tewari and Arora (2014) isolated an EPS producing *Pseudomonas aeruginosa* PF23 that was able to tolerate salinity stress upto 2000 mM NaCl concentration. The study reported that inoculation of PF23 enhanced growth and yield of sunflower under saline stress conditions and also inhibited the incidence of charcoal rot disease of sunflower.

Other important activities root associated bacteria use in plant growth stimulation and stress tolerance are modulation of proline synthesis, activation of plant defense mechanisms reducing the toxicity of reactive oxygen species (Batool et al. 2019). Inoculation of plant with *B. subtilis* increased osmoprotectants such as glycine, betaine and proline, and reduced the oxidative damage through modulation of antioxidant enzymes system (SOD, CAT, POD, GR, APX) (Hashem et al. 2016). Haidar et al. (2016) reported an increased CAT, APX, and GPX activity in plant tissue of basil by *Pseudomonas* sp. Similar observation was demonstrated with rice plant inoculated with *Trichoderma asperellum* and *P. fluorescens* whereas enzymes such as SOD, POD CAT, APX levels were increased under salt stress (Singh et al. 2020).

Plant hormones play an important role in plant physiology, such as seed germination, root formation, root elongation, and blossom formation. IAA is the most abundant naturally occurring auxin with a well-documented ability to regulate many aspects of plant development some of them include the differentiation of vascular tissues, elongation growth, apical dominance, lateral root initiation, fruit setting, and ripening (Woodward and Bartel 2005). There are many reports that demonstrate phytohormone production by root associated bacteria, e.g., IAA by *Enterobacter* sp. (Sorty et al. 2016) and *Ochrobactrum* spp. (Mishra et al. 2016), ABA, IAA, GA3, and jasmonic acid by *Arthrobacter koreensis* (Piccoli et al. 2011). Inoculation of *Sulla carnosa* with *Pseudomonas* sp. improved root system and shoot growth through IAA production under saline soil conditions (Hidri et al. 2016). In other study ABA producing bacteria *Bacillus amyloliquefaciens* stimulated plant growth of rice under saline conditions (Shahzad et al. 2016). An increased root and shoot biomass, nodule formation by rhizobial bacteria synthesizing auxins, cytokinins, and abscisic acids were observed (Hayat et al. 2010). Similar observation was reported by Kim et al. (2017), *Bacillus amyloliquefaciens* with gibberellins and abscisic acid production ability stimulated the root and shoot biomass of Chinese cabbage, radish, tomato, and mustard plants under salt stress conditions.

Plant growth promoting microorganisms from extreme conditions show diverse mechanisms of survival and abilities to improve the soil productivity and growth in adverse conditions. These microbes need to be explored and utilized in agro-ecosystems, particularly those facing the impact of abiotic stresses.

1.5 Conclusion

Soil salinity has potential negative effect on soil and plant biodiversity, agricultural crops, other plants and soil health. Although much of the development has been made in improvement of soil health under stress and stress tolerance of plants by microbial technologies, there are still gaps in achieving the sustainability in soil and improving the agricultural productivity. Salt-tolerant microorganisms able to surivive in harsh environment can contribute to recover soil and plant health under saline soils. They modulate plant physiological processes, protect plants from various soil-borne pathogens and increase tolerance to abiotic stresses and play an essential role in nutrient acquisition. To this, it becomes vital to explore novel characteristics of the plant beneficial microbes and mechanisms of their action in plant stress tolerance. These stress tolerant microbes have a great biotechnological potential to improve soil productivity and plant health of saline soils under arid conditions. Understanding the stress responsive genes in plants is important to reveal their action boosting plant defense systems in achieving stress tolerance under extreme environment.

References

Abdullah AS, Moffat CS, Lopez-Ruiz FJ et al (2017) Host–multi–pathogen warfare: pathogen interactions in coinfected plants. Front Plant Sci 8:1–12

Ahmad F, Ahmad I, Khan MS (2008) Screening of free-living rhizospheric bacteria for their multiple plant growth promoting activities. Microb Res 163:173–181

Ahmad P, Alyemeni MN, Ahanger MA et al (2018) Salicylic acid (SA) induced alterations in growth, biochemical attributes and antioxidant enzyme activity in faba bean (*Vicia faba* L.) seedlings under NACL toxicity. Russ J Plant Physiol 65(1):104–114

Akhtar SS, Amby DB, Hegelund JN, Fimognari L, Großkinsky DK, Westergaard JC, Müller R, Moelbak L, Liu F, Roitsch T (2020) Bacillus licheniformis FMCH001 increases water use efficiency via growth stimulation in both normal and drought conditions. Front Plant Sci 11:297

Akram NA, Shafiq F, Ashraf M et al (2020) Advances in salt tolerance of some major fiber crops through classical and advanced biotechnological tools: a review. J Plant Growth Regul. https://doi.org/10.1007/s00344-020-10158-5

Amna XY, Farooq MA, Javed NT (2020) Multi-stress tolerant PGPR Bacillus xiamenensis PM14 activating sugarcane (Saccharum officinarum L.) red rot disease resistance. Plant Physiol Biochem 151:640–649

Ansari M, Shekari F, Mohammadi MH et al (2019) Salt-tolerant plant growth-promoting bacteria enhanced salinity tolerance of salt-tolerant alfalfa (Medicago sativa L.) cultivars at high salinity. Acta Physiol Plant 41:1–13

Arora NK, Fatima T, Mishra J et al (2020) Halo-tolerant plant growth promoting rhizobacteria for improving productivity and remediation of saline soils. J Adv Res. https://doi.org/10.1016/j.jare.2020.07.003

Ashraf M, Hasnain S, Berge O, Mahmood T (2004) Inoculating wheatseedlings with exopolysaccharide-producing bacteria restricts sodium uptake and stimulates plant growth under salt stress. Biol Fertil Soils 40(3):157–162

Backer R, Rokem JS, Ilangumaran GJ et al (2018) Plant growth-promoting rhizobacteria: context, mechanisms of action, and roadmap to commercialization of biostimulants for sustainable agriculture. Front Plant Sci 9:1473

Batool R, Rehman SU, Rafique M et al (2019) Biocontrol potential of Bacillus gibsonii and Brevibacterium frigoritolerans in suppression of Fusarium stalk rot of maize: a sustainable approach. Asian J Agric Biol 7:320–333

Berg G, Alavi M, Schmidt CS et al (2013) Biocontrol and osmoprotection for plants under salinated conditions. In: de Bruijn FJ (ed) Molecular microbial ecology of the rhizosphere. John Wiley & Sons, Inc., Hoboken

Canfora L, Bacci G, Pinzari F et al (2014) Salinity and bacterial diversity: To what extent does the concentration of salt affect the bacterial community in a saline soil? PLoS ONE 9(9):e106662

Cardinale M, Grube M, Erlacher A, Quehenberger J, Berg G (2015) Bacterial networks and co-occurrence relationships in the lettuce root microbiota. Environ Microbiol 7(1):239–252

Castro D, Torres M, Sampedro I et al (2020) Biological control of Verticillium wilt on olive trees by the salt-tolerant strain Bacillus velezensis XT1. Microorganisms 8:1080

Chaves MM, Flexas J, Pinheiro C (2009) Photosynthesis under drought and salt stress. Regulation mechanisms from whole plant to cell. Ann Bot 103(2009):551–560. https://doi.org/10.1093/aob/mcn125

Chen XY, Zhang YY, Fu XC, Li Y, Wang Q (2016) Isolation and characterization of Bacillus amyloliquefaciens PG12 for the biological control of apple ring rot. Postharvest Biol Technol 115:113–121

Cho ST, Chang HH, Egamberdieva D et al (2015) Genome analysis of Pseudomonas fluorescens PCL1751: a rhizobacterium that controls root diseases and alleviates salt stress for its plant host. PLoS One. https://doi.org/10.1371/journal.pone.0140231

Choudhary DK, Kasotia A, Jain S, Vaishnav A, Kumari S, Sharma KP, Varma A (2016) Bacterial-mediated tolerance and resistance to plants under abiotic and biotic stresses. J Plant Growth Regul 35(1):276–300

Danish S, Zafar-ul-Hye M, Mohsin F, Hussain M (2020) ACC-deaminase producing plant growth promoting rhizobacteria and biochar mitigate adverse effects of drought stress on maize growth. PLoS ONE 15(4):e0230615

Deketelaere S, Tyvaert L, França SC et al (2017) Desirable traits of a good biocontrol agent against Verticillium wilt. Front Microbiol 8:1–23

Dodd IC, Perez-Alfocea F (2012) Microbial amelioration of crop salinity stress. J Exp Bot 63:3415–3428

Egamberdieva D (2012a) Pseudomonas chlororaphis: a salt tolerant bacterial inoculant for plant growth stimulation under saline soil conditions. Acta Physiol Plant 34:751–756

Egamberdieva D (2012b) Colonisation of Mycobacterium phlei in the rhizosphere of wheat grown under saline condition. Turk J Biol 36:487–492

Egamberdieva D, Renella G, Wirth S, Islam R (2010) Secondary salinity effects on soil microbial biomass. Biol Fertil Soils 46(5):445–449

Egamberdieva D, Kucharova Z, Davranov K et al (2011) Bacteria able to control foot and root rot and to promote growth of cucumber in salinated soils. Biol Fertil Soils 47:197–205

Egamberdieva D, Berg G, Lindström K, Räsänen LA (2013) Alleviation of salt stress of symbiotic *Galega officinalis* L. (goat's rue) by co-inoculation of rhizobium with root colonising Pseudomonas. Plant Soil 369(1):453–465

Egamberdieva D, Wirth S, Shurigin V et al (2017) Endophytic bacteria improve plant growth, symbiotic performance of chickpea (*Cicer arietinum* L.) and induce suppression of root rot caused by Fusarium solani under salt stress. Front Microbiol 28(8):1887

Egamberdieva D, Wirth S, Dorothea Bellingrath-Kimura S et al (2019) Salt-tolerant plant growth promoting rhizobacteria for enhancing crop productivity of saline soils. Front Microbiol 10:1–18

Egamberdiyeva D (2005) Characterization of Pseudomonas species isolated from the rhizosphere of plants grown in serozem soil, semi arid region of Uzbekistan. Sci World J 5:501–509

Egamberdiyeva D, Höflich G (2003a) Influence of growth promoting bacteria on the growth of wheat at different soils and temperatures. Soil Biol Biochem 35:973–978

Egamberdiyeva D, Höflich G (2003b) The effect of associative bacteria from different climates on plant growth of pea at different soils and temperatures. Arch Agron Soil Sci 49(2):203–213

Egamberdiyeva D, Mamiev M, Poberejskaya S (2001) The influence of mineral fertilizer combined with a nitrification inhibitor on microbial populations and activities in calcareous Uzbekistanian soil under cotton cultivation. Sci World J 1(S2):108–113

Egamberdiyeva D, Qarshieva D, Davranov K (2004) Growth and yield of soybean inoculated with Bradyrhizobium spp in calcareous soil. Biol Fertil Soils 4:144–146

El-Esawi MA, Al-Ghamdi AA, Ali HM, Alayafi AA (2019) Azospirillum lipoferum FK1 confers improved salt tolerance in chickpea (*Cicer arietinum* L.) by modulating osmolytes, antioxidant machinery and stress related genes expression. Environ Exp Bot 159:55–65

Etesami H (2018) Can interaction between silicon and plant growth promoting rhizobacteria benefit in alleviating abiotic and biotic stresses in crop plants? Agric Ecosyst Environ 253:98–112

Fatima T, Mishra I, Verma R, Arora NK (2020) Mechanisms of halotolerant plant growth promoting Alcaligenes sp. involved in salt tolerance and enhancement of the growth of rice under salinity stress. 3 Biotech 10:361

Garcia MM, Pereira LC, Braccini AL et al (2017) Effects of Azospirillum brasilense on growth and yield compounds of maize grown at nitrogen limiting conditions. Rev Ciênc Agrár 40 (2):353–362

Glick BR (2014) Bacteria with ACC deaminase can promote plant growth and help to feed the world. Microbiol Res 169:30–39

Gond SK, Torres MS, Bergen MS et al (2015) Induction of salt tolerance and up-regulation of aquaporin genes in tropical corn by rhizobacterium Pantoea agglomerans. Lett Appl Microbiol 60:392–399

Haas D, Défago G (2005) Biological control of soil-borne pathogens by fluorescent pseudomonads. Nat Rev Microbiol 3:307–319. https://doi.org/10.1038/nrmicro1129

Hahm MS, Son JS, Hwang YJ, Kwon DK, Ghim SY (2017) Alleviation of salt stress in pepper (*Capsicum annum* L.) plants by plant growth-promoting rhizobacteria. J Microbiol Biotechnol 27(10):1790–1797

Haidar R, Fermaud M, Calvo-garrido C et al (2016) Modes of action for biological control of *Botrytis cinerea* by antagonistic bacteria. Phytopathol Mediterr 55:301–322

Hashem A, Abd Allah EF, Alqarawi A et al (2016) The interaction between arbuscular mycorrhizal fungi and endophytic bacteria enhances plant growth of Acacia gerrardii under salt stress. Front Plant Sci 7:1089

Hayat R, Ali S, Amara U et al (2010) Soil beneficial bacteria and their role in plant growth promotion: a review. Ann Microbiol 60:579–598. https://doi.org/10.1007/s13213-010-0117-1

Hidri R, Barea JM, Mahmoud OM, Abdelly C, Azcón R (2016) Impact of microbial inoculation on biomass accumulation by Sulla carnosa provenances, and in regulating nutrition, physiological and antioxidant activities of this species under non-saline and saline conditions. J Plant Physiol 201:28–41

Irwin JA (2010) Sustainable wheat (*Triticum aestivum* L.) production in saline fields: a review. Crit Rev Biotechnol 39:99–1014

Islam F, Yasmeen T, Arif MS et al (2016) Plant growth promoting bacteria confer salt tolerance in *Vigna radiata* by up-regulating antioxidant defense and biological soil fertility. Plant Growth Regul 80:23–36

Jiang H, Li W, Birkeland NK, Egamberdieva D (2019) Thermophilic and halophilic extremophiles in Eurasian environments. Front Microbiol. https://doi.org/10.3389/fmicb.2019.00379

Kamilova F, Validov S, Azarova T, Mulders I, Lugtenberg B (2005) Enrichment for enhanced competitive plant root tip colonizers selects for a new class of biocontrol bacteria. Environ Microbiol 7:1809–1817

Kamran M, Parveen A, Ahmar S et al (2019) An overview of hazardous impacts of soil salinity in crops, tolerance mechanisms, and amelioration through selenium supplementation. Int J Mol Sci 21:1–27

Kang SM, Shahzad R, Bilal S et al (2019) Indole-3-acetic-acid and ACC deaminase producing *Leclercia adecarboxylata* MO1 improves Solanum lycopersicum L. growth and salinity stress tolerance by endogenous secondary metabolites regulation. BMC Microbiol 19(1):80

Kapoor R, Gupta MK, Kumar N et al (2017) Analysis of nhaA gene from salt tolerant and plant growth promoting *Enterobacter ludwigii*. Rhizosphere 4:62–69

Kaymakanova M (2009) Effect of salinity on germination and seed physiology in bean (*Phaseolus vulgaris* L.). Biotech Biotechnol Equip 23(1):326–329

Khan AL, Hamayun M, Kim YH et al (2011) Gibberellins producing endophytic *Aspergillus fumigatus* sp. LH02 influenced endogenous phytohormonal levels, plant growth and isoflavone biosynthesis in soybean under salt stress. Process Biochem 46:440–447

Khan MA, Asaf S, Khan AL et al (2019) Halotolerant rhizobacterial strains mitigate the adverse effects of NaCl stress in soybean seedlings. Biomed Res Int 2019:1–15

Kim M, Radhakrishnan R, Kang S et al (2017) Plant growth promoting effect of Bacillus amyloliquefaciens H-2-5 on crop plants and influence on physiological changes in soybean under soil salinity. Physiol Mol Biol Plants 23:571–580

Lastochkina O, Pusenkova L, Yuldashev R et al (2017) Effects of Bacillus subtilis on some physiological and biochemical parameters of *Triticum aestivum* L. (wheat) under salinity. Plant Physiol Biochem 121:80–88

Latef AAH, Hashem A, Rasool S et al (2016) Arbuscular mycorrhizal symbiosis and abiotic stress in plants: a review. J Plant Biol 59:407–426

Li Y, Li X, Liu Y, Wang ET et al (2016) Genetic diversity and community structure of rhizobia nodulating *Sesbania cannabina* in saline–alkaline soils. Syst Appl Microbiol 39(3):195–202
Li H, Lei P, Pang X et al (2017) Enhanced tolerance to salt stress in canola (*Brassica napus* L.) seedlings inoculated with the halotolerant Enterobacter cloacae HSNJ4. Appl Soil Ecol 119:26–34
Li-Hua C, Jin-Hai Z, Xiao-Hou S et al (2016) Effects of *Trichoderma harzianum* T83 on Suaeda salsa L. in coastal saline soil. Ecol Eng 91:58–64
Lijuan C, Changsheng L, Qi F et al (2017) Shifts in soil microbial metabolic activities and community structures along a salinity gradient of irrigation water in a typical arid region of China. Sci Total Environ 598:64–70
Litalien A, Zeeb B (2020) Curing the earth: A review of anthropogenic soil salinization and plant-based strategies for sustainable mitigation. Sci Total Environ 698:1–15
Liu YH, Guo JW, Li L et al (2017) Endophytic bacteria associated with endangered plant Ferula sinkiangensis KM Shen in an arid land: diversity and plant growth-promoting traits. J Arid Land 9(3):432–445
Liu Y, Mohamad OAA, Salam N et al (2019) Diversity, community distribution and growth promotion activities of endophytes associated with halophyte Lycium ruthenicum Murr. Biotech 9:144
Lugtenberg BJJ, Dekkers L, Bloemberg GV (2001) Molecular determinants of rhizosphere colonization by Pseudomonas. Annu Rev Phytopathol 39:461–490
Lugtenberg BJ, Malfanova N, Kamilova F, Berg G (2013) Plant growth promotion by microbes. In: de Bruijn FJ (ed) Mol microb ecol rhizosphere. John Wiley & Sons, New York, pp 559–573
Machado RMA, Serralheiro RP (2017) Soil salinity: effect on vegetable crop growth management practices to prevent and mitigate soil salinization. Horticulturae 3:1–13
Marasco R, Rolli E, Vigani G et al (2013) Are drought-resistance promoting bacteria cross compatible with different plant models? Plant Signal Behav 8:e26741
Meena KK, Sorty AM, Bitla UM, Choudhary K, Gupta P, Pareek A et al (2017) Abiotic stress responses and microbe-mediated mitigation in plants: theomics strategies. Front Plant Sci 8:172. https://doi.org/10.3389/fpls.2017.00172
Mendes R, Garbeva P, Raaijmakers JM (2013) The rhizosphere microbiome: significance of plant beneficial, plant pathogenic, and human pathogenic microorganisms. FEMS Microbiol Rev 37:634–663
Merino N, Aronson HS, Bojanova DP et al (2019) Living at the extremes: extremophiles and the limits of life in a planetary context. Front Microbiol 10:780. https://doi.org/10.3389/fmicb.2019.00780
Mishra BK, Meena KK, Dubey PN et al (2016) Influence on yield and quality of fennel (Foeniculum vulgare Mill.) grown under semi-arid saline soil, due to application of native phosphate solubilizing rhizobacteria isolates. Ecol Eng 97:327–333
Mwando E, Angessa TT, Han Y et al (2020) Salinity tolerance in barley during germination—homologs and potential genes. Biomed Biotechnol 21:93–121
Naseem H, Ahsan M, Shahid MA, Khan N (2018) Exopolysaccharides producing rhizobacteria and their role in plant growth and drought tolerance. J Basic Microbiol 58(12):1009–1022
Nelson DR, Mele PM (2007) Subtle changes in the rhizosphere microbial community 482 structure in response to increased boron and sodium chloride concentrations. Soil Biol Biochem 39:340–351
Pahari A, Dangar TK, Mishra B (2016) Siderophore quantification of bacteria from sundarban and its effect on growt of brinjal (*Solanum Melongena* L.). Bioscan 11(4):2147–2151
Panosyan H, Hakobyan A, Birkeland NK, Trchounian A (2018) Bacilli community of saline-alkaline soils from the Ararat Plain (Armenia) assessed by molecular and culture-based methods. Syst Appl Microbiol 41(3):232–240
Piccoli P, Travaglia C, Cohen A, Sosa L, Cornejo P, Masuelli R, Bottini R (2011) An endophytic bacterium isolated from roots of the halophyte Prosopis strombulifera produces ABA, IAA,

gibberellins A1 and A3 and jasmonic acid in chemically-defined culture medium. Plant Growth Regul 64:207–210

Pitman MG, Lauchli A (2002) Global impact of salinity and agriculture ecosystems. In: Lauchi A, Luttge U (eds) Salinity: environment-plants-molecules. Springer, Dordrecht, pp 3–20

Qessaoui R, Bouharroud R, Furze JN et al (2019) Applications of New Rhizobacteria Pseudomonas isolates in agroecology via fundamental processes complementing plant growth. Sci Rep 9:12832

Qurashi AW, Sabri AN (2012) Bacterial exopolysaccharide and biofilm formation stimulate chickpea growth and soil aggregation under salt stress. Braz J Microbiol 43(3):1183–1191

Rajput L, Imran A, Mubeen F et al (2013) Salt-tolerant PGPR strain Planococcus rifietcensis promotes the growth and yield of wheat (Triticum aestivum L.) cultivated in saline soil. Pak J Bot 45:1955–1962

Salwan R, Sharma A, Sharma V (2019) Microbes mediated plant stress tolerance in saline agricultural ecosystem. Plant Soil 442:1–22

Sandhya V, Ali SZ, Grover M et al (2010) Effect of plant growth promoting Pseudomonas spp. on compatible solutes, antioxidant status and plant growth of maize under drought stress. Plant Growth Regul 62(1):21–30

Saralov AI (2019) Adaptively of archaeal and bacterial extremophiles. Microbiology 88:379–401

Sarkar A, Ghosh PK, Pramanik K et al (2018) A halotolerant *Enterobacter* sp. displaying ACC deaminase activity promotes rice seedling growth under salt stress. Microbiol Res 169:20–32

Sarwar S, Khaliq A, Yousra M et al (2020) Screening of siderophore-producing PGPRs isolated from groundnut (*Arachis hypogaea* L.) rhizosphere and their influence on iron release in soil. Com Soil Sci Plant Anal 51:1680–1692

Schmidt R, Cordovez V, de Boer W, Raaijmakers J, Garbeva P (2015) Volatile affairs in microbial interactions. ISME J 9(11):2329–2335

Shah G, Jan M, Afreen M, Anees M, Rehman S, Daud MK, Malook I, Jamil M (2017) Halophilic bacteria mediated phytoremediation of salt-affected soils cultivated with rice. J Geochem Explor 174:59–65

Shahzad R, Waqas M, Khan AL, Asaf S, Khan MA, Kang SM et al (2016) Seed-borne endophytic Bacillus amyloliquefaciens RWL-1 produces gibberellins and regulates endogenous phytohormones of Oryza sativa. Plant Physiol Biochem 106:236–243. https://doi.org/10.1016/j.plaphy.2016.05.006

Shoda M (2000) Bacterial control of plant diseases. J Biosci Bioeng 89:515–521

Shrivastava P, Kumar R (2015) Soil salinity: a serious environmental issue and plant growth promoting bacteria as one of the tools for its alleviation. Saudi J Biol Sci 22:123–131

Shurigin V, Egamberdieva D, Li L et al (2019) Endophytic bacteria associated with halophyte *Seidlitzia rosmarinus* Ehrenb. ex Boiss. – diversity and plant beneficial traits. J Arid Land. https://doi.org/10.1007/s40333-020-0019-4

Singh RP, Jha PN (2016) The multifarious PGPR *Serratia marcescens* CDP-13 augments induced systemic resistance and enhanced salinity tolerance of wheat (*Triticum aestivum* L.). PLoS One 11:e0155026

Singh DP, Singh V, Gupta VK et al (2020) Microbial inoculation in rice regulates antioxidative reactions and defense related genes to mitigate drought stress. Sci Rep 10:4818

Sorty AM, Meena KK, Choudhary K, Bitla UM, Minhas PS, Krishnani KK (2016) Effect of plant growth promoting bacteria associated with halophytic weed (*Psoralea corylifolia* L.) on germination and seedling growth of wheat under saline conditions. Appl Biochem Biotechnol 180:872–882

Tanveer M, Shah AN (2017) An insight into salt stress tolerance mechanisms of *Chenopodium album*. Environ Sci Pol 24:16531–16535

Tewari S, Arora NK (2014) Multifunctional exopolysaccharides from Pseudomonas aeruginosa PF23 involved in plant growth stimulation, biocontrol and stress amelioration in sunflower under saline conditions. Curr Microbiol 69:484

Tewari S, Arora NK (2016) Fluorescent *Pseudomonas* sp. PF17 as an efficient plant growth regulator and biocontrol agent for sunflower crop under saline conditions. Symbiosis 68 (1–3):99–108

Tewari S, Arora NK (2018) Role of salicylic acid from *Pseudomonas aeruginosa* PF23EPS^{+} in growth promotion of sunflower in saline soils infested with phytopathogen *Macrophomina phaseolina*. Environ Sustain 1:49–59

Upadhyay SK, Singh JS, Saxena AK, Singh DP (2012) Impact of PGPR inoculation on growth and antioxidant status of wheat under saline conditions. Plant Biol 14:605–611

Verma M, Singh A, Dwivedi H, Arora NKA (2020) Zinc and phosphate solubilizing *Rhizobium radiobacter* (LB2) for enhancing quality and yield of loose leaf lettuce in saline soil. Environ Sustain 3:209–218

Wang W, Wu Z, He Y, Huang Y, Li X (2018) Plant growth promotion and alleviation of salinity stress in *Capsicum annuum* L. by Bacillus isolated from saline soil in Xinjiang. Ecotox Environ Safe 164:520–529

Wani SH, Kumar V, Khare T et al (2020) Engineering salinity tolerance in plants: progress and prospects. Planta 251:1–29

Woodward AW, Bartel B (2005) Auxin: regulation, action, and interaction. Ann Bot 95:707–735

Yang H, Hu J, Long X, Liu Z, Rengel Z (2016) Salinity altered root distribution and increased diversity of bacterial communities in the rhizosphere soil of Jerusalem artichoke. Sci Rep 6:20687

Zhang R, Vivanco JM, Shen Q (2017) The unseen rhizosphereroot–soil–microbe interactions for crop production. Curr Opin Microbiol 37:8–14

Zhang K, Shi Y, Cui X et al (2019) Salinity is a key determinant for soil microbial communities in a desert ecosystem. Ecol Evol Sci 4:e00225

Chapter 2
Insights into the Microbial Diversity in Saline-Alkaline Soils of China

Shuang Wang, Lei Sun, Manik Prabhu Narsing Rao, Lihua Wang, Yue Wang, and Wen-Jun Li

Abstract There are many types of saline-alkali soils and lakes on the earth. In some unique saline-alkali environments, the salinity can reach saturation and the alkaline pH value even can reach 11. Despite these two drastic circumstances, there is a kind of special life-haloalkaliphiles inhabiting and breeding in this habitat. To resolve such extreme conditions, these haloalkaliphiles have unique adaptabilities. This chapter mainly introduces the distribution and sources of some types of saline-alkali soil in China. The focus is on understanding their microbial diversity, including culture-dependent and independent approaches. This chapter also focuses on listing the new strains and their characteristics reported in China from different saline-alkaline soils.

Keywords Saline-alkali soils · Distribution · Halophiles · Alkaliphiles · Microbial diversity

---

S. Wang · L. Sun · L. Wang
Institute of Soil Fertilizer and Environment Resources, Heilongjiang Academy of Agricultural Sciences, Harbin, People's Republic of China

M. P. Narsing Rao
State Key Laboratory of Biocontrol, Guangdong Provincial Key Laboratory of Plant Resources and Southern Marine Science and Engineering Guangdong Laboratory (Zhuhai), School of Life Sciences, School of Life Sciences, Sun Yat-Sen University, Guangzhou, People's Republic of China

Y. Wang
Plant Protection and Resource Utilization, College of Modern Agriculture and Ecological Environment, Heilongjiang University, Harbin, People's Republic of China

W.-J. Li (✉)
State Key Laboratory of Biocontrol, Guangdong Provincial Key Laboratory of Plant Resources and Southern Marine Science and Engineering Guangdong Laboratory (Zhuhai), School of Life Sciences, School of Life Sciences, Sun Yat-Sen University, Guangzhou, People's Republic of China

State Key Laboratory of Desert and Oasis Ecology, Xinjiang Institute of Ecology and Geography, Chinese Academy of Sciences, Urumqi, People's Republic of China
e-mail: liwenjun3@mail.sysu.edu.cn

D. Egamberdieva et al. (eds.), *Microbial Communities and their Interactions in the Extreme Environment*, Microorganisms for Sustainability 32,
https://doi.org/10.1007/978-981-16-3731-5_2

2.1 Introduction

Salinization is the result of free salt accumulation to the point that it causes severe degradation of vegetation and soils (Rengasamy 2006). Salinization is a major environmental issue with a negative effect on sustainable agricultural production (Shrivastava and Kumar 2015). Salt-affected soils are distributed widely across the Earth, from cold and temperate zones to subtropical and tropical zones. It is estimated that about 1 billion hectares of soil worldwide are impaired by salinity (Rengasamy 2006; Khan et al. 2020). The distribution is shown in Table 2.1.

Generally speaking, salt-affected soils are characterized by a specific electrical conductivity (EC) (above 4 dS m^{-1}), and the pH is usually less than 8.5 (Richards 1954; Bazilevich and Pankova 1972; Zhang 2014). The major problem with saline soils is the presence of soluble salts, cations: Na^+, Ca^{2+}, Mg^{2+}, K^+ and the anions: Cl^-, SO_4^{2-}, HCO_3^-, CO_3^{2-}, and nitrate NO_3^-. It may also contain boron, selenium, strontium, lithium, silica, rubidium, fluorine, molybdenum, manganese, barium, and aluminum (Sparks 2003; Shahid et al. 2018). When alkalinity takes place, the high pH level does not permit plant growth. Excess sodium on the exchange complex destroys the soil. It is hard for the plants to live in such harsh soil environments, but there are few species, halophytes that thrive in such conditions (Hasanuzzaman et al. 2014). Apart from halophytes, some microbial communities also survive (Ventosa 2006; Wang et al. 2020a). The unique ecological environment with the abundant saline-alkali soil provides a good natural environment for the growth of unique and diverse microbial resources (Yu et al. 2018a; Li et al. 2020a; Zhang et al. 2019).

Microorganisms inhabiting hypersaline environments are designated as halophiles and must cope not only with high ion composition but also with other environmental factors such as alkaline pH values, low oxygen availability, high or low temperatures, the presence of certain ions or toxic compounds (Oren 2002). The two classical groups of halophiles from hypersaline environments are the extremely aerobic halophilic Archaea (haloarchaea) and the moderately halophilic Bacteria, represented by a limited number of species (De la Haba et al. 2011). Another significant group of extremophiles is alkaliphiles, which originated in deep-ocean

Table 2.1 Global distribution of salt-affected soils (table from reference Zhang 2014)

Region	Area (1000 ha)	Percentage (%)
North America	15,755	1.65
Mexico and Central America	1965	0.21
South America	129,163	13.53
Africa	80,538	8.43
South Asia	87,608	9.17
North and Central Asia	211,686	22.17
Southeast Asia	19,983	2.09
Australasia	357,330	37.42
Europe	50,804	5.32
Total	954,832	

alkaline hydrothermal vent systems billions of years ago and are thought to be the earliest life forms on earth (Herschy et al. 2014; Sojo et al. 2016). Alkaliphiles have adapted to thrive in high pH environments that are often lethal to other forms of life (Wang et al. 2020b). However, in saline-alkali soil, salt and alkali often accompany each other, a large number of haloalkaliphiles have evolved to cope with double stress. These extremophiles play an important role in the biogeochemical process or in the ecological cycle of saline-alkaline soils, and it is of great importance to dig the resources and functions of haloalkaliphiles. In general, given that microorganisms are rapidly affected by the change of their environment (Jiang et al. 2012), some soil biological properties such as microbial diversity, structure, and composition are often considered as sensitive and early soil ecological stress indicators or dynamic environmental changes (Li et al. 2011; Liu and Kang 2014). Therefore, in this chapter, we will discuss the distribution and biotechnological importance of halophiles and alkaliphiles in saline-alkaline soils of China.

2.2 The Characteristics and Distribution of Saline Soils in China

China is one of the countries with large areas of saline-alkali land, ranking third among the top 10 countries in the world (Zhang 2014). According to the natural geographical conditions and the process of soil formation, the saline soils include five distribution areas in China: (1) Semiarid and semi-humid lowland saline land in Northeast China. (2) Northwest inland saline land. (3) Semiarid saline land in the middle reaches of the Yellow River. (4) Semiarid and arid lowland saline lands in the Yellow-Huai-Hai River Plain. (5) Coastal semi-humid saline land (Fig. 2.1).

2.2.1 Distribution of Saline-Alkali Soil in Northwest China

Many locations in Northwest China are located in arid and semiarid areas, where rainfall is scarce, evaporation is large, long, dry and windy sunshine duration, which provides favorable conditions for salt accumulation on the surface of the soil (Feng et al. 2019). This area covers most of Xinjiang, Chaidamu Basin in Qinghai, the Hexi Corridor in Gansu, and west inner Mongolia. The area is characterized by a terrestrial climate, annual precipitation of 100–300 mm, and a water table of 3–10 m (1–2 m in some areas). The groundwater mineral content is 3–5 g/L, with a maximum of 10 g/L. The major anions present are Cl^- and SO_4^{2-}, and the total soil salinity is 1–4% in the topsoil (Zhang 2014). The distribution of salinization degree of cultivated land in the northwest irrigation area is shown in Table 2.2.

There are obvious contrasts in soil salinization among different provinces in northwest China. The accumulation of salt in the soil is the main characteristic of

Fig. 2.1 China's saline-alkali land distribution

saline-alkali soil in Gansu Province, especially in 80–100 cm soil layer, most of which are more than 2000 μs·cm^{-1}. There are great differences in salinity of soils in Ningxia province, the lowest of electrical conductivity is 608 μs·cm^{-1} (Xidatan) the highest is 9980 μs·cm^{-1} (Miaotai country), with a difference of nearly 16 times (Li et al. 2020a, b). Chloride-type dry solonchaks showed the highest soil total salt (40.65%) in Hexi Corridor of northwestern Gansu province, China, the lowest total salt (6.36%) was found in sulfate-chloride-type meadow solonchaks (Nan et al. 2020). Saline soil in Gansu Province can be classified into five subtypes: meadow solonchaks, orthic solonchaks, bog solonchaks, alkalized solonchaks, and dry solonchaks (Nan et al. 2020).

The Xinjiang Province, located in the hinterland of Eurasia, is another concentrated area for the distribution of salt-alkali soil in northwest China. It is China's driest place, with less precipitation and severe evaporation. The saline-alkali soil area is 2.8 to 105 km^2, representing 28.3% of China's total saline-alkali soil area. (Cai et al. 2009). Here, the salt-affected soils have been widely distributed and the types of salinization and alkalization were most diverse. The most significant characteristics of these saline-alkali soils were the surface hardening and have the poor structural composition (Stuyt et al. 2000). In the vertical spatial distribution of salinity, the salt content in the 0–30 cm which is higher than that in other horizons, and its variation range was significantly higher than that in 30–100 cm soil horizon.

Table 2.2 Distribution of salinization degree of cultivated land in Northwest irrigation area

Distribution area	Saline-alkali cultivated land area ($\times 10^4$ Mu, 1 Mu = 1/15 ha)						
	Severe salinization	Proportion (%)	Moderate salinization	Proportion (%)	Mild salinization	Proportion (%)	Total
Hetao irrigation district in inner Mongolia	5.296	16.30	9.919	30.60	17.216	53.10	32.431
Hexi corridor and yellow river diversion irrigation area in Gansu Province	5.015	15.50	11.367	35.10	16.030	49.40	32.411
Irrigation area of diverting water from the Yellow River in Ningxia	2.513	16.50	3.799	24.90	8.934	58.60	15.246
Xinjiang Province	7.407	4.50	31.909	19.60	123.504	75.90	162.820

However, in horizontal spatial distribution, the salinity increases from southwest to northeast and is closely related to topographic (Hua et al. 2018).

2.2.2 Distribution of Saline-Alkali Soil in Northeast China

This region includes the Songnen Plain and inner Mongolia district. Songnen Plain (42°30′–51°20′N and 121°40′–128°30′E) is the second-largest plain in China after the Yellow–Huai–Hai River Plain in central China. It is an important agricultural and animal husbandry base in northern China and one of the main salt-alkali soil distribution areas in China (Wang et al. 2009a). These saline-alkali soils are mainly distributed in the west part of Jilin and Heilongjiang provinces, which show slow permeability to freshwater because of their large content of montmorillonite clay and sodium bicarbonate. The groundwater mineral content is 2–5 g/L, with a maximum of 10 g/L, and the major anions present are CO_3^{2-} and HCO_3^-. The soil salinity is about 0.3%. The soil is strongly alkaline with a pH value as high as 8.5, and it can reach 9 to 10.5 m in severe areas. These soils possess poor physical and chemical properties that affect the growth of most crops (Qadir and Schubert 2002). Here, the plant community and productivity of Songnen grassland have been driven by soil salinization and alkalinity. There is a wide variety of salt- and alkali-tolerant plants. *Leymus chinensis* grows in a large area of grassland, *Phragmites australis* is distributed in some wetland environments. In some of the more serious salinization of bare land around the growth of *Chloris virgata* Sw., *Suaeda glauca, Carex tristachya, Artemisia anethifolia, Tripolium vulgare* Nees., *Imperata cylindrica L,* etc. (Fig. 2.2). The slow growth of these typical vegetation communities is mainly due to the effects of ion toxicity and osmotic adjustment. Furthermore, as the vegetation biomass decreases, the surface water evaporation increases, and the surface soil moisture decreases. Under the joint action of human and natural factors, the degradation of salinized grassland is more and more serious, resulting in large areas of alkali spots and saline bare land (Zhang 2016).

The annual rainfall exceeds the potential evaporation due to poor management such as overgrazing and irrational utilization (Gao and Liu 2010; Liu et al. 2011). These natural factors and human disturbance accelerate the soil salinization process. Thus, the affected alkali-saline area has been increasing in size by 20,000 ha per year (Zhang et al. 2015a). Under increase aridity and soil alkalization, large areas of croplands have been abandoned (Shang et al. 2003) and large proportions of grasslands have degraded seriously to unprecedented levels (Li et al. 2014).

Fig. 2.2 Some representative plants in saline-alkali soil of Songnen Plain

2.3 Culture-Independent Microbial Diversity and Its Related Influencing Factors in Saline Soils

A number of investigations have been conducted on soil biological attributes in saline soils (Zahran et al. 1992; Sarig et al. 1996; Zahran 1997; Sardinha et al. 2003; Yuan et al. 2007), and the effects of soil salinization on microbial communities. Several recent studies on the composition of soil microbial diversity in saline soils revealed that soil salinization had negative effects not merely on soil biochemical properties, but also the structure of microbial communities (Foti et al. 2007; Yuan et al. 2007; Hidri et al. 2013; Wang et al. 2014; Zhang et al. 2015a). In already salinized soil, however, it is presumably unknown whether an increase in salinization would induce a change in soil microbial diversity because the soil microbe has already developed tolerance mechanisms (Yan and Marschner 2012).

2.3.1 Bacterial Diversity in Saline-Alkali Soil of Northeast China

Soil salinization is a serious environmental problem in arid or semiarid grasslands in northeast China. Zhang et al. (2015a) studied the response of soil properties, bacterial community composition, and metabolic diversity of soil salinization in semiarid grassland in the Horqin region of inner Mongolia of China. They found that grassland salinization has a significant negative effect not only on vegetation but also on the physical, chemical, and biological properties of soil, as well as on the metabolic diversity of the soil bacterial community.

To explore the soil factors that affect bacterial community composition, a total of 29 saline soil samples (with a mean depth of 0–15 cm) across Jilin and Heilongjiang provinces were randomly collected (Fig. 2.3). Except for one soil with a pH 8.84, the other samples had a pH range from 9.5 to 11.0. The concentration of the eight ions and the electrical conductivity among the samples changed, especially in the EC value, the highest EC was 26.3 times higher than the lowest value (Wang et al. 2020a). The Illumina MiSeq sequencing method was practiced to investigate the bacterial diversity and composition in saline-alkali soil. The dominant bacterial phyla reported were *Firmicutes*, *Bacteroidetes*, *Actinobacteria*, *Proteobacteria*,

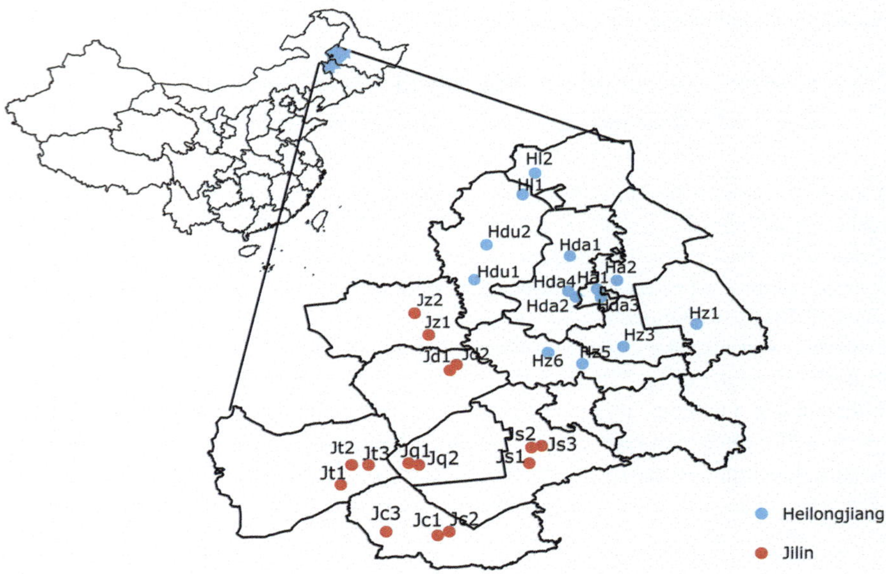

Fig. 2.3 Sampling positions in Jilin province and Heilongjiang province from Songnen plain

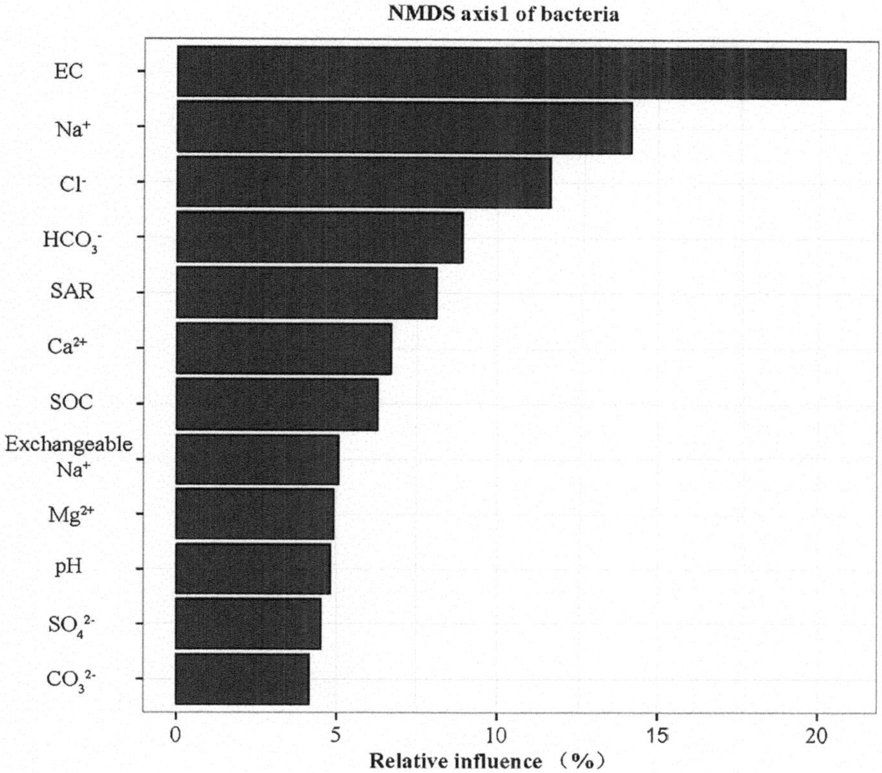

Fig. 2.4 Relative variable importance plot (%) of environmental driving factors for composition of bacteria

Planctomycetes, *Chloroflexi*, *Acidobacteria*, and *Nitrospirae*. However, the proportion of these taxa in different saline soil samples varies greatly. For example, the relative abundance of OTUs of *Firmicutes* ranged from 1.87% to 47.88%, while that of *Bacteroidetes*, *Proteobacteria*, *Planctomycetes* and *Actinobacteria* ranged between 3.34–36.42%, 10.98–30.00%, 0.94–20.32%, and 0.33–15.58%, respectively. There was no significant difference in the alpha diversity index among the samples. The bacterial β-diversity and community structure correlate with the salt gradient. It was reported that the total dissolved salt content of soils was the most important driving force for microbial composition (20.83%), and the second most influencing factor was Na^+ content (14.17%) (Fig. 2.4).

Similarly, Ma and Gong (2013) reported *Proteobacteria*, *Actinobacteria*, *Firmicutes*, *Acidobacteria*, and *Bacteroidetes* as a dominant phylum in saline soils. *Proteobacteria* has been reported as salinity related phylum in many studies (Valenzuela-Encinas et al. 2009; Yang et al. 2018).

2.3.2 Arbuscular Mycorrhizal Fungal Diversity in Saline-Alkali Soil of Northeast China

Natural arbuscular mycorrhizal fungal (AMF) diversity plays a key role in the formation of group effects and the release of vegetation restoration potential. AMF diversity is influenced by abiotic and biotic factors and intrinsic species-specific characteristics (Hazard et al. 2013; Alguacil et al. 2015). Zhang et al. (2020) studied the AMF diversity by high-throughput sequencing technology based on Miseq platform in typical Songnen plain salt-alkali soil on a regional scale. Their results showed that soil pH and EC were the most important drivers for the variation of AMF diversity in saline-alkali soils. This conclusion was consistent with the results of Bencherif et al. (2015), Krishnamoorthy et al. (2014), Dumbrell et al. (2010) and Hazard et al. (2013). AMF diversity was negatively or positively affected by soil salinity or nutrients but was weakly affected by climatic variables at the regional scale. However, in coastal ecosystems, the soil EC tends to be the dominant factor driving the soil AMF diversity (Ramos et al. 2011).

2.3.3 Archaeal Diversity in Saline-Alkali Soil of Northwest China

Salinity may reduce soil respiration (Asghar et al. 2012) and strongly affects the microbial community composition, specifically favoring archaea (Rousk et al. 2011). Nan et al. (2020) studied the diversity of the archaeal community in different saline-alkali soil types in the arid area of Northwest China based on high-throughput sequencing technology and found that *Euryarchaeota* phylum as the most microflora followed by *Crenarchaeota* phylum. High salinity resulted in enrichment of the archaeal community and the pH value of saline soils plays a key role in altering microbial communities (Nan et al. 2020). Significant differences in the archaeal community in different soil types have been noticed. These variations can be explained by the differences in soil organic matter, total salt, sulfate ion content and pH value. Similarly, Auguet et al. (2010) found salinity to be the most important environmental factor controlling the archaeal distribution worldwide.

2.4 Culture-Dependent Haloalkaliphilic Resources from Saline Soils

At the time of writing this chapter, the type strains of cultivated haloalkaliphilic bacteria reported in all references were *Firmicutes* (about 50%), *Proteobacteria* (37%), *Actinobacteria* (9%), the rest are found in *Spirochaetes* and *Bacteroidetes* phylum. At the class level, they were *Bacilli* (about 40%), *Gammaproteobacteria* (γ-*Proteobacteria*, 30%), *Clostridia* (11%), *Deltaproteobacteria* (δ-*Proteobacteria*, 6%), and *Actinobacteria* (6%). The most abundant family was *Bacillaceae* (about 31%), followed by *Halomonadaceae* (about 11%), *Desulfobacteraceae* (about 6%) and *Ectothiorhodospiraceae* (about 6%). However, the proportion of *Lactobacillus* family, *Staphylococcaceae* family and *Alteromonadaceae* accounts for less than 5% (Zhao and Li 2017).

2.4.1 Haloalkaliphilic Microorganisms in Northwest Saline-Alkali Soil

Halophilic and alkaliphilic microorganism resources are very important in the construction of the saline-alkali soil ecosystem in northwest China. Recent studies on hypersaline soils in northwest China have revealed the presence of considerable diversity of microorganisms, constituting moderately halophilic as well as a halotolerant microbial taxon (Li et al. 2005a, b). Li et al. (2020b) used Illumina HiSeq high-throughput sequencing technology to reveal the distribution characteristics of bacterial community diversity in saline soil in different areas of Ningxia, and found that *Proteobacteria*, *Bacteroidetes*, and *Actinobacteria* were the dominant phylum. At the genus level, *Bacillus* was dominant in all sampling sites. Although the soil ecosystem from northwest China was under the pressure of salinity and alkalinity, it also breeds some unique and diverse resources of halophilic and alkaliphilic microorganisms. In recent years, many novel strains, *Phytoactinopolyspora halophila* (Ding et al. 2019), *Planococcus salinus* (Gan et al. 2018a), *Ornithinibacillus salinisoli* (Gan et al. 2018b), *Sinococcus qinghaiensis* (Li et al. 2006) were isolated from saline-alkali soil in Gansu, Xinjiang, Qinghai, Ningxia provinces of northwest China. Other new taxa from the northwest saline-alkali soil are listed in Table 2.3.

2.4.2 Haloalkaliphilic Microorganisms in Northeast Saline-Alkali Soil

Regarding the soil microbial community in saline soils of Songnen Plain, some representative new species of halophilic and alkaliphilic bacteria have been reported by pure culture methods, including halophilic bacterium *Halomonas daqingensis* isolated from an oilfield soil (Wu et al. 2008), *Streptomyces daqingensis* isolated from saline-alkaline soil (Pan et al. 2016). Besides, some novel species of haloalkaliphilic bacteria, such as *Bacillus daqingensis* (Wang et al. 2014), *Halomonas alkalitolerans* (Wang et al. 2011), and *Nesterenkonia haasae* (Wang et al. 2020b) were reported by the author of this chapter isolated from northeast saline-alkali soil. These isolates were moderately halophilic but can tolerate the pH

Table 2.3 List of novel microorganisms isolated from saline and (or) alkali habitat of Northwest China

No.	Name	Category	Taxa level	Habitat	Characteristics	References
1	*Gracilibacillus salitolerans*	Bacterium	Novel species	Saline soil of northwest of China	Moderately halophilic	Gan et al. (2020)
2	*Phytoactinopolyspora halophila*	Bacterium	Novel species	Soil from the edge of a saline lake in Xinjiang	Moderately halophilic	Ding et al. (2019)
3	*Marinimicrobium alkaliphilum*	Bacterium	Novel species	Soil sample near a soda lake From Shanxi	Alkaliphilic	Song et al. (2019)
4	*Planococcus salinus*	Bacterium	Novel species	Saline-alkali soil from Gansu	Halophilic and alkaliphilic	Gan et al. (2018a)
5	*Ornithinibacillus salinisoli*	Bacterium	Novel species	Saline-alkali soil from Gansu	Halophilic and alkaliphilic	Gan et al. (2018b)
6	*Halomonas saliphila*	Bacterium	Novel species	Saline-alkali soil from Gansu	Halophilic and alkali-tolerant	Gan et al. (2018c)
7	*Natribaculum breve*	Archaeon	Novel genus	Saline soil from Xinjiang	Halophilic	Liu et al. (2015)
8	*Natribaculum longum*	Archaeon	Novel species	Saline soil from Xinjiang	Halophilic	
9	*Falsirhodobacter deserti*	Bacterium	Novel species	Sandy soil from Xinjiang	Halotolerant	Wang et al. (2015a)
10	*Halorussus amylolyticus*	Archaeon	Novel species	An inland salt lake from Shanxi	Halophilic	Yuan et al. (2015)
11	*Rheinheimera tuosuensis*	Bacterium	Novel species	Saline lake from Qinghai	Halophilic and alkaliphilic	Zhong et al. (2014a)
12	*Idiomarina planktonica*	Bacterium	Novel species	Saline lake of Qinghai	Halophilic	Zhong et al. (2014b)
13	*Streptomyces sparsus*	Bacterium	Novel species	Saline and alkaline soil from Qinghai	Salt- and alkali-tolerant	Jiang et al. (2011)
14	*Virgibacillus subterraneus*	Bacterium	Novel species	Saline soil from Qinghai	Moderately halophilic	Wang et al. (2010a)

15	*Paracoccus saliphilus*	Bacterium	Novel species	Salt lake from Xinjiang	Moderately halophilic	Wang et al. (2009b)
16	*Bacillus solisalsi*	Bacterium	Novel species	Saline soil from Shanxi	Halotolerant, alkaliphilic	Liu et al. (2009)
17	*Kocuria halotolerans*	Bacterium	Novel species	Saline soil from Xinjiang	Halotolerant	Tang et al. (2009)
18	*Salinicoccus salitudinis*	Bacterium	Novel species	Hypersaline soil from Qaidam basin	Moderately halophilic	Chen et al. (2008a)
19	*Salinimicrobium xinjiangense*	Bacterium	Novel species	Saline lake from Xinjiang	Moderately halophilic	Lim et al. (2008)
20	*Nesterenkonia halophila*	Bacterium	Novel species	Saline soil from Xinjiang	Moderately halophilic, alkali-tolerant	Li et al. (2008)
21	*Brevibacterium album*	Bacterium	Novel species	Saline soil from Xinjiang	Moderately halophilic	Tang et al. (2008)
22	*Salinicoccus halodurans*	Bacterium	Novel species	Saline soil from Qinghai	Moderately halophilic	Wang et al. (2008)
23	*Salinimicrobium terrae*	Bacterium	Novel species	Saline soil from Qinghai	Slightly halophilic	Chen et al. (2008b)
24	*Sinococcus qinghaiensis*	Bacterium	Novel genus	Hypersaline soil from Qinghai	Halophilic and alkaliphilic	Li et al. (2006)
25	*Paenibacillus xinjiangensis*	Bacterium	Novel species	Alkaline soil from Xinjiang	Alkali-tolerant, halotolerant	Lim et al. (2006)
26	*Tenuibacillus multivorans*	Bacterium	Novel genus	Soil sample from a neutral salt lake from Xinjiang	Moderately halophilic	Ren and Zhou (2005)
27	*Lentibacillus salarius*	Bacterium	Novel species	Soil sediment of a salt lake from Xinjiang	Moderately halophilic	Jeon et al. (2005)
28	*Microbacterium halotolerans*	Bacterium	Novel species	Saline soil from Qinghai	Moderately halophilic	Li et al. (2005a)
29	*Isoptericola halotolerans*	Bacterium	Novel species	Saline soil from Qinghai	Moderately halophilic	Zhang et al. (2005)

(continued)

Table 2.3 (continued)

No.	Name	Category	Taxa level	Habitat	Characteristics	References
30	*Yania halotolerans*	Bacterium	Novel genus	Saline soil from Xinjiang	Halotolerant	Li et al. (2004a)
31	*Nocardiopsis salina*	Bacterium	Novel species	Saline soil from Xinjiang	Moderately halophilic	Li et al. (2004b)
32	*Nesterenkonia halotolerans*	Bacterium	Novel species	Hypersaline soils from Xinjiang	–	Li et al. (2004c)
33	*Nesterenkonia xinjiangensis*	Bacterium	Novel species	Hypersaline soils from Xinjiang	–	
34	*Saccharomonospora paurometabolica*	Bacterium	Novel species	Soil from Xinjiang	Moderately halophilic	Li et al. (2003a)
35	*Prauserella halophila*	Bacterium	Novel species	Soil from Xinjiang	Moderately halophilic	Li et al. (2003b)
36	*Prauserella alba*	Bacterium	Novel species	Soil from Xinjiang	Moderately halophilic	
37	*Nocardiopsis xinjiangensis*	Bacterium	Novel species	Saline soil from Xinjiang	Halophilic	Li et al. (2003c)

above 10.0. The optimal pH for growth was 10.0, 9.5, and 9.0, respectively. Other new taxa from the Northeast saline-alkali soil are listed in Table 2.4.

Archaea are microorganisms that are genetically suited to living in habitats with saline soils (Petrova et al. 2010). The number and diversity of cultivated haloalkaliphilic archaea were much less than that of halophilic bacteria. All of the reported type strains of cultivated haloalkaliphilic archaea have been classified to *Euryarchaeota* according to their phylogenetic status. At the taxonomic level of class, order and family, about 92% of the haloalkaliphilic archaea belong to *Halobacteria*, *Halobacteriales*, and *Halobacteriaceae* family, while only about 8% belong to *Methanomicrobia* class, *Methanomicrobiales* order, and *Methanocalculaceae* family. The most abundant genus is *Halorubrum* (about 20%), followed by *Natrialba*, *Natronoccus*, *Natronnolibius*, and *Natronorubrum*, each accounting for about 12%. However, *Halostagnicola*, *Haloterrigena*, *Methanocalculus*, *Methanosalsum*, *Natronomonas*, and *Halalkaliccus* all accounted for about 4% of the total (Zhao and Li 2017).

More than 90% of the isolates of haloalkaliphilic archaea were suitable for growing at salinity higher than 1.7 mol/L (approximately 10% (w/v) NaCl), so the salt-dependent and salt-tolerant ability of the haloalkaliphilic archaea was much higher than that haloalkaliphilic bacteria. However, the pH value suitable for the growth of haloalkaliphilic archaea rarely up to 10. Therefore, the dependence and tolerance of haloalkaliphilic archaea to pH were lower than that of haloalkaliphilic bacteria. Zhao et al. (2018) reported an aerobic extremely halophilic alkali-thermophilic archaeon, which was isolated from the sediments of Wadi An Natrun in Egypt. The strain had a wide tolerance range to salinity and alkalinity and grew in the presence of 2.6 M to saturating Na^{+} (optimum 3.3–4.6 M) and $pH^{55°C}$ 7.5–10.5 (optimum $pH^{55°C}$ 9.0–9.5). The author of this chapter also isolated an extremely haloalkaliphilic archaea (*Haloterrigena daqingensis*) from saline-alkaline soil in northeast China and its optimum growth was reported at 2.0–2.5 M NaCl and pH 10.0 (Wang et al. 2010b).

2.5 Halophiles and Alkaliphiles in Biotechnology

Affected by salinity and alkalinity, microorganisms have evolved many unique functions. (Loredana et al. 2015). In recent years, attention has specifically been paid to the biotechnological potential of microorganisms in saline soils (Canfora et al. 2014; Li et al. 2015).

The prominent existing applications of halophiles and alkaliphiles are related to their active enzymes which are used in detergent formulation, textile and leather processes, and pulp and paper production (Demirjian et al. 2001). For example, halophilic archaea itself has high salt tolerance, wide pH and temperature and adaptability. Enzymes from these strains have the advantages of salt tolerance, high-temperature tolerance, and so on. Therefore, the unique halophilic and halotolerant enzymes produced by halophilic archaea have important application

Table 2.4 List of novel taxa isolated from saline and (or) alkali habitat of Northeast China

No.	Name	Category	Taxa level	Habitat	Characteristics	References
1	*Phytoactinopolyspora limicola*	Bacterium	Novel species	Soda alkali-saline soil from Heilongjiang	Alkaliphilic	Wei et al. (2021)
2	*Nesterenkonia haasae*	Bacterium	Novel species	Soda alkali-saline soil from Heilongjiang	Alkaliphilic	Wang et al. (2020b)
3	*Streptomyces durbertensis*	Bacterium	Novel species	Saline-alkali soil from Heilongjiang	Alkaliphilic	Yu et al. (2018b)
4	*Streptomyces daqingensis*	Bacterium	Novel species	Saline-alkaline soil from Heilongjiang	Halophilic and alkaliphilic	Pan et al. (2016)
5	*Pseudomonas zhaodongensis*	Bacterium	Novel species	Saline and alkaline soil from Heilongjiang	–	Zhang et al. (2015b)
6	*Halomonas heilongjiangensis*	Bacterium	Novel species	Saline and alkaline soil from Heilongjiang	Moderately halophilic	Dou et al. (2015)
7	*Bacillus lindianensis*	Bacterium	Novel species	Saline and alkaline soil from Heilongjiang	Halophilic	Dou et al. (2016)
8	*Kocuria dechangensis*	Bacterium	Novel species	Saline and alkaline soil from Heilongjiang	–	Wang et al. (2015b)
9	*Streptomonospora halotolerans*	Bacterium	Novel species	Muddy soil collected from a riverbank in Jilin	Halotolerant	Zhao et al. (2015)
10	*Halobacillus andaensis*	Bacterium	Novel species	Saline and alkaline soil from Heilongjiang	Moderately halophilic	Wang et al. (2015c)
11	*Planococcus dechangensis*	Bacterium	Novel species	Saline and alkaline soil from Heilongjiang	Moderately halophilic	Wang et al. (2015d)
12	*Bacillus daqingensis*	Bacterium	Novel species	Saline and alkaline soil from Heilongjiang	Halophilic, alkaliphilic	Wang et al. (2014)
13	*Halomonas songnenensis*	Bacterium	Novel species	Saline and alkaline soil from Heilongjiang	Moderately halophilic	Jiang et al. (2014)
14	*Halomonas zhaodongensis*	Bacterium	Novel species	Saline and alkaline soil from Heilongjiang	slightly halophilic	Jiang et al. (2013)

15	*Anditalea andensis*	Bacterium	Novel genus	Extreme alkali–saline soil from Heilongjiang	Alkaliphilic, halotolerant	Shi et al. (2012)
16	*Halomonas alkalitolerans*	Bacterium	Novel species	Soda meadow saline soil from Heilongjiang	Moderately halophilic	Wang et al. (2011)
17	*Haloterrigena daqingensis*	Archaeon	Novel species	Saline and alkaline soil from Heilongjiang	Extremely haloalkaliphilic	Wang et al. (2010b)
18	*Halomonas daqingensis*	Bacterium	Novel species	Soil sample contaminated with crude oil from Heilongjiang	Moderately halophilic	Wu et al. (2008)

value in biotransformation or biocatalysis of saline system, which cannot be replaced by conventional enzymes (Cui 2016). Besides, bacteriorhodopsin produced by halophilic archaea can be used to develop nano biomaterials (Tayier et al. 2019). Carotenoids from them can be used as food additives, and polyhydroxyalkanoates (PHAs) (Quillaguanán et al. 2010; Bhattacharyya et al. 2014; Koller 2015; Koller et al. 2015) and Poly (β-hydroxybutyrate) (PHB) (Quillaguanán et al. 2005) can be used to make biodegradable plastics.

In addition to its carotenoids, organic acids, enzymes, and exopolysaccharides, alkaliphile biomass has been used directly as a food and feed supplement. For example, alkaliphilic cyanobacteria (*Arthrospira platensis*, which is previously known as spirulina) has been used as food and feed (Plavsic et al. 2004; Prasanna et al. 2010). Among alkaliphiles, some strains have been proven as biosurfactant and siderophores producers (Khalikova et al. 2020). Halophiles and alkaliphiles are playing crucial roles in mining and oil drilling, in waste and effluents management (Somee et al. 2021; Khalikova et al. 2020).

As far as pharmaceutically important metabolites, several interesting bioactive compounds from alkaliphiles have already been reported. For example, the *Bacillus halodurans* produces haloduricin (antimicrobial agent) (Danesh et al. 2011; Lawton et al. 2007), *Bacillus alkalophilshaggy* JY-827 produces aminoglycoside antibiotic (Chun et al. 2002), *Streptomyces tanashiensis* produces antifungal compound (Singh et al. 2009), *Nocardiopsis* sp. produces naphthospironone (antimicrobial agent) (Ding et al. 2010), anticancer (Ding et al. 2010), Griseusin D (Höltzel et al. 2003), *Nocardiopsis alkaliphila* YIM-80379 produces nocardiopyrones A and B, *Streptomyces castaneoglobisporus* AJ9 produces 2-Hydroxy-4-methoxy-6-methyl-methyl ester (Adlin et al. 2018), etc.

2.6 Conclusions and Future Perspectives

Soil salinization and alkalization are one of the most devastating environmental problems, threatening the sustainable development of agriculture. A variety of halophytes, including grasses, shrubs, and trees can remove the salt from different kinds of salt-affected areas. Apart from halophytes, halophilic, and alkaliphilic microorganisms play an important role in overcoming salt and alkali stress. They also produce valuable products that are industrially important. The culture-dependent analysis showed that these areas harbor many novel strains. However, culture-independent analysis shows still many strains remained uncultivated and plenty of room for research.

References

Adlin J, Remya R, Velmurugan S, Babu M, Citarasu T (2018) *Streptomyces castaneoglobisporus* AJ9, a haloalkaliphilic actinomycetes isolated from solar salt works in southern India and its pharmacological properties. Indian J Mar Sci 47:475–488

Alguacil MM, Torrecillas E, Lozano Z, Roldán A (2015) Arbuscular mycorrhizal fungi communities in a coral cay system (Morrocoy, Venezuela) and their relationships with environmental variables. Sci Total Environ 505:805–813

Asghar HN, Setia R, Marschner P (2012) Community composition and activity of microbes from saline soils and non-saline soils respond similarly to changes in salinity. Soil Biol Biochem 47:175–178

Auguet JC, Barberan A, Casamayor EO (2010) Global ecological patterns in uncultured archaea. ISME J 4:182–190

Bazilevich NI, Pankova EI (1972) Soil classification by the content of soluble salts and ions. Bull VV Dokuchaev Soil Sci Inst 5:36–40

Bencherif K, Boutekrabt A, Fontaine J, Laruelle F, Dalpè Y, Sahraoui ALH (2015) Impact of soil salinity on arbuscular mycorrhizal fungi biodiversity and microflora biomass associated with *Tamarix articulate* Vahll rhizosphere in arid and semi-arid Algerian areas. Sci Total Environ 533:488–494

Bhattacharyya A, Saha J, Haldar S, Bhowmic A, Mukhopadhyay UK, Mukherjee J (2014) Production of poly-3-(hydroxybutyrate-co-hydroxyvalerate) by *Haloferax mediterranei* using rice-based ethanol stillage with simultaneous recovery and re-use of medium salts. Extremophiles 18:463–470

Cai Y, Xue Q, Chen Z, Zhang R, Lai H-X (2009) Classification and salt-tolerance of actinomycetes in the Qinghai lake water and lakeside saline soil. J Sustainable Dev. https://doi.org/10.5539/jsd.v2n1P107

Canfora L, Bacci G, Pinzari F, Papa GL, Dazzi C, Benedetti A (2014) Salinity and bacterial diversity: to what extent does the concentration of salt affect the bacterial community in a saline soil? PLoS One 9(9):e106662. https://doi.org/10.1371/journal.pone.0106662

Chen Y-G, Cui X-L, Li W-J, Xu L-H, Wen M-L, Peng Q, Jiang C-L (2008a) *Salinicoccus salitudinis* sp. nov., a new moderately halophilic bacterium isolated from a saline soil sample. Extremophiles 12:197–203

Chen Y-G, Cui X-L, Zhang Y-Q, Li W-J, Wang Y-X, Kim C-J, Lim J-M, Xu L-H, Jiang C-L (2008b) *Salinimicrobium terrae* sp. nov., isolated from saline soil, and emended description of the genus *Salinimicrobium*. Int J Syst Evol Microbiol 58:2501–2504

Chun JY, Ryu IH, Park JS, Lee KS (2002) Anticaries activity of antimicrobial material from *Bacillus alkalophilshaggy* JY-827. Microb Biotechnol 12:18–24

Cui H-L (2016) Taxonomy of halophilic archaea: a review. Microbiology 43(5):1113–1122

Danesh A, Mamo G, Mattiasson B (2011) Production of haloduracin by *Bacillus halodurans* using solid-state fermentation. Biotechnol Lett 33:1339–1344

De la Haba RR, Sanchez-Porro C, Marque MC (2011) Taxonomy of halophiles. In: Horikoshi K, Antranikian G, Bull AT, Robb FT, Stetter KO (eds) Extremophiles handbook. Springer, Cham, pp 256–308

Demirjian DC, Morís-Varas F, Cassidy CS (2001) Enzymes from extremophiles. Curr Opin Chem Biol 5(2):144–151

Ding ZG, Li MG, Zhao JY, Ren J, Huang R, Xie MJ et al (2010) Naphthospironone A: an unprecedented and highly functionalized polycyclic metabolite from an alkaline mine waste extremophile. Chemistry 16:3902–3905

Ding Z, Ji Y, Yin M, Zhao Y-R, Feng Y-Z, Chunyu W-X, Tang S-K (2019) *Phytoactinopolyspora halophila* sp. nov., a halophilic actinomycete isolated from a saline soil. Int J Syst Evol Microbiol 69:384–389

Dou G, He W, Ma Y (2015) *Halomonas heilongjiangensis* sp. nov., a novel moderately halophilic bacterium isolated from saline and alkaline soil. Antonie Van Leeuwenhoek 108:403–413

Dou G, Liu H, He W, Ma Y (2016) *Bacillus lindianensis* sp. nov., a novel alkaliphilic and moderately halotolerant bacterium isolated from saline and alkaline soils. Antonie Van Leeuwenhoek 109:149–158

Dumbrell AJ, Nelson M, Helgason T, Dytham C, Fitter AH (2010) Relative roles of niche and neutral processes in structuring a soil microbial community. ISME J 4(3):337–345

Feng B-Q, Cui J, Wu D, Guan X-Y, Wang S-L (2019) Preliminary studies on causes of salinization and alkalinization in irrigation districts of northwest China and countermeasures. China Water Resour 9:43–46

Foti MJ, Sorokin DY, Lomans B, Mussman M, Zacharova EE, Pimenov NV, Kuenen JG, Muyzer G (2007) Diversity, activity, and abundance of sulfate-reducing bacteria in saline and hypersaline soda lakes. Appl Environ Microbiol 73:2093–2100

Gan L, Zhang H, Tian J, Li X, Long X, Zhang Y, Dai Y, Tian Y (2018a) *Planococcus salinus* sp. nov., a moderately halophilic bacterium isolated from a saline-alkali soil. Int J Syst Evol Microbiol 68:589–595

Gan L, Zhang H, Long X, Tian J, Wang Z, Zhang Y, Dai Y, Tian Y (2018b) *Ornithinibacillus salinisoli* sp. nov., a moderately halophilic bacterium isolated from a saline-alkali soil. Int J Syst Evol Microbiol 68:769–775

Gan L, Long X, Zhang H, Hou Y, Tian J, Zhang Y, Tian Y (2018c) *Halomonas saliphila* sp. nov., a moderately halophilic bacterium isolated from a saline soil. Int J Syst Evol Microbiol 68:1153–1159

Gan L, Li X, Chen J, Zhang S, Zhang R, Peng B, Tian Y (2020) *Gracilibacillus salitolerans* sp. nov., a moderate halophile isolated from saline soil in Northwest China. IJSEM 70(6):4224. https://doi.org/10.1099/ijsem.0.004224

Gao J, Liu Y (2010) Determination of land degradation causes in Tongyucounty, Northeast China via land cover change detection. Int J Appl Earth Obs Geoinf 12:9–16

Hasanuzzaman M, Nahar K, Alam Md M, Bhowmik PC, Hossain Md A, Rahman MM, Prasad MNV, Ozturk M, Fujita M (2014) Potential use of halophytes to remediate saline soils. Biomed Res Int 2014:589341. https://doi.org/10.1155/2014/589341

Hazard C, Gosling P, Van der GCJ, Mitchell DT, Doohan FM, Bending GD (2013) The role of local environment and geographical distance in determining community composition of arbuscular mycorrhizal fungi at the landscape scale. ISME J 7(3):498–508

Herschy B, Whicher A, Camprubí C, Eloi WC, Dartnell L, Ward J, Evans JRG, Lane N (2014) An origin-of-life reactor to simulate alkaline hydrothermal vents. J Mol Evol 79:213–227

Hidri DE, Guesmi A, Najjari A, Cherif H, Ettoumi B, Hamdi C, Boudabous A, Cherif A (2013) Cultivation-dependent assessment, diversity, and ecology of haloalkaliphilic bacteria in arid saline systems of southern Tunisia. Biomed Res Int 10:15. https://doi.org/10.1155/2013/648141

Höltzel A, Dieter A, Schmid DG, Brown R, Beil W, Jung G, Fiedler H-P (2003) Lactonamycin Z, an antibiotic and antitumor compound produced by *Streptomyces sanglieri* strain AK 623. J Antibiot 56:1058–1061

Hua D, Yi J, Li J (2018) Study on reclamation of saline-alkali land in Akto Plain Region in Xinjiang. IOP Conf Ser Earth Environ Sci 170(2):022146. https://doi.org/10.1088/1755-1315/170/2/022146

Jeon CO, Lim J-M, Lee J-C, Lee G-S, Lee J-M, Xu L-H, Jiang C-L, Kim C-J (2005) *Lentibacillus salarius* sp. nov., isolated from saline sediment in China, and emended description of the genus *Lentibacillus*. Int J Syst Evol Microbiol 55:1339–1343

Jiang Y, Cao Y-R, Wiese J, Tang S-K, Xu L-H, Imhoff JF, Jiang C-L (2011) *Streptomyces sparsus* sp. nov., isolated from a saline and alkaline soil. Int J Syst Evol Microbiol 61:1601–1605

Jiang Y, Cheng C, Xu Z, Liu Y (2012) Effects of single and mixed species forest ecosystems on diversity and function of soil microbial community in subtropical China. J Soils Sediments 12:228–240

Jiang J, Pan Y, Meng L, Hu S, Zhang X, Hu B, Meng J, Li C, Huang H, Wang K, Su T (2013) *Halomonas zhaodongensis* sp. nov., a slightly halophilic bacterium isolated from saline–alkaline soils in Zhaodong, China. Antonie Van Leeuwenhoek 104:685–694

Jiang J, Pan Y, Hu S, Zhang X, Hu B, Huang H, Hong S, Meng J, Li C, Wang K (2014) *Halomonas songnenensis* sp. nov., a moderately halophilic bacterium isolated from saline and alkaline soils. Int J Syst Evol Microbiol 64:1662–1669

Khalikova E, Somersalo S, Korpela T (2020) Metabolites produced by alkaliphiles with potential biotechnological applications. In: Mamo G, Mattiasson B (eds) Alkaliphiles in biotechnology. Springer, Cham, pp 157–194

Khan W, Tanveer M, Shaukat R, Ali M, Pirdad F (2020) An overview of salinity tolerance mechanism in plants. In: Hasanuzzaman M, Tanveer M (eds) Salt and drought stress tolerance in plants. Springer, Cham, pp 1–16

Koller M (2015) Recycling of waste streams of the biotechnological poly (hydroxyalkanoate) production by *Haloferax mediterranei* on whey. Int J Polym Sci 2015:370164. https://doi.org/10.1155/2015/370164

Koller M, Chiellini E, Braunegg G (2015) Study on the production and re-use of poly (3-hydroxybutyrate-*co*-3-hydroxyvalerate) and extracellular polysaccharide by the archaeon *Haloferax mediterranei* strain DSM 1411. Chem Biochem Eng Q 29:87–98

Krishnamoorthy R, Kim K, Kim C, Sa T (2014) Changes of arbuscular mycorrhizal traits and community structure with respect to soil salinity in a coastal reclamation land. Soil Biol Biochem 72:1–10

Lawton EM, Cotter PD, Hill C, Ross RP (2007) Identification of a novel two-peptide lantibiotic, Haloduracin, produced by the alkaliphile *Bacillus halodurans* C-125. FEMS Microbiol Lett 267:64–71

Li W-J, Tang S-K, Stackebrandt E, Kroppenstedt RM, Schumann P, Xu L-H, Jiang C-L (2003a) *Saccharomonospora paurometabolica* sp. nov., a moderately halophilic actinomycete isolated from soil in China. Int J Syst Evol Microbiol 53:1591–1594

Li W-J, Xu P, Tang S-K, Xu L-H, Kroppenstedt RM, Stackebrandt E, Jiang C-L (2003b) *Prauserella halophila* sp. nov. and *Prauserella alba* sp. nov., moderately halophilic actinomycetes from saline soil. Int J Syst Evol Microbiol 53:1545–1549

Li M-G, Li W-J, Xu P, Cui X-L, Xu L-H, Jiang C-L (2003c) *Nocardiopsis xinjiangensis* sp. nov., a halophilic actinomycete isolated from a saline soil sample in China. Int J Syst Evol Microbiol 53:317–321

Li W-J, Chen H-H, Xu P, Zhang Y-Q, Schumann P, Tang S-K, Xu L-H, Jiang C-L (2004a) *Yania halotolerans* gen. nov., sp. nov., a novel member of the suborder *Micrococcineae* from saline soil in China. Int J Syst Evol Microbiol 54:525–531

Li W-J, Park D-J, Tang S-K, Wang D, Lee J-C, Xu L-H, Kim C-J, Jiang C-L (2004b) *Nocardiopsis salina* sp. nov., a novel halophilic actinomycete isolated from saline soil in China. Int J Syst Evol Microbiol 54:1805–1809

Li W-J, Chen H-H, Zhang Y-Q, Schumann P, Stackebrandt E, Xu L-H, Jiang C-L (2004c) *Nesterenkonia halotolerans* sp. nov. and *Nesterenkonia xinjiangensis* sp. nov., actinobacteria from saline soils in the west of China. Int J Syst Evol Microbiol 54:837–841

Li W-J, Chen H-H, Kim C-J, Park D-J, Tang S-K, Lee J-C, Xu L-H, Jiang C-L (2005a) *Microbacterium halotolerans* sp. nov., isolated from a saline soil in the west of China. Int J Syst Evol Microbiol 55:67–70

Li W-J, Schumann P, Zhang Y-Q, Xu P, Chen G-Z, Xu L-H, Stackebrandt E, Jiang C-L (2005b) Proposal of *Yaniaceae* fam. nov. and *Yania flava* sp. nov. and emended description of the genus *Yania*. Int J Syst Evol Microbiol 55:1933–1938

Li W-J, Zhang Y-Q, Schumann P, Tian X-P, Zhang Y-Q, Xu L-H, Jiang C-L (2006) *Sinococcus qinghaiensis* gen. nov. sp. nov. a novel member of the order *Bacillales* from a saline soil in China. Int J Syst Evol Microbiol 56:1189–1192

Li W-J, Zhang Y-Q, Schumann P, Liu H-Y, Yu L-Y, Zhang Y-Q, Stackebrandt E, Xu L-H, Jiang C-L (2008) *Nesterenkonia halophila* sp. nov., a moderately halophilic, alkalitolerant actinobacterium isolated from a saline soil. Int J Syst Evol Microbiol 58:1359–1363

Li X, Liu B, Cui J, Liu D, Ding S, Gilna B, Luo J, Fang Z, Cao W, Han Z (2011) No evidence of persistent effects of continuously planted transgenic insect-resistant cotton on soil microorganisms. Plant Soil 339:247–257

Li Q, Yu P, Chen X, Li G, Zhou D, Zheng W (2014) Facilitative and inhibitory effect of litter on seedling emergence and early growth of six herbaceous species in an early successional old field ecosystem. Sci World J 2014:101860. https://doi.org/10.1155/2014/101860

Li HY, Niu SQ, Kong WB, Zhu XT, Zhang AM, Da WY, Han CH, Yan WR, Geng H (2015) Diversity of actinomycetes at the saline-alkali soils in downstream area of shiyang river of Hexi corridor-illustrated by the case of minqin county. China Environ Sci 35:1805–1813

Li X, Zhang HH, Sun ML, Xu N, Sun GY, Zhao MC (2020a) Land use change from upland to paddy field in Mollisols drives soil aggregation and associated microbial communities. Appl Soil Ecol 146:103351. https://doi.org/10.1016/j.apsoil.2019.09.001

Li M, Bi JT, Wang J (2020b) Bacterial community structure and key influence factors in saline soil of different sites in Ningxia. Acta Ecol Sin 40(4):1316–1330

Lim J-M, Jeon CO, Park D-J, Xu L-H, Jiang C-L, Kim C-J (2006) *Paenibacillus xinjiangensis* sp. nov., isolated from Xinjiang province in China. Int J Syst Evol Microbiol 56:2579–2582

Lim J-M, Jeon CO, Lee SS, Park D-J, Xu L-H, Jiang C-L, Kim C-J (2008) Reclassification of *Salegentibacter catena* Ying et al. 2007 as *Salinimicrobium catena* gen. nov., comb. nov. and description of *Salinimicrobium xinjiangense* sp. nov., a halophilic bacterium isolated from Xinjiang province in China. Int J Syst Evol Microbiol 58:438–442

Liu S, Kang Y (2014) Changes of soil microbial characteristics in saline-sodic soils under drip irrigation. J Siol Sci Plant Nutr 14:139–150

Liu H, Zhou Y, Liu R, Zhang K-Y, Lai R (2009) *Bacillus solisalsi* sp. nov., a halotolerant, alkaliphilic bacterium isolated from soil around a salt lake. Int J Syst Evol Microbiol 59:1460–1464

Liu S, Kang Y, Wan S, Wang Z, Liang Z, Sun X (2011) Water and salt regulation and its effects on Leymuschinensis growth under drip irrigation in saline-sodic soils of the Songnen plain. Agric Water Manag 98:1469–1476

Liu Q, Ren M, Li Z (2015) *Natribaculum breve* gen. nov., sp. nov. And Natribaculum longum sp. nov., halophilic archaea isolated from saline soil. Int J Syst Evol Microbiol 65:604–608

Loredana C, Giuseppe LP, Livia VA, Giuseppe B, Carmelo D, Anna B (2015) Spatial microbial community structure and biodiversity analysis in "extreme" hypersaline soils of a semiarid Mediterranean area. Appl Soil Ecol 93:120–129

Ma B, Gong J (2013) A meta-analysis of the publicly available bacterial and archaeal sequence diversity in saline soils. World J Microbiol Biotechnol 29:2325–2334

Nan L, Guo Q, Cao S (2020) Archaeal community diversity in different types of saline-alkali soil in arid regions of Northwest China. J Biosci Bioeng 130(4):382–389

Oren A (2002) Halophilic microorganisms and their environments. Kluwer Academic Publishers, New York, pp 3–14

Pan T, He H, Li C, Zhao J, Zhang Y, Li J, Wang X, Liu C, Zhang J, Xiang W (2016) *Streptomyces daqingensis* sp. nov., isolated from saline-alkaline soil. Int J Syst Evol Microbiol 66:1358–1363

Petrova SN, Andronov EE, Pinaev AG, Pershina EV (2010) Prospects for using the methods of molecular genetic analysis in soil ecology. Vestn Orlovsk Gos Agrarn Univ 26:45–48

Plavsic M, Terzic S, Ahel M, Van Den Berg CMG (2004) Folic acid in coastal waters of the Adriatic Sea. Mar Freshw Res 53:1245–1252

Prasanna R, Sood A, Jaiswal P, Nayak S, Gupta V, Chaudhary V, Joshi M, Natarajan C (2010) Rediscovering cyanobacteria as valuable sources of bioactive compounds (review). Appl Biochem Microbiol 46:119–134

Qadir M, Schubert S (2002) Degradation processes and nutrient constraints in sodic soils. Land Degrad Dev 13:275–294

Quillaguanán J, Hashim S, Bento F, Mattiasson B, Hatti-Kaul R (2005) Poly (β-hydroxybutyrate) production by a moderate halophile, *Halomonas boliviensis* LC1 using starch hydrolysate as substrate. J Appl Microbiol 99:151–157

Quillaguanán J, Guzmán H, Van-Thuoc D, Hatti-Kaul R (2010) Synthesis and production of polyhydroxyalkanoates by halophiles: current potential and future prospects. Appl Microbiol Biotechnol 85:1687–1696

Ramos AC, Facanha AR, Palma LM, Okorokov LA, Cruz ZMA, Silva AG, Siqueira AF, Bertolazi AA, Canton GC, Melo J, Santos WO, Schimitberger VMB, Okorokova-Facanha AL (2011) An outlook on ion signaling and ionome of mycorrhizal symbiosis. Braz J Plant Physiol 23 (1):79–89

Ren P-G, Zhou P-J (2005) *Tenuibacillus multivorans* gen. nov., sp. nov., a moderately halophilic bacterium isolated from saline soil in Xin-Jiang, China. Int J Syst Evol Microbiol 55:95–99

Rengasamy P (2006) World salinization with emphasis on Australia. J Exp Bot 57:1017–1023

Richards LA (1954) Diagnosis and improvement of saline and alkaline soils: USDA agriculture handbook. No. 60, p 160

Rousk J, Elyaagubi FK, Jones DL, Godbold DL (2011) Bacterial salt tolerance is unrelated to soil salinity across an arid agroecosystem salinity gradient. Soil Biol Biochem 43:1881–1887

Sardinha M, Müller T, Schmeisky H, Joergensen RG (2003) Microbial performance in soils along a salinity gradient under acidic conditions. Appl Soil Ecol 23:237–244

Sarig S, Fliessbach A, Steinberger Y (1996) Microbial biomass reflects the nitrogen and phosphorus economy of halophytes grown in salty desert soil. Biol Fertil Soils 21:128–130

Shahid SA, Zaman M, Heng L (2018) Soil salinity: historical perspectives and a world overview of the problem. In: Zaman M, Shahid SA, Heng L (eds) Guideline for salinity assessment, mitigation and adaptation using nuclear and related techniques. Springer, Cham, pp 43–53

Shang Z, Gao Q, Dong M (2003) Impacts of grazing on the alkalinized-salinized meadow steppe ecosystem in the Songnen plain, China- a simulation study. Plant Soil 249:237–251

Shi W, Takano T, Liu S (2012) *Anditalea andensis* gen. nov., sp. nov., an alkaliphilic, halotolerant bacterium isolated from extreme alkali–saline soil. Antonie Van Leeuwenhoek 102:703–710

Shrivastava P, Kumar R (2015) Soil salinity: a serious environmental issue and plant growth promoting bacteria as one of the tools for its alleviation. Saudi J Biol Sci 22(2):123–131

Singh LS, Mazumder S, Bora TC (2009) Optimisation of process parameters for growth and bioactive metabolite produced by a salt-tolerant and alkaliphilic actinomycete, *Streptomyces tanashiensis* strain A2D. J Mycol Med 19:225–233

Sojo V, Herschy B, Whicher A, Camprubí E, Lane N (2016) The origin of life in alkaline hydrothermal vents. Astrobiology 16:181–197

Somee MR, Dastgheib SMM, Shavandi M, Zolfaghar M, Zamani N, Ventosa A, Amoozegar MA (2021) Chapter 11- Halophiles in bioremediation of petroleum contaminants: challenges and prospects. In: Kumar V, Saxena G, Shah MP (eds) Bioremediation for environmental sustainability. Springer, Cham, pp 251–291

Song L, Liu H, Cai S, Zhou Y (2019) *Marinimicrobium alkaliphilum* sp. nov., an alkaliphilic bacterium isolated from soil and emended description of the genus *Marinimicrobium*. Int J Syst Evol Microbiol 69:481–485

Sparks DL (2003) The chemistry of saline and sodic soils. Environmental soil chemistry, 2nd edn. Elsevier, London. https://doi.org/10.1016/B978-0-12-656446-4.X5000-2

Stuyt LCPM, Dierickx W, Beltrán JM (2000) Materials for subsurface land drainage systems. FAO, Rome, pp 21–75

Tang S-K, Wang Y, Schumann P, Stackebrandt E, Lou K, Jiang C-L, Xu L-H, Li W-J (2008) Brevibacterium album sp. nov., a novel actinobacterium isolated from a saline soil in China. Int J Syst Evol Microbiol 58:574–577

Tang S-K, Wang Y, Lou K, Mao P-H, Xu L-H, Jiang C-L, Kim C-J, Li W-J (2009) *Kocuria halotolerans* sp. nov., an actinobacterium isolated from a saline soil in China. Int J Syst Evol Microbiol 59:1316–1320

Tayier B, Deng Z, Wang Y, Wang W, Mu Y, Yan F (2019) Biosynthetic nanobubbles for targeted gene delivery by focused ultrasound. Nanoscale 11:14757–14768

Valenzuela-Encinas C, Neria-González I, Alcántara-Hernández RJ, Estrada-Alvarado I, Zavala-Díaz de la Serna FJ, Dendooven L, Marsch R (2009) Changes in the bacterial populations of the

highly alkaline saline soil of the former Lake Texcoco (Mexico) following flooding. Extremophiles 13:609–621
Ventosa A (2006) Unusual micro-organisms from unusual habitats: hypersaline environments. In: Logan NA, Lappin-Scott HM, Oyston PCF (eds) Prokaryotic diversity: mechanisms and significance. Cambridge University Press, Cambridge, pp 223–253
Wang X, Xue Y, Yuan S, Zhou C, Ma Y (2008) *Salinicoccus halodurans* sp. nov., a moderate halophile from saline soil in China. Int J Syst Evol Microbiol 58:1537–1541
Wang Y, Tang S-K, Lou K, Mao P-H, Jin X, Jiang C-L, Xu L-H, Li W-J (2009a) *Paracoccus saliphilus* sp. nov., a halophilic bacterium isolated from a saline soil. Int J Syst Evol Microbiol 59:1924–1928
Wang L, Seki K, Miyazaki T, Ishihama Y (2009b) The causes of soil alkalinization in the Songnen plain of Northeast China. Paddy Water Environ 7:259–270
Wang X, Xue Y, Ma Y (2010a) *Virgibacillus subterraneus* sp. nov., a moderately halophilic Gram-positive bacterium isolated from subsurface saline soil. Int J Syst Evol Microbiol 60:2763–2767
Wang S, Yang Q, Liu Z-H, Sun L, Wei D, Zhang J-Z, Song J-Z, Yuang H-F (2010b) *Haloterrigena daqingensis* sp. nov., an extremely haloalkaliphilic archaeon isolated from a saline-alkaline soils. Int J Syst Evol Microbiol 60:2267–2271
Wang S, Yang Q, Liu ZH, Sun L, Wei D, Zhang JZ, Song JZ, Wang Y, Song J, Fan JX (2011) *Halomonas alkalitolerans* sp. nov., a novel moderately halophilic bacteriun isolated from soda meadow saline soil in Daqing, China. J Microbiol 49:4–28
Wang S, Sun L, Wei D, Zhou B, Zhang J, Gu X, Zhang L, Liu Y, Li Y, Guo W, Jiang S, Pan Y, Wang Y (2014) *Bacillus daqingensis* sp. nov., a halophilic, alkaliphilic bacterium isolated from saline-sodic soil in Daqing, China. J Microbiol 52:548–553
Wang L, Zhou Z, Wu G, Chen M, Lin M, Zhang W, Chen W (2015a) *Falsirhodobacter deserti* sp. nov., isolated from sandy soil. Int J Syst Evol Microbiol 65:650–655
Wang K, Zhang L, Liu Y, Pan Y, Meng L, Xu T, Zhang C, Liu H, Hong S, Huang H, Jiang J (2015b) *Kocuria dechangensis* sp. nov., an actinobacterium isolated from saline and alkaline soils. Int J Syst Evol Microbiol 65:3024–3030
Wang K, Zhang L, Yang Y, Pan Y, Meng L, Liu H, Hong S, Huang H, Jiang J (2015c) *Halobacillus andaensis* sp. nov., a moderately halophilic bacterium isolated from saline and alkaline soil. Int J Syst Evol Microbiol 65:1908–1914
Wang K, Zhang L, Li J, Pan Y, Meng L, Xu T, Zhang C, Liu H, Hong S, Huang H, Jiang J (2015d) *Planococcus dechangensis* sp. nov., a moderately halophilic bacterium isolated from saline and alkaline soils in Dechang Township, Zhaodong City, China. Antonie Van Leeuwenhoek 107:1075–1083
Wang S, Sun L, Ling N, Zhu C, Chi F, Li W, Hao X, Zhang W, Bian J, Chen L, Wei D (2020a) Exploring the diversity of bacteria along a soil salinity gradient in Songnen Plain of Northeast China. Front Microbiol 10:10. https://doi.org/10.1007/s00203-020-02073-w
Wang S, Sun L, Wei D, Salam N, Fang B-Z, Dong Z-Y, Hao X-Y, Zhang M, Zhang Z, Li W-J (2020b) *Nesterenkonia haasae* sp. nov., an alkaliphilic actinobacterium isolated from a degraded pasture in Songnen plain. Arch Microbiol. https://doi.org/10.1007/s00203-020-02073-w
Wei W, Wang K, Hu X, Yang S, Zeng C, Hou Z, Liu S, Cui H, Zhu L (2021) *Phytoactinopolyspora limicola* sp. nov., an alkaliphilic actinomycete isolated from a soda alkali-saline soil. Arch Microbiol. https://doi.org/10.1007/s00203-020-02134-0
Wu G, Wu XQ, Wang YN, Chi CQ, Tang YQ, Kida K, Wu XL, Luan ZK (2008) *Halomonas daqingensis* sp. nov., a moderately halophilic bacterium isolated from an oilfield soil. Int J Syst Evol Microbiol 58:2859–2865
Yan N, Marschner P (2012) Response of soil respiration and microbial biomass to changing EC in saline soils. Soil Biol Biochem 65:322–328
Yang J, Zhan C, Li Y, Zhou D, Yu Y, Yu J (2018) Effect of salinity on soil respiration in relation to dissolved organic carbon and microbial characteristics of a wetland in the Liaohe River estuary, Northeast China. Sci Total Environ 642:946–953

Yu BQ, Yang S, Zhu L, Gao F, Zeng CL, Liu SK, Cui HL, Wei W (2018a) Diversity characteristic of archaeal community responding to soil anion in saline-alkali soil. China Environ Sci 38:2731–2739

Yu Y, Fu Y, Guo X, Yan R, Wang H, Zhao J, Wang X, Zhang J, Xiang W (2018b) *Streptomyces durbertensis* sp. nov., isolated from saline–alkali soil. Int J Syst Evol Microbiol 68:3635–3640

Yuan BC, Li ZZ, Liu H, Gao M, Zhang YY (2007) Microbial biomass and activity in salt affected soils under arid conditions. Appl Soil Ecol 35:319–328

Yuan P-P, Ye W-T, Pan J-X, Han D, Zhang W-J, Cui H-L (2015) *Halorussus amylolyticus* sp. nov., isolated from an inland salt lake. Int J Syst Evol Microbiol 65:3734–3788

Zahran HH (1997) Diversity, adaptation and activity of the bacterial flora in saline environments. Biol Fertil Soils 25:211–223

Zahran HH, Moharram AM, Mohammad HA (1992) Some ecological and physiological studies on bacteria isolated from salt affected soils of Egypt. J Basic Microbiol 32:405–413

Zhang J (2014) Coastal saline soil rehabilitation and utilization based on forestry approaches in China. Springer, Berlin, pp 4–7. https://doi.org/10.1007/978-3-642-39915-2

Zhang H (2016) The soil microbial characteristics in different vegetation grasslands and the relation between plant and soil microbes in Songnen Plain. Jilin Agricultural University, pp 1–6

Zhang Y-Q, Schumann P, Li W-J, Chen G-Z, Tian X-P, Stackebrandt E, Xu L-H, Jiang C-L (2005) *Isoptericola halotolerans* sp. nov., a novel actinobacterium isolated from saline soil from Qinghai Province, north-west China. Int J Syst Evol Microbiol 55:1867–1870

Zhang Y, Cao C, Guo L, Wu Q, Cui Z (2015a) Soil properties, bacterial community composition, and metabolic diversity responses to soil salinization of a semiarid grassland in Northeast China. J Soil Water Conserv 70:110–120

Zhang L, Pan Y, Wang K, Zhang X, Zhang C, Zhang S, Fu X, Jiang J (2015b) *Pseudomonas zhaodongensis* sp. nov., isolated from saline and alkaline soils. Int J Syst Evol Microbiol 65:1022–1030

Zhang MM, Wang N, Zhang JY, Hu YB, Cai DJ, Guo JH, Wu D, Sun GY (2019) Soil physico-chemical properties and the rhizosphere soil fungal community in a mulberry (*Morus alba* L.)/alfalfa (*Medicago sativa* L.) intercropping system. Forests 10:167. https://doi.org/10.3390/f10020167

Zhang Z, Wang H, Song X, Liang Z, Tang Z (2020) Arbuscular mycorrhizal fungal diversity is affected by soil salinity and soil nutrients in typical saline-sodic grasslands dominated by *Leymus chinensis*. Arid Land Res Manag 34:68–82

Zhao B, Li J (2017) Biodiversity of culture-dependent haloalkaliphilic microorganisms. Acta Microbiol Sin 57(9):1409–1420

Zhao J, Guo L, Liu C, Sun P, Li J, Li W-J, Xiang W, Wang X (2015) *Streptomonospora halotolerans* sp. nov., an actinomycete isolated from soil. Int J Syst Evol Microbiol 65:3183–3189

Zhao B, Hu Q, Guo X, Liao Z, Sarmiento F, Mesbah NM, Yan Y, Li J, Wiegel J (2018) *Natronolimnobius aegyptiacus* sp. nov., an extremely halophilic alkalithermophilic archaeon isolated from the athalassohaline Wadi An Natrun, Egypt. Int J Syst Evol Microbiol 68:498–506

Zhong Z-P, Liu Y, Liu L-Z, Wang F, Zhou Y-G, Liu Z-P (2014a) *Rheinheimera tuosuensis* sp. nov., isolated from a saline lake. Int J Syst Evol Microbiol 64:1142–1148

Zhong Z-P, Liu Y, Liu H-C, Wang F, Song L, Liu Z-P (2014b) *Idiomarina planktonica* sp. nov., isolated from a saline lake. Int J Syst Evol Microbiol 64:3411–3416

Chapter 3
Microbial Diversity of High-Altitude Geothermal Springs in Tajikistan

M. M. Dzhuraeva, A. A. Margaryan, C. M. Ratnadevi, H. H. Panosyan, N.-K. Birkeland, and Kh. I. Bobodzhanova

Abstract Thermophiles are a group of heat loving microbes thriving in various hot ecological niches including geothermal springs. Many high-altitude geothermal springs are located on the territory of Tajikistan, microbial diversity of which have not yet been explored.

Over the last decades, thermophilic microbes have fascinated researches in many fields due to their ability to withstand and function under extreme conditions. Thermophilic microbes are a great source of thermostable enzymes, which show unique features that can be used in biotechnological processes at elevated temperatures and under other harsh conditions. Unexplored geothermal springs in Tajikistan are promising sources for isolation of thermophilic microbes with useful enzymatic activities.

In this regard, a total of 21 thermophilic aerobic bacteria and one thermophilic anaerobic bacterium were isolated from three high-altitude mineralized Tajik geothermal springs with temperatures ranging from 50.5 to 89 °C. The isolates were phylogenetically identified and studied for their potential to produce extracellular hydrolytic enzymes (such as protease, amylase, lipase, and cellulase). The identification of isolates based on 16S rRNA gene sequences revealed relationships to members of more than 10 different species belonging to 5 genera, namely

M. M. Dzhuraeva
Center of Biotechnology, Tajik National University, Dushanbe, Tajikistan

Department of Biochemistry, Microbiology and Biotechnology, Yerevan State University, Yerevan, Armenia

Department of Biological Sciences, University of Bergen, Bergen, Norway

A. A. Margaryan · H. H. Panosyan
Department of Biochemistry, Microbiology and Biotechnology, Yerevan State University, Yerevan, Armenia

C. M. Ratnadevi · N.-K. Birkeland
Department of Biological Sciences, University of Bergen, Bergen, Norway

K. I. Bobodzhanova (✉)
Center of Biotechnology, Tajik National University, Dushanbe, Tajikistan

D. Egamberdieva et al. (eds.), *Microbial Communities and their Interactions in the Extreme Environment*, Microorganisms for Sustainability 32,
https://doi.org/10.1007/978-981-16-3731-5_3

Aeribacillus, *Anoxybacillus*, *Geobacillus*, *Parageobacillus*, and *Thermotoga*. *Geobacillus* spp. were found to be the most abundant (cultivable) aerobic species in the studied geothermal springs. Some of the isolated bacilli shared less than 96% sequence similarity with their closest match in GenBank, indicating that Tajik geothermal springs harbor potentially novel bacilli species. More than 20 isolates actively produced one or more extracellular hydrolases like proteases, amylases, or lipases, which makes them a potential source for thermostable hydrolases for use in biotechnological processes.

Keywords Thermophiles · Geothermal springs · Tajikistan · 16S rRNA genes · Hydrolases

3.1 Introduction

It has long been known that microorganisms play an important role in natural environments, especially in extreme conditions such as hot springs. Although Baas-Becking's postulate of microbial biogeography, "Everything is everywhere, but the environment selects" is accepted by many microbiologists, it has been challenged by others including studies using the hyperthermophilic archaeon, *Sulfolobus*, and thermoacidophilic methanotrophic Verrucomicrobia as model microorganisms to show that geographical distribution patterns exist in geothermal environments on a global scale (Whitaker et al. 2003; Erikstad et al. 2019). Therefore, enhanced knowledge of microorganisms that thrive in terrestrial hot springs would enhance our understanding of both thermophilic microbial diversity in nature and the evolutionary mechanisms that shape it. Hot springs are modern analogs of ancient hydrothermal systems where life may have emerged and evolved. They are distributed throughout the world and are formed as a result of the movement of hot water from the Earth's crust via faults created by tectonic movement or volcanic eruption, and represent favorable natural habitats for thermophilic organisms (Stetter 1999). The temperature of the hot springs in geothermal areas is usually rather constant, not varying more than 1–2 °C over many years even though the temperature between the hot springs can vary greatly, from 20 °C to above 100 °C, depending on location and pressure (Salmond and Whittenbury 1985). In addition to relatively stable temperatures, geothermal areas offer wide ranges of acidity, oxidation/reduction states, solute concentrations, gas compositions, mineralogy and nutrients to support growth of both chemolithotrophs and chemoorganotrophs. This variety in growth parameters results in enormous genetic and metabolic diversity (Amend and Shock 2001).

Studies of such extreme environments are necessary because different parameters (e.g., physicochemical and geological properties) shape microbial diversities differently in each specific environment (Lau et al. 2009). The Great plate count anomaly illustrates that less than 1% of the existing microorganisms in nature are cultivable.

Thus, culture-independent approaches are also of great importance for the exploration of microbial diversity in different habitats (Hou et al. 2013).

Consistent efforts are still on to decipher the ecological role of microbial communities in thermal springs. These include community structure determination on Tibetan and Tengchong thermal springs (China), Nakabusa hot springs (Japan), Siloam hot water springs (South Africa), Andean Mountain hot water springs (Colombia), Solfataric Fields (Iceland), Great Basin Hot springs, and Yellowstone National Park (USA) (Liu et al. 2016). While some of these studies were performed to understand the geochemistry, geomicrobiology, and bioenergetics of biogeothermal systems (Inskeep et al. 2005), others were done to determine the biotechnological potential of the microbes (Liu et al. 2016). Interestingly, it was found that the chemical type of hot springs influence the organization of microbial community (Berelson et al. 2011). In sulfur rich environments with anaerobic zones, sulfur bacteria were responsible for energy production which in turn helps in maintenance of the community structure (Zaigham et al. 2009).

Diverse microorganisms were identified in Tibetan hot springs, including members of the archaeal phyla *Crenarchaeota*, *Euryarchaeota*, and *Thaumarchaeota*, and members of various bacterial phyla such as *Cyanobacteria*, *Chloroflexi*, *Chlorobi*, *Proteobacteria*, and *Firmicutes* (Wang et al. 2014). Although similarities can be found between different geographical areas, microbial composition usually differs due to the influence of physical and chemical conditions such as temperature and pH (Cole et al. 2013; Hou et al. 2013). Therefore, unique microbial communities are often shaped by distinct niches (Tang et al. 2018). Hot springs have a range of temperature (47–96 °C) and pH (3.2–8.6) conditions. *Proteobacteria*, *Aquificae*, *Firmicutes*, *Deinococcus-Thermus*, and *Bacteroidetes* comprised a large fraction of the bacterial communities in acidic hot springs in Yunnan. Hot springs (both in Yunnan and Tibet) harbor more and variable bacterial phyla than acidic springs; the major phyla of Tibetan hot springs were similar to the Yunnan nonacidic samples but showed different relative abundances. For example, *Bacteroidetes* in Tibetan nonacidic hot springs shows higher abundance than Yunnan. *Desulfurococcales* and unclassified *Crenarchaeota* were the dominating groups in archaeal populations from most of the nonacidic hot springs, whereas the archaeal community structure in acidic hot springs was simpler and dominated by *Sulfolobales* and *Thermoplasmata*. The phylogenetic analyses showed that *Aquificae* and *Crenarchaeota* were predominant in the investigated springs and possessed many phylogenetic lineages that had never been found in other hot springs in the world (Song et al. 2013; Xian et al. 2018).

Few investigations have been done to explore the diversity in natural hot springs/thermal lakes in The Indian Himalayas. Himalayas have unique hot springs at high elevations, such as Soldhar and Ringigad hot springs, Uttaranchal Himalaya (Kumar et al. 2004a, b); Chumathang hot spring, NW Indian Himalayas (Yadav 2015); Vashisht, Khirganga and Kasol hot springs (Shirkot and Verma 2015) and Tattapani hot spring in NW Indian Himalayas (Priya et al. 2016).

Thermophilic microbes living in the hot springs, are producing thermophilic enzymes that have many industrial implications (Mirete et al. 2016). Since the discovery of thermophilic microorganisms and novel enzymes, such as *Taq* polymerase from *Thermus aquaticus*, a large number of thermophiles have been isolated and studied for industrial applications. Thus, thermophilic microbes are important sources of thermostable enzymes such as proteases, amylase, lipase, xylanase, cellulase, DNA restriction enzymes, etc. (Herbert and Sharp 1992; Schallmey et al. 2004).

Thermophilic bacilli belonging to genera *Aeribacillus*, *Anoxybacillus*, *Geobacillus*, and *Parageobacillus* constitute the most abundant thermophiles isolated from geothermal springs and have huge biotechnological potential as producers of thermozymes (Sharma et al. 2019).

Many high-altitude geothermal springs have been found on the territory of Tajikistan, the microbial diversity of which has not been explored yet. In this chapter we provide geochemical and geophysical profiles of three geothermal springs located in Khodja-Obi-Garm, Tamdykul, and Obigarm. The geology of the region where Tajikistan is located is complex due to ongoing tectonic activity and volcanism resulting in the existence of numerous geothermal springs with different geochemical properties. Thermal springs found in Tajikistan are not investigated yet from microbiological and biotechnological stands of view. The purpose of this chapter is to summarize initial findings of microbiological studies of several geothermal springs of Tajikistan.

3.2 Geothermal Springs in Tajikistan and their Geochemical Profiling

More than 200 deposits of mineral water springs have been explored on the territory of Tajikistan, the temperature range of which varied from 5 °C (Novobedak) to 89.0 °C (Khodja-Obi-Garm).

In this chapter we provide geochemical and geophysical profiles of three geothermal springs located in Khodja-Obi-Garm, Tamdykul, and Obigarm (Fig. 3.1).

The physiochemical characteristics of studied geothermal springs are presented on Table 3.1.

The Khodja-Obi-Garm field is located on the Bank of the Varzob river, 50 km North of Dushanbe, at an altitude of 1780–1870 m (Fig. 3.2) and the physiochemical characteristics of studied geothermal springs are presented on Table 3.1. Geothermal water from the Khodja-Obi-Garm field is used for hot water supply and heating of the entire Spa complex (Razykov 2007; Normatov 2010).

Obigarm mineral waters come to the surface of the Obigarm river 100 km East of Dushanbe, at an altitude of 1300–1400 m (Fig. 3.2). The physiochemical characteristics of these geothermal springs are presented on Table 3.1. The baths of the field Obigarm have two types of water—siliceous and without "specific" components and properties. The composition of mineral waters is chloride-sulfate calcium-sodium (Razykov 2007; Normatov 2010).

Fig. 3.1 The map of Tajikistan showing the studied geothermal springs. The geographical locations are indicated by red marks. Close up photographs of geothermal springs are also shown: (1) Khodja-Obi-Garm, (2) Obigarm, (3) Tamdykul. The source of the map is https://av.wikipedia.org/wiki/%D0%A4%D0%B0%D0%B9%D0%BB:Tajikistan_adm_location_map.svg

The Tamdykul field is located 25 km North-Northwest of the Jirgital district center, in the upper part of the valley, of the Tamdykul river, at an altitude of 2198 m (Fig. 3.2). The physiochemical characteristics of these geothermal springs are presented on Table 3.1. Based on the Tamdykul field, a local seasonal water treatment facility operates (Razykov 2007; Normatov 2010).

3.3 Bacterial Diversity of Geothermal Springs in Tajikistan Based on Cultivation-Dependent and Molecular Studies

In total 22 thermophilic (Tamdykul—15 isolates, Khodja-Obi-Garm—6 isolates, Obigarm—1 isolate) aerobic and anaerobic bacilli strains were isolated from hot spring sediment samples and identified based on phenotypic and phylogenetic characteristics. The optimal growth temperature of the isolates varied from 55 to 75 °C and the optimal pH from 7.4 to 10. The isolates were catalase and oxidase

Table 3.1 Geographical location, physiochemical profiling, and brief characteristics of main geothermal springs distributed on the territory of Tajikistan

Thermal mineral spring	Spring GPS location/altitude (m above sea level	Conductivity (μS/cm)	Temperature and pH in the outlet (°C/pH)	Description
Khodja-Obi-Garm	38°54′10″N, 68°48′3″E/1835	4378.3	93/8.5	This type of water is associated with crack-core water pressure systems. Siliceous high sulfate-chloride–hydrocarbonate sodium waters
Tamdykul	39°25′102,0″N, 071°13′103″E/2198	1019	88/7.4	The current regional sanatorium and greenhouse are used
Obigarm	38°42′27,24″N, 69°41′40,84″E/1333	797	50.5/8.2	Balneotherapy procedures are provided by a bath spa. Geothermal water from the Obigarm field is also used for hot water supply in the village of Obigarm

positive, and were able to reduce nitrate. 16S rRNA gene sequence analyses showed that the majority of the isolates shared 83–99% similarities with representatives of genera *Aeribacillus*, *Anoxybacillus*, *Geobacillus* (G. *vulcani*, *G. caldoxylosilyticus*, *G. thermocatenulatus*, *G. jurassicus*, and *G. stearothermophilus*), and *Parageobacillus* (*P. caldoxylosilyticus*). Some isolates shared less than 96% sequence identity with their closest match in GenBank (Table 3.2), indicating that Khodja-Obi-Garm, Obigarm, and Tamdykul hot springs harbor novel bacilli, at least at the species level.

The *Geobacillus* genus contains Gram-positive, rod-shaped, spore-forming bacteria that have an optimum growth temperature between 55 and 65 °C (Nazina et al. 2001). Members of the *Geobacillus* genus were originally classified in Group 5 of the *Bacillus* genus (Ash et al. 1991). In 2001, based on a combination of 16S rRNA sequence analysis, fatty acid composition and DNA-DNA hybridization (DDH), some members of Group 5 were reclassified into the new genus *Geobacillus*, with the word *Geobacillus* meaning "soil or earth small rod" (Nazina et al. 2001). Recently it was proposed that the *Geobacillus* genus should be separated into two genera based on a comparative genomics analysis (Aliyu et al. 2016). There is an extensive interest in the members of *Geobacillus* genus for biotechnological purposes such as for bioremediation, production of thermostable enzymes, and biofuels (Bhalla et al. 2013; Wiegand et al. 2013). In addition, *Geobacillus* spp. are common spoilage organisms in food manufacturing plants and products (Donk 1920; Postollec et al. 2012). *Geobacillus* spp. have been isolated from temperate as well as hot environments including hot springs, oilfields, deep sea sediments, sugar refineries, canned foods, dehydrated vegetables and dairy factories. The type species *G. stearothermophilus* was first described in 1920 and was isolated from canned

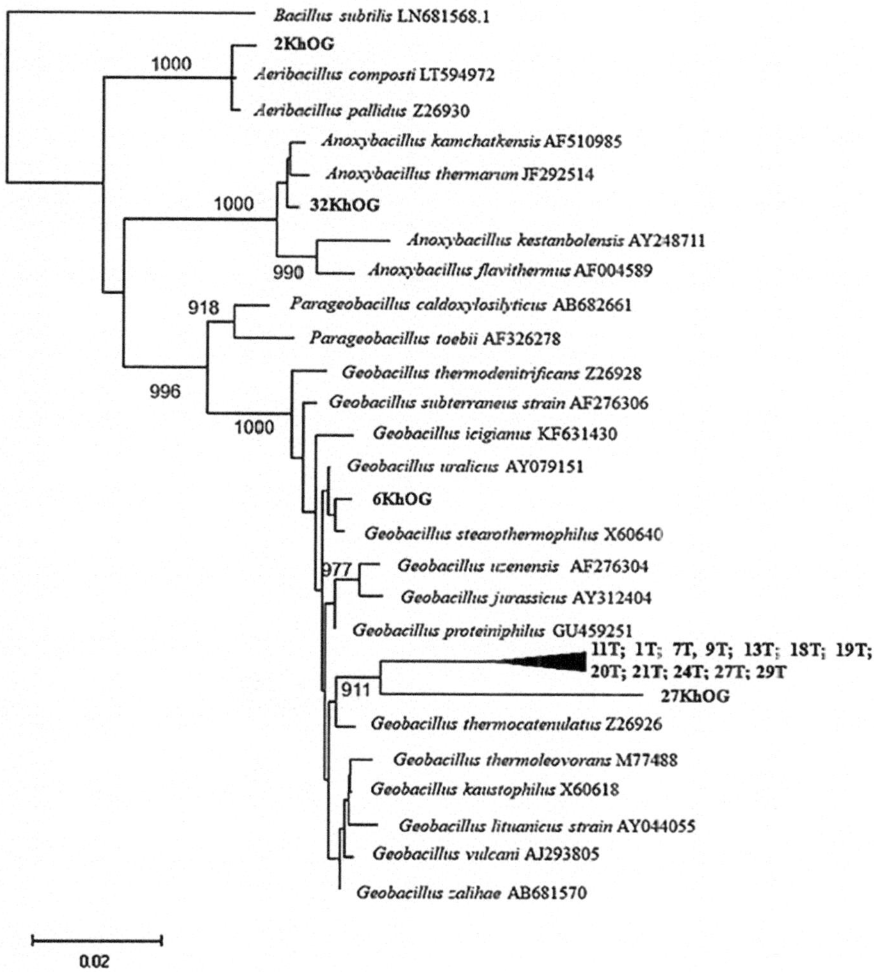

Fig. 3.2 Neighbor-Joining phylogenetic tree showing the phylogenetic positions of the bacterial isolates (in bold) and selected close relatives based on 16S rRNA gene sequences. The lineage containing 12 isolates is collapsed. Database accession numbers of reference strains are given in bold. Bootstrap values ≥911 are indicated at branch nodes and based on 1000 iterations. Positions containing gaps or missing data were excluded from the analysis. The tree was rooted using *Bacillus subtilis* as outgroup. The bar indicates the number of base substitutions per site

cream-style corn (Donk 1920). *G. stearothermophilus* is a common contaminant of dairy products, particularly milk powder and has also been isolated from dried soups and vegetables. Until the 1980s, *G. stearothermophilus* was regarded as the only known obligate thermophile of the *Bacillus* genus (Suzuki et al. 1983; Zarilla and Perry 1987). According to the List of Prokaryotic names with Standing in Nomenclature (https://www.bacterio.net/) (Parte 2018), as of April 2018, there were sixteen

Table 3.2 BLAST results of 16S rRNA gene sequences of thermophilic bacilli isolates and accession numbers

Geothermal springs, no isolates	Closest match taxonomic affiliation	Similarity to closest match (%)	Accession number
Khodja-Obi-Garm			
2KhOG	*Aeribacillus pallidus strain* BTPS-2	99.03	MT801094
6KhOG	*Geobacillus stearothermophilus* P4	99.45	MT801095
25 KhOG	*Geobacillus. vulcani* Manikaran-107	91.75	MT801096
27KhOG	*Geobacillus thermocatenulatus* BGSC 93A1	95.24	MT801097
32KhOG	*Anoxybacillus* sp. DR04	99.66	MT801098
KhT	*Thermotoga caldifontis* AZM44c09	99.86	MT809042
Obigarm			
14OG	*Parageobacillus caldoxylosilyticus*	94.03	MT804630
Tamdykul			
1T	*Geobacillus stearothermophilus* strain G1017_C12	99.77	MT808162
4T	*Geobacillus caldoxylosilyticus* BGSC W98A1	96.48	MT808163
6T	*Geobacillus* sp. *TC-Y1*	83.93	MT808164
7T	*Geobacillus jurassicus* WSUCF-022A	99.27	MT808165
9T	*Geobacillus stearothermophilus* G1017_C12	99.89	MT808166
11T	*Geobacillus stearothermophilus* strain ARM 1	96.02	MT808167
13T	*Geobacillus jurassicus* WSUCF-022A	99.77	MT808168
18T	*Geobacillus stearothermophilus* ARM 1	94.90	MT808170
19T	*Geobacillus stearothermophilus* H-2	99.43	MT808169
20T	*Geobacillus stearothermophilus* ARM 1	99.89	MT808171
21T	*Geobacillus stearothermophilus* H-2	99.43	MT808172
24T	*Geobacillus jurassicus* WSUCF-022A	99.40	MT808173
26T	*Geobacillus stearothermophilus* AD24	98.09	MT808174
27T	*Geobacillus stearothermophilus* AD24	99.26	MT808175
29T	*Geobacillus stearothermophilus* H-2	99.73	MT808176

Geobacillus species (*G. caldoxylosilyticus*, *G. galactosidasius*, *G. icigianus*, *G. jurassicus*, *G. kaustophilus*, *G. lituanicus*, *G. stearothermophilus*, *G. subterraneus*, *G. thermantarcticus*, *G. thermocatenulatus*, *G. thermodenitrificans*, *G. thermoglucosidasius*, *G. thermoleovorans*, *G. toebii*, *G. uzenensis*, and *G. vulcani*) (Margaryan et al. 2018) described with validly published names (Ahmad et al. 2000; Nazina et al. 2001, 2004). However, the classification of many of these species remains uncertain. To date over 60 *Geobacillus* genomes have been sequenced, mainly to identify genes that could be used in different biotechnological applications (Aliyu et al. 2016). There are eleven species for which the type strain have been genome sequenced; *G. caldoxylosilyticus*, NBRC 10776; *G. icigianus* DSM 28325, *G. jurassicus* DSM 15726, *G. kaustophilus* NBRC 102445, *G. stearothermophilus* ATCC 12980, *G. subterraneus* DSM 13552, *G. thermantarcticus* M1, *G. thermodenitrificans* DSM 465 *G. thermoglucosidasius* NBRC 107763, *G. thermoleovorans* DSM 5366, and *G. toebii* DSM 14590 (Aliyu et al. 2016; Yao et al. 2013; Bryanskaya et al. 2014). Recent studies have shown that it is possible by using a comparative genomics approach to resolve the taxonomy of this important genus (Aliyu et al. 2016; Studholme 2015).

The family *Bacillaceae* (Skerman et al. 1980; Fischer 1895) is one of the largest bacterial families and currently consists of 57 genera. The *Bacillaceae* are either rod-shaped (bacilli) or spherical (cocci) Gram-positive bacteria, the majority of which produce endospores (Goh et al. 2013).

Anoxybacillus is one of the genera within the *Bacillaceae* (Skerman et al. 1980; Fischer 1895), classified within the phylum *Firmicutes* (Gibbons and Murray 1978), class *Bacilli* (Ludwig et al. 2009; Validation List no. 132 2010) and order *Bacillales* (Skerman et al. 1980; Prévot et al. 1953). *Anoxybacillus* spp. are alkalo-thermophiles with optimum growth at temperatures between 50 °C and 65 °C and at pH 5.6–9.7 (Goh et al. 2013). Most of the *Anoxybacillus* spp. are found in hot springs (Goh et al. 2013; Margaryan et al. 2018), but *Anoxybacillus* has also been found in animal manure, contaminated diary and meat products (Goh et al. 2013), animals (i.e., fish gut) (Goh et al. 2013), insects (i.e., glassy-winged sharpshooter and spiraling whitefly) (Skerman et al. 1980), and plants (i.e., Indian mulberry) (Skerman et al. 1980; Rogers and Backus 2014). To date, a total of 22 species and two subspecies of *Anoxybacillus* have been described (Goh et al. 2013; Cihan et al. 2014).

Almost all members of the *Bacillaceae* are excellent industrial enzyme producers (Goh et al. 2013). Members of the genus *Anoxybacillus* exhibit the additional advantage of thermostability compared to the mesophilic *Bacillaceae*. It has been reported that enzymes from *Anoxybacillus* spp. can degrade various substrates such as starch, cellulose, fats, and proteins (Goh et al. 2013). Many carbohydrase-encoding genes have been identified in *Anoxybacillus* spp. genomes, and some of the well-studied starch-degrading enzymes are α-amylase (Skerman et al. 1980; Chai et al. 2012), pullulanase (Skerman et al. 1980), amylopullulanase (Skerman et al. 1980), CDase (Skerman et al. 1980; Turner et al. 2005), and xylose-isomerase. In addition, xylanolytic enzymes such as xylanase (Wang et al. 2010) and α-L-arabinofuranosidase (Canakci et al. 2008) have been characterized from

Anoxybacillus spp. Apart from their hydrolytic capabilities, *Anoxybacillus* spp. have been proposed as agents for bioremediation of Hg^{2+}, Cr^{2+}, Al^{3+}, As^{3+} ions (Goh et al. 2013; Beris et al. 2011; Jiang et al. 2016), and nitrogen oxide (Chen et al. 2015), and as possible candidates for biohydrogen production (Goh et al. 2013). Among the members of the family *Bacillaceae*, intensive genome sequencing efforts have been undertaken for *Geobacillus* (Nazina et al. 2001) (>80 projects) and *Bacillus* (Skerman et al. 1980; Cohn 1872) (>1500 projects), which have been registered in the NCBI BioProject database. In contrast, genomic studies on *Anoxybacillus* are rather limited, with only 12 *Anoxybacillus* species with complete genome sequence in NCBI. Therefore, the genomic study of *Anoxybacillus* spp. is essential not only to fully understand their biochemical networks, but also to discover their potential applicability in industrial processes (Belduz et al. 2015).

The genus *Aeribacillus* was first proposed by (Miñana-Galbis et al. 2010) when *Geobacillus pallidus* (Banat et al. 2004) was reclassified in a novel genus, as *Aeribacillus pallidus*. This genus belongs to the phylum *Firmicutes*, order *Bacillales* and family *Bacillaceae*, and is most closely related to the genera *Geobacillus* and *Anoxybacillus* (Nazina et al. 2001). A Gram-positive, aerobic, endospore-forming, thermophilic bacterium, strain N.8T, was isolated from the curing step of an olive mill pomace compost sample, collected at the Composting Experimental Centre (CESCO, Salerno, Italy) (Finore et al. 2017).

A Neighbor-joining evolutionary distance tree based on 16S rRNA gene sequences of the isolated strains and selected reference sequences was constructed (Fig. 3.2).The sequences were aligned using CLUSTAL X, version 2.0 (Larkin et al. 2007) and the tree constructed using the neighbor-joining method (NJ) implemented in MEGA X (Kumar et al. 2018). The partial 16S rRNA gene sequences of 22 strains were submitted to NCBI GenBank and Accession Numbers are presented on Table 3.2. 16S rRNA gene sequencing and phylogenetic analysis of selective isolates from thermal springs of Tajikistan, revealed that all the isolates showed 83 to 99% identity with the sequences within the GenBank database. The phylogenetic tree was constructed to determine the phylogenetic affiliations. Analysis of the 16S rRNA sequences from Khodja-Obi-Garm hot spring revealed that tree strains belonged to *Geobacillus* sp., one strain to *Aeribacillus* sp. and one strain to *Anoxybacillus* sp. From Obigarm, one isolate belonging to *P. caldoxylosilyticus* was obtained. Tamdykul hot spring samples yielded 12 strains belonging to *Geobacillus*. Overall, all identified bacteria belonged to 21 species of 5 different genera namely, *Anoxybacillus*, *Geobacillus*, *Aeribacillus*, *Parageobacillus* and *Thermotoga*, with members of *Geobacillus* as the most frequently isolated representatives.

Collected biomass from Khodja-Obi-Garm was used to enrich anaerobic bacteria in a basal mineral medium supplemented with yeast extract and peptone. Flasks were incubated for 14 days at 55 °C. Growth was monitored by measuring culture turbidity at 590 nm. Following several transfers using the same medium a pure culture was obtained. BLAST results of the 16S rRNA gene sequence of this isolate shared 99.86% similarity with *Thermotoga caldifontis* (Table 3.2) (Abdusamadzoda et al. 2016).

The genus *Thermotoga* comprises chemoheterotrophs that are strictly anaerobic, non-spore-forming, thermophilic bacteria and have a characteristic sheath-like structure called a "toga." These characteristics are common to most genera in the family *Thermotogaceae*, although some species grow under moderately thermophilic or mesophilic conditions (Ben Hania et al. 2013). The first described species of this genus was *Thermotoga maritima*, which was isolated from geothermally heated marine sediments (Huber et al. 1986). Eight additional species have so far been described: *Thermotoga elfii*, *T. hypogea*, *T. lettingae*, *T. naphthophila*, *T. neapolitana*, *T. petrophila*, *T. subterranea*, and *T. thermarum* (Balk et al. 2002). In addition to marine thermal environments, members of the genus *Thermotoga* have been isolated from oil reservoirs and terrestrial hot springs in Japan (Mori et al. 2014).

3.4 Biotechnological Potential of Thermophilic Isolates Obtained from Tajik Geothermal Springs

Extreme environments represent a major source of untapped microbes with potential for resources for biotechnological processes and products. The discovery of thermostable DNA polymerase (*Taq* polymerase) from the thermophilic bacterium, *Thermus aquaticus* isolated from Yellowstone National Park, has been used world-wide in Polymerase Chain Reaction (Podar and Reysenbach 2006) and triggered interest in the study of microbial diversity in hot springs (Amanuel and Nancy 2018).

One of the most attractive attributes of thermophiles is that they produce enzymes capable of catalyzing biochemical reactions at temperatures higher than those of mesophilic organisms (Demirjian et al. 2001). The higher thermal stability and tolerance to many chemical denaturants e.g., organic solvents, enable them to resist harsh process conditions. They also show high catalytic activity at elevated temperatures and longer shelf-life as commercial products (Aguilar et al. 1998). Elevated temperatures in biotechnological processes can influence the bioavailability and solubility of organic compounds such as polyaromatic, aliphatic hydrocarbons, and polymeric substances. The elevated temperature is accompanied by a decrease in viscosity and an increase in the diffusion coefficient of reactants. In biotechnological processes where high operating temperatures are required, the risk of contamination by other organisms also become substantially reduced (Adams and Kelly 1998). Furthermore, in large-scale fermentations with heat sensitive microorganisms, extensive efforts must be made for cooling the fermentation process and as much as 10% of the energy cost of a microbial fermentation may be for heat transfer. Thermophilic fermentations, on the other hand, do not need to be cooled (Niehaus et al. 1999).

Extracellular proteases are naturally produced by microorganisms mainly to degrade proteins and large polypeptides in the medium into peptides and amino acids before cellular uptake. Man has commercially exploited such enzymes to assist in protein breakdown in various industrial processes. Protease enzymes constitute one of the most important groups of industrial enzymes which are extensively used in the food, pharmaceutical, protein hydrolysis, detergent, cheese-making, brewing, photographic, baking, meat and leather industries and inclusions in animal and human food as digestive aids. About 75% of the world sales of industrial enzymes are hydrolytic enzymes of which proteolytic enzymes constitute about 60% (Chu 2007; Ningthoujam et al. 2009; Ninghoujam and Kshetri 2010; Rai et al. 2010). Proteases are produced commercially from plants, animals and microbial sources. Microorganisms offer an attractive source of proteases because they can be cultured in large quantities in a short period of time using established fermentation techniques (Gupta et al. 2002; Tambekar et al. 2009; Dabananda and Kshetri 2010). In addition, the protein products they produce are more stable than those from plants and animals. Approximately 40% of these enzymes are of microbial origin. A crude alkaline protease from a thermophile has been used for the acceleration of the gelatin layer hydrolysis of X-ray films (Fujiwara et al. 1991). In food industry, especially in meat tenderization, thermophilic proteases have been preferred as they possess the ability to hydrolyze connective tissue proteins as well as muscle fiber proteins. It was found that the most suitable enzymes for solubilization of meat hydrolysates are alkaliphilic, with optimal activity at pH 8.5 and temperature 55–60 °C (Kumar et al. 1999). In detergent industry, alkaline proteases added to laundry detergents enable the release of proteinaceous materials from stains (Wilson and Remigio 2012). Ideally, the enzymes used in detergent formulations should have high activity and stability over a broad range of pH and temperature. However, not all the processes require the same thermal stability and lower wash temperatures are also sometimes preferred (Kumar et al. 1999).

Proteases are arguably the most important group of industrial enzymes and certainly form a major portion of world-wide enzyme sales. Thermophilic proteases, with their high specific activities and their superior chemical and physical stability, would seem to be good candidates for current and future biotechnological applications.

Starch is composed of α-glucose units that are linked by α-1,4- or α-1,6-glycosidic bonds. Enzymes involved in the hydrolysis of starch can simply be classified into two groups, endo-acting and exo-acting enzymes. Endo-acting enzymes such as α-amylase hydrolyze linkages in the interior of the starch in a random fashion leading to the formation of linear and branched oligosaccharides. Exo-acting enzymes include β-amylase, glucoamylase, and α-glucosidase. These enzymes attack the substrate from the non-reducing end, producing oligo and/or monosaccharides. Enzymes capable of hydrolyzing α-1,6-bonds in pullulan and amylopectin are defined as debranching enzymes. Cyclodextrin glycosyltransferases produce a series of non-reducing cyclic dextrins from starch, amylose and other polysaccharides (Niehaus et al. 1999); (Bertoldo and Antranikian 2002). Such enzymes can be derived from mesophilic enzymes by chemical modifications or mutagenesis, but natural thermophilic amylases exist. Alpha-amylases from *B. stearothermophilus*, *B. licheniformis*, *Thermus* spp. etc. are well characterized and widely used in starch processing. Thermophilic amylases are also found in thermophilic and hyperthermophilic archaea (Eichler 2001; Bertoldo and Antranikian 2002). Amylase from the hyperthermophile, *Pyrococcus woesei*, which is important in starch industry, has been purified and characterized and its amylase encoding gene was expressed in the mesophilic hosts *Bacillus subtilis* and *Escherichia coli* (Aguilar et al. 1998).

Lipases are a versatile group of enzymes and often express other activities like phospholipase, isophospholipase, cholesterol esterase, cutinase, amidase, and other esterase type of activities (Svendsen 2000). They have a number of unique characteristics, including substrate specificity, stereospecificity, regiospecificity, and ability to catalyze a heterogeneous reaction at the interface of water soluble and water insoluble systems (Sharma et al. 2001). The esters produced play an important role in the food industry as flavor and aroma constituents (Gandhi et al. 1995; Pandey et al. 1999). Whereas long chain methyl and ethyl esters of carboxylic acid moieties provide valuable oleo- chemical species that may function as fuel for diesel engines, esters of long chain carboxylic acid and alcohol moieties (waxes) have applications as lubricants and additives in cosmetic formulations (Linko et al. 1994). Other applications include the removal of the pitch from pulp produced in the paper industry, hydrolysis of milk fat in the dairy industry, removal of non-cellulosic impurities from raw cotton before further processing into dyed and finished products, drug formulations in the pharmaceutical industry and in the removal of subcutaneous fat in the leather industry (Pandey et al. 1999; Traore and Buschle-Diller 2000). A biodiesel was derived from vegetable oils using immobilized *Candida antarctica* lipase (Shimada et al. 1999). Lipases are of widespread occurrence throughout the earth's flora and fauna. More abundantly, however, they are found in bacteria, fungi and yeasts (Wu et al. 1996; Sharma et al. 2013).

Cellulose is the most abundant and renewable natural polymer on earth. Cellulose compounds are structurally heterogeneous and have both amorphous and highly ordered crystalline forms. Thus, they require a multitude of endo- and exoglucanases that must act synergistically to achieve the desired hydrolysis (Niehaus et al. 1999).

The major applications of cellulases are in textile industry for bio-polishing of fabrics and in house-hold laundry detergents for improving fabric softness and brightness. They are also used in animal feed, processing of fruit juice, in baking and deinking of recycled paper (Mawadza et al. 2000). Industrial ethanol production is based on corn-starch which is liquefied and saccharified. The oligosaccharide syrup is then used for ethanol fermentation. It has been described that the use of cellulases during starch saccharification and liquefaction increase the yield (Vieille and Zeikus 2001). Since these steps are performed at high temperatures, thermophilic endoglucanases are important for this process. In the current industrial processes, cellulolytic enzymes are employed in the color extraction of juices, in detergents causing color brightening and softening, in the biostoning of jeans, in the pretreatment of biomass that contains cellulose to improve nutritional quality of forage and in the pretreatment of industrial wastes (Buchert et al. 1997; Niehaus et al. 1999; Bhat 2000; Nakamura et al. 2001; Van et al. 2001; Zhou et al. 2001).

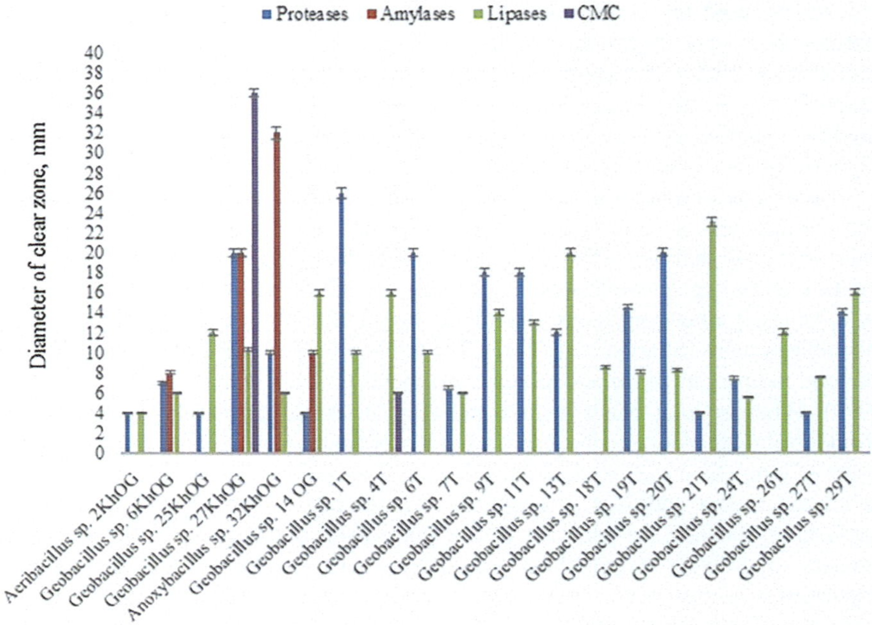

Fig. 3.3 Production of hydrolases (protease, amylase, lipase, and cellulose) by 21 thermophilic aerobic bacilli isolated from geothermal springs in Tajikistan. The ability of the strains to produce hydrolases was determined at 65 °C, i.e., at their optimal growth temperature

To find new sources of thermostable enzymes, bacterial strains isolated from Tajik geothermal springs were screened for amylolytic, proteolytic, lipolytic, and cellulolytic properties. To determine amylolytic strains a starch hydrolysis test on starch agar plates was used. The strains were streaked on the plates, followed by incubation at 65 °C, for 24 h. After incubation, 1% iodine solution (Lugol solution from Gram's staining) was flooded on the starch agar plate. A clear zone of hydrolysis on starch (after addition of iodine), around bacterial growth, is an indication of amylase production (Singh et al. 2016) (Fig. 3.3).

Protease activity was assessed on LB agar containing 3% skimmed milk. Plates were streaked with test strains followed by incubation at 37 °C for 24 h. The presence of a transparent zone around the colonies indicated caseinase activity (Burke et al. 1991) (Fig. 3.3).

Lipase activity was observed by the appearance of a turbid halo around the inoculate on Tryptic soy agar plates supplemented with 1% Tween 80 as explained by Rollof et al. (1987) (Fig. 3.3).

The screening of cellulase-producing bacteria was conducted in medium containing carboxymethylcellulose sodium salt (CMC) 1.0%, agar 1.5%, and pH 7.0. The isolates were transferred to Carboxymethylcellulose (CMC) agar plates and incubated for 24–48 h. For the detection of cellulase expression, 1 mL of grams

Iodine solution was added to the CMC plate after completion of growth for 30 min and excess iodine was removed. The strains forming clear halo zones on CMC media were selected as cellulase-producing strains (Fig. 3.3).

Among 21 aerobic, thermophilic strains isolated from Tamdykul, Khodja-Obi-Garm and Obigarm geothermal springs, 18 bacterial isolates produced one or more extracellular hydrolytic enzyme (protease, amylase, lipase or cellulose) (Fig. 3.3). Among the selected isolates, only one bacterial strain (27KhOG) was efficient producer of all four types of thermostable enzymes. There are many reports on the production of thermostable cellulase and xylanase from bacterial isolates from thermal springs/high temperature habitats (Gerasimova and Kuisiene 2012; Kumar et al. 2014; Pandey et al. 2013; Suman et al. 2015; Graham et al. 2011). In general, the cellulases are of special interest due to abundance of cellulose and hemicellulose in nature. In the present investigation, strains capable of producing thermostable proteases and amylases were identified. The thermophilic bacteria produced thermo-active enzymes, which is of great interest for both fundamental research and industrial applications. Different thermo-active enzyme production by microbes, in this study, may find applications in various industries (Pandey et al. 2015).

The physical-chemical analysis showed specific variations among the sediments and water samples collected from Tamdykul, Khodja-Obi-Garm and Obigarm hot springs (Table 3.1). The water temperature of Tamdykul and Khodja-Obi-Garm springs reached 88–93 °C, while in the Obigarm geothermal springs the maximum temperature was 50.5 °C. In total, two nutrient media were used for extraction of cultured bacteria from three geothermal hot springs. A total of 22 bacteria were isolated from Tamdykul, Khodja-Obi-Garm, and Obigarm hot springs. Similarly, recent reports on other natural extreme habitats provided thermophiles with a significant impact on biotechnology (Suman et al. 2015; Huang et al. 2011; Khiyami et al. 2012; Sahoo et al., 2015; Sen and Maiti 2014; Stefanova et al. 2016; Yadav et al. 2015, Sharma et al. 2009; Arya et al. 2015; Graham et al. 2011).

3.5 Conclusions and Future Perspectives

This study demonstrated that diverse and novel thermophilic bacterial populations with biotechnological potentials thrive in high-altitude geothermal springs in Tajikistan. Most of the microbial isolates are still mostly uncharacterized and awaits further exploration. Environmental factors play an important role in structuring microbial communities, and hence these factors should be considered in future analyses.

Overall, this study contributes to the global knowledge on bacterial communities by comprehensively profiling culture-based bacterial diversity in the hot springs of Tajikistan. Further studies are required for investigating these isolates by genome sequence analysis.

In conclusion, heat-adapted bacteria have attracted the attention of the scientific community due to their ability to produce thermostable extracellular hydrolytic enzymes. Microbial extracellular enzymes with optimal activity at high temperature provide opportunities to study the adaptation of life in thermal/high temperatures habitats and the potential for biotechnological exploitation. Diverse thermo-active enzymes detected in this study may be used to serve various industrial, agricultural and medicinal purposes.

References

Abdusamadzoda D, Panosyan H, Birkeland N-K, Bobodzhanova KI (2016) Anaerobic Thermophiles from geothermal springs Khodja-Obi-Garm, Tajikistan. Materials of the Republican scientific and theoretical conference of the teaching staff and students of TNU, dedicated to the 25th anniversary of the State Independence of the Republic of Tajikistan. Dushanbe, pp 587–588a

Adams MW, Kelly RM (1998) Finding and using hyperthermophilic enzymes. Trends Biotechnol 16(8):329–332

Aguilar A, Ingemansson T, Magnien E (1998) Extremophile microorganisms as cell factories: support from the European union. Extremophiles 2(3):367–373

Ahmad S, Scopes RK, Rees GN, Patel BKC (2000) *Saccharococcus caldoxylosilyticus* sp nov., an obligately thermophilic, xylose-utilizing, endospore-forming bacterium. Int J Syst Evol Microbiol 50:517–523

Aliyu H, Lebre P, Blom J, Cowan D, De Maayer P (2016) Phylogenomic reassessment of the thermophilic genus *Geobacillus*. Syst Appl Microbiol 39:527–533

Amanuel MG, Nancy LMB (2018) Thermotolerant bacteria of biotechnological potential from hot springs in Eritrea. Afr J Microbiol Res 12:512–524

Amend JP, Shock EL (2001) Energetics of overall metabolic reactions of thermophilic and hyperthermophilic archaea and bacteria. FEMS Microbiol Rev 25(2):175–243

Arya M, Joshi GK, Gupta AK, Kumar A, Raturi A (2015) Isolation and characterization of thermophilic bacterial strains from Soldhar (Tapovan) hot spring in Central Himalayan region, India. Ann Microbiol 65(3):1457–1464

Ash C, Farrow JAE, Wallbanks S, Collins MD (1991) Phylogenetic heterogeneity of the genus *Bacillus* revealed by comparative-analysis of small-subunit-ribosomal RNA sequences. Lett Appl Microbiol 13(4):202–206

Balk M, Weijma J, Stams AJM (2002) *Thermotoga lettingae* sp nov., a novel thermophilic, methanol-degrading bacterium isolated from a thermophilic anaerobic reactor. Int J Syst Evol Microbiol 52:1361–1368

Banat IM, Marchant R, Rahman TJ (2004) *Geobacillus debilis* sp. nov., a novel obligately thermophilic bacterium isolated from a cool soil environment, and reassignment of Bacillus pallidus to *Geobacillus pallidus* comb. nov. Int J Syst Evol Microbiol 54:2197–2201

Belduz AO, Canakci S, Chan K-G, Kahar UM, Chan CS, Yaakop AS, Goh KM (2015) Genome sequence of *Anoxybacillus ayderensis* AB04T isolated from the Ayder hot spring in Turkey. Stand Genomic Sci 10:70. https://doi.org/10.1186/s40793-015-0065-2

Ben Hania W, Postec A, Aullo T, Ranchou-Peyruse A, Erauso G, Brochier-Armanet C, Hamdi M, Ollivier B, Saint-Laurent S, Magot M, Fardeau ML (2013) *Mesotoga infera* sp nov., a mesophilic member of the order *Thermotogales*, isolated from an underground gas storage aquifer. Int J Syst Evol Microbiol 63:3003–3008

Berelson WM, Corsetti FA, Pepe-Ranney C, Hammond DE, Beaumont W, Spear JR (2011) Hot spring siliceous stromatolites from Yellowstone National Park: assessing growth rate and laminae formation. Geobiology 9(5):411–424

Beris FS, De Smet L, Karaoglu H, Canakci S, Van Beeumen J, Belduz AO (2011) The atpase activity of the g2alt gene encoding an aluminium tolerance protein from *Anoxybacillus gonensis* G2. J Microbiol 49(4):641–650
Bertoldo C, Antranikian G (2002) Starch-hydrolyzing enzymes from thermophilic Archaea and Bacteria. Curr Opin Chem Biol 6(2):151–160
Bhalla A, Kainth AS, Sani RK (2013) Draft genome sequence of lignocellulose-degrading thermophilic bacterium *Geobacillus* sp. Strain wsucf1. Genome Announc 1:4
Bhat MK (2000) Cellulases and related enzymes in biotechnology. Biotechnol Adv 18(5):355–383
Bryanskaya AV, Rozanov AS, Logacheva MD, Kotenko AV, Peltek SE (2014) Draft genome sequence of *Geobacillus icigianus* strain g1w1t isolated from hot springs in the valley of Geysers, Kamchatka (Russian Federation). Genome Announc 2:5
Buchert J, Pere J, Oijusluoma L, Rahkamo L, Viikari L (1997) Cellulases–tools for modification of cellulosic materials. Int J Life Sci 6:284–290
Burke V, Robinson JO, Richardson CJ, Bundell CS (1991) Longitudinal studies of virulence factors of *Pseudomonas aeruginosa* in cystic fibrosis. Pathology 23(2):145–148
Canakci S, Kacagan M, Inan K, Belduz AO, Saha BC (2008) Cloning, purification, and characterization of a thermostable alpha-l-arabinofuranosidase from *Anoxybacillus kestanbolensis* ac26sari. Appl Microbiol Biotechnol 81(1):61–68
Chai YY, Rahman RN, Illias RM, Goh KM (2012) Cloning and characterization of two new thermostable and alkalitolerant alpha-amylases from the *Anoxybacillus* species that produce high levels of maltose. J Ind Microbiol Biotechnol 39(5):731–741
Chen J, Li Y, Hao HH, Zheng J, Chen JM (2015) Fe(ii)edta-no reduction by a newly isolated thermophilic *Anoxybacillus* sp. Ha from a rotating drum biofilter for nox removal. J Microbiol Methods 109:129–133
Chu WH (2007) Optimization of extracellular alkaline protease production from species of *Bacillus*. J Ind Microbiol Biotechnol 34(3):241–245
Cihan AC, Cokmus C, Koc M, Ozcan B (2014) *Anoxybacillus calidus* sp. Nov., a thermophilic bacterium isolated from soil near a thermal power plant. Int J Syst Evol Microbiol 64 (Pt1):211–219
Cohn F (1872) Untersuchungen über bakterien. Beiträge Biol Pflanzen 1:127–224
Cole JK, Peacock JP, Dodsworth JA, Williams AJ, Thompson DB, Dong HL, Wu G, Hedlund BP (2013) Sediment microbial communities in great boiling spring are controlled by temperature and distinct from water communities. ISME J 7(4):718–729
Dabananda SN, Kshetri P (2010) A thermostable alkaline protease from a moderately halo-alkalithermotolerant *Bacillus subtilis* strain SH1. Aust J Basic Appl Sci 10:5126–5134
Demirjian DC, Moris-Varas F, Cassidy CS (2001) Enzymes from extremophiles. Curr Opin Chem Biol 5(2):144–151
Donk PJ (1920) A highly resistant thermophilic organism. J Bacteriol 5(4):373–374
Eichler J (2001) Biotechnological uses of archaeal extremozymes. Biotechnol Adv 19(4):261–278
Erikstad HA, Ceballos RM, Smestad NB, Birkeland NK (2019) Global biogeographic distribution patterns of thermoacidophilic *Verrucomicrobia methanotrophs* suggest allopatric evolution. Front Microbiol 10:1129
Finore I, Gioiello A, Leone L, Orlando P, Romano I, Nicolaus B, Poli A (2017) *Aeribacillus composti* sp nov., a thermophilic *Bacillus* isolated from olive mill pomace compost. Int J Syst Evol Microbiol 67(11):4830–4835
Fischer A (1895) Untersuchungen über bakterien. Jahrbücher Wiss Bot 27:1–163
Fujiwara N, Yamamoto K, Masui A (1991) Utilization of a thermostable alkaline protease from an alkalophilic thermophile for the recovery of silver from used X-ray-film. J Ferment Bioeng 72 (4):306–308
Gandhi N, Sawant S, Joshi J (1995) Studies on the lipozymecatalyzed synthesis of butyl laurate. Biotechnol Bioeng 46:1–12
Gerasimova J, Kuisiene N (2012) Characterization of the novel xylanase from the thermophilic *Geobacillus thermodenitrificans* JK1. Microbiology 81:418–424

Gibbons NE, Murray RGE (1978) Proposals concerning the higher taxa of bacteria. Int J Syst Bacteriol 28:1–6

Goh KM, Kahar UM, Chai YY, Chong CS, Chai KP, Ranjani V et al (2013) Recent discoveries and applications of *Anoxybacillus*. Appl Microbiol Biotechnol 97:1475–1488

Graham JE, Clark ME, Nadler DC, Huffer S, Chokhawala HA, Rowland SE, Blanch HW, Clark DS, Robb FT (2011) Identification and characterization of a multidomain hyperthermophilic cellulase from an archaeal enrichment. Nat Commun 2:375

Gupta R, Beg QK, Khan S, Chauhan B (2002) An overview on fermentation, downstream processing and properties of microbial alkaline proteases. Appl Microbiol Biotechnol 60:381–395

Herbert RA, Sharp RJ (1992) Molecular biology and biotechnology of extremophiles. Chapman and Hall, New York

Hou W, Wang S, Dong H, Jiang H, Briggs BR, Peacock JP, Huang Q, Huang L, Wu G, Zhi X, Li W, Dodsworth JA, Hedlund BP, Zhang C, Hartnett HE, Dijkstra P, Hungate BA (2013) A comprehensive census of microbial diversity in hot springs of Tengchong, Yunnan province China using 16S rRNA gene pyrosequencing. PLoS One 8(1):e53350

Huber R, Langworthy TA, Konig H, Thomm M, Woese CR, Sleytr UB, Stetter KO (1986) *Thermotoga maritima* sp-nov represents a new genus of unique extremely thermophilic Eubacteria growing up to 90-degrees-c. Arch Microbiol 144(4):324–333

Inskeep WP, McDermott TR, Thermal Biology Institute (2005) Geothermal biology and geochemistry in Yellowstone National Park. In: Proceeding of the Thermal biology Institute Workshop, 1st edn. Montana State University Publications, Bozeman, MT

Jiang Z, Li P, Jiang DW, Dai XY, Zhang R, Wang YH, Wang YX (2016) Microbial community structure and arsenic biogeochemistry in an acid vapor-formed spring in Tengchong geothermal area, China. PLoS One 11(1):e0146331

Khiyami MA, Serour EA, Shehata MM, Bahklia AH (2012) *Thermoaerobic* bacteria from geothermal springs in Saudi Arabia. Afr J Biotechnol 11:4053–4062

Kumar CG, Tiwari MP, Jany KD (1999) Novel alkaline serine proteases from alkalophilic *Bacillus* spp.: purification and some properties. Process Biochem 34(5):441–449

Kumar B, Trivedi P, Mishra AK, Pandey A, Palni LMS (2004a) Microbial diversity of soil from two hot springs in Uttaranchal Himalaya. Microbiol Res 159:141–146

Kumar B, Trivedi P, Mishra AK, Pandey A, Palni LMS (2004b) Microbial diversity of soil from two hot springs in Uttaranchal Himalaya. Microbiol Res 159:141–146

Kumar M, Yadav AN, Tiwari R, Prasanna R, Saxena AK (2014) Deciphering the diversity of culturable thermotolerant bacteria from Manikaran hot springs. Ann Microbiol 64:741–751

Kumar S, Stecher G, Li M, Knyaz C, Tamura K (2018) Mega X: molecular evolutionary genetics analysis across computing platforms. Mol Biol Evol 35(6):1547–1549

Larkin MA, Blackshields G, Brown NP, Chenna R, McGettigan PA, McWilliam H, Valentin F, Wallace IM, Wilm A, Lopez R, Thompson JD, Gibson TJ, Higgins DG (2007) Clustal W and Clustal X version 2.0. Bioinformatics 23(21):2947–2948

Lau MCY, Aitchison JC, Pointing SB (2009) Bacterial community composition in thermophilic microbial mats from five hot springs in Central Tibet. Extremophiles 13(1):139–149

Linko Y, Lamsa M, Huhatala A, Linko P (1994) Lipase catalyzed transesterification of rapeseed oil snf 2-ethyl-1-hexanol. JAOCS 71:1411–1414

Liu L, Salam N, Jiao JY, Jiang HC, Zhou EM, Yin YR, Ming H, Li WJ (2016) Diversity of culturable thermophilic actinobacteria in hot springs in Tengchong, China and studies of their biosynthetic gene profiles. Microb Ecol 72(1):150–162

Ludwig W, Schleifer K-H, Whitman WB et al (2009) Class I. Bacilli class nov. In: De Vos P, Garrity G, Jones D, Krieg NR, Ludwig W, Rainey FA (eds) Bergey's manual of systematic bacteriology, vol 3, 2nd edn. Springer, New York, pp 19–20

Margaryan A, Shahinyan G, Hovhannisyan P, Panosyan H, Birkeland N-K, Trchounian A (2018) *Geobacillus* and *Anoxybacillus* spp. from terrestrial geothermal springs worldwide: diversity and biotechnological applications. In: Microorganisms for sustainability: Extremophiles in Eurasian ecosystems: Ecology, diversity, and applications. Springer, Singapore, pp 119–166

Mawadza C, Hatti-Kaul R, Zvauya R, Mattiasson B (2000) Purification and characterization of cellulases produced by two *Bacillus* strains. J Biotechnol 83(3):177–187

Miñana-Galbis D, Pinzón DL, Loren JG, Manresa A, Oliart-Ros RM (2010) Reclassification of *Geobacillus pallidus* (Scholz et al. 1988) Banat et al. (2004) as *Aeribacillus pallidus* gen. nov., comb. nov. Int J Syst Evol Microbiol 60:1600–1604

Mirete S, Morgante V, Gonzalez-Pastor JE (2016) Functional metagenomics of extreme environments. Curr Opin Biotechnol 38:143–149

Mori K, Yamazoe A, Hosoyama A, Ohji S, Fujita N, Ishibashi JI, Kimura H, Suzuki KI (2014) *Thermotoga profunda* sp. Nov. And *Thermotoga caldifontis* sp. nov., anaerobic thermophilic bacteria isolated from terrestrial hot springs. Int J Syst Evol Microbiol 64(6):2128–2136

Nakamura H, Kubota H, Kono T, Isogai A and Onabe F (2001) Modification of pulp properties by cellulase treatment and application of cellulase to wastepaper deinking and mechanical pulp refining. In: 68th pulp and paper research conference, 18–19/06, Japan, pp 2–5

Nazina TN, Tourova TP, Poltaraus AB, Novikova EV, Grigoryan AA, Ivanova AE, Lysenko AM, Petrunyaka VV, Osipov GA, Belyaev SS, Ivanov MV (2001) Taxonomic study of aerobic thermophilic bacilli: Descriptions of *Geobacillus subterraneus* gen. Nov., sp. Nov. and *Geobacillus uzenensis* sp. Nov. From petroleum reservoirs and transfer of *Bacillus stearothermophilus*, *Bacillus thermocatenulatus*, *Bacillus thermoleovorans*, *Bacillus kaustophilus*, *Bacillus thermodenitrificans to Geobacillus* as the new combinations G. *stearothermophilus*, g. Th. Int J Syst Evol Microbiol 51:433–446

Nazina TN, Lebedeva EV, Poltaraus AB, Tourova TP, Grigoryan AA, Sokolova DS, Lysenko AM, Osipov GA (2004) *Geobacillus gargensis* sp nov., a novel thermophile from a hot spring, and the reclassification of *Bacillus vulcani* as *Geobacillus vulcani* comb. Nov. Int J Syst Evol Microbiol 54:2019–2024

Niehaus F, Bertoldo C, Kahler M, Antranikian G (1999) Extremophiles as a source of novel enzymes for industrial application. Appl Microbiol Biotechnol 51(6):711–729

Ninghoujam DS, Kshetri P (2010) A thermostable alkaline protease from a moderately haloalkalithermotolerant *Bacillus subtilis* strain SH1. Australian J Basic Appl Sci 4:5126–5134

Ningthoujam DS, Kshetri P, Sanasam S, Nimaichand S (2009) Screening, identification of best producers and optimization of extracellular proteases from moderately halophilic alkalithermotolerant indigenous *Actinomycetes*. World Appl Sci J 7:907–916

Normatov I (2010) Geothermal water resources of the Republic of Tajikistan and a perspective on their use. In: Proceedings World Geothermal Congress 2010 Bali, Indonesia, pp 25–29

Pandey A, Benjamin S, Soccol CR, Nigam P, Krieger N, Soccol VT (1999) The realm of microbial lipases in biotechnology. Biotechnol Appl Biochem 29(2):119–131

Pandey S, Singh S, Yadav AN, Nain L, Saxena AK (2013) Phylogenetic diversity and characterization of novel and efficient cellulase producing bacterial isolates from various extreme environments. Biosci Biotechnol Biochem 77:1474–1480

Pandey A, Dhakar K, Sharma A, Priti P, Sati P, Kumar B (2015) Thermophilic bacteria that tolerate a wide temperature and pH range colonize the Soldhar (95 a degrees C) and ringigad (80 a degrees C) hot springs of Uttarakhand, India. Ann Microbiol 65(2):809–816

Parte AC (2018) LPSN - list of prokaryotic names with standing in nomenclature (bacterio.Net), 20 years on. Int J Syst Evol Microbiol 68(6):1825–1829

Podar M, Reysenbach AL (2006) New opportunities revealed by biotechnological explorations of extremophiles. Curr Opin Biotechnol 17(3):250–255

Postollec F, Mathot AG, Bernard M, Divanac'h ML, Pavan S, Sohier D (2012) Tracking spore-forming bacteria in food: from natural biodiversity to selection by processes. Int J Food Microbiol 158(1):1–8

Prévot AR, Hauderoy P, Ehringer G, Guillot G, Magrou J, Prévot AR, Rosset D et al (1953) Dictionnaire des bactéries pathogènes, 2nd edn. Masson et Cie, Paris, pp 1–692

Priya I, Dhar M, Bajaj B, Koul S, Vakhlu J (2016) Cellulolytic activity of thermophilic *Bacilli* isolated from Tattapani hot spring sediment in North West Himalayas. Indian J Microbiol 56:228–231

Rai SK, Roy JK, Mukherjee AK (2010) Characterization of a detergentstable alkaline protease from a novel thermophilic strain *Paenibacillus tezpurensis* sp. Nov. AS-S24-II. Appl Micrbiol Biotechnol 85:1437–1450

Razykov BH (2007) Features of placement and economic assessment of mineral waters of Tajikistan. Dushanbe 99:15–18

Rogers EE, Backus EA (2014) Anterior foregut microbiota of the glassy-winged sharpshooter explored using deep 16S rRNA gene sequencing from individual insects. PLoS ONE 9:e106215

Rollof J, Hedstrom SA, Nilsson-Ehle P (1987) Lipolytic activity of *Staphylococcus aureus* strains from disseminated and localized infections. Acta Pathol Microbiol Immunol Scand B 95 (2):109–113

Salmond GPC, Whittenbury R (1985) Biology of microorganisms. Nature 314(6006):49–49

Schallmey M, Singh A, Ward OP (2004) Developments in the use of *Bacillus* species for industrial production. Can J Microbiol 50(1):1–17

Sen R, Maiti N (2014) Genomic and functional diversity of bacteria isolated from hot springs in Odisha, India. Geomicrobiol J 31:541–550

Sharma R, Chisti Y, Banerjee UC (2001) Production, purification, characterization, and applications of lipases. Biotechnol Adv 19(8):627–662

Sharma A, Pandey A, Shouche YS, Kumar B, Kulkarni G (2009) Characterization and identification of *Geobacillus* spp. isolated from Soldhar hot spring site of Garhwal Himalaya. J Basic Microb 49(2):187–194

Sharma R, Thakur V, Sharma M, Birkeland N-K (2013) Biocatalysis through thermostable lipases: adding flavor to chemistry. Biotechnology. https://doi.org/10.1007/978-94-007-5899-5

Sharma P, Gupta S, Sourirajan A, Baumler DJ, Dev K (2019) Draft genome sequence of hyperthermophilic, halotolerant *Parageobacillus toebii* PW12, isolated from the Tattapani hot spring, Northwest Himalayas. Microbiol Resour Announc 8:4

Shimada Y, Watanabe Y, Samukawa T, Sugihara A, Noda H, Fukuda H, Tominaga Y (1999) Conversion of vegetable oil to biodiesel using immobilized *Candida antarctica* lipase. J Am Oil Chem Soc 76(7):789–793

Shirkot P, Verma A (2015) Assessment of thermophilic bacterial diversity of thermal springs of Himachal Pradesh. ENVIS Bull Himal Ecol 23:27–34

Singh NR, Bahuguna A, Chauhan P, Sharma VK, Kaur S, Singh SK, Khan A (2016) Production, purification and characterization of thermostable α-amylase from soil isolate *Bacillus* sp. strain B-10. Biosci Biotechnol 5(1):37–43

Skerman VBD, McGowan V, Sneath PHA (1980) Approved lists of bacterial names. Int J Syst Bacteriol 30:225–420

Song ZQ, Wang FP, Zhi XY, Chen JQ, Zhou EM, Liang F, Xiao X, Tang SK, Jiang HC, Zhang CL, Dong H, Li WJ (2013) Bacterial and archaeal diversities in Yunnan and Tibetan hot springs, China. Environ Microbiol 15(4):1160–1175

Stefanova K, Tomova I, Tomova A, Radchenkova N, Atanassov I, Kambourova M (2016) Archaeal and bacterial diversity in two hot springs from geothermal regions in Bulgaria as demonstrated by 16S rRNA and GH-57 genes. Int Microbiol 18:217–223

Stetter KO (1999) Extremophiles and their adaptation to hot environments. FEBS Lett 452 (1-2):22–25

Studholme DJ (2015) Some (*Bacilli*) like it hot: Genomics of *Geobacillus* species. Microb Biotechnol 8(1):40–48

Suman A, Verma P, Yadav AN, Saxena AK (2015) Bioprospecting for extracellular hydrolytic enzymes from culturable thermotolerant bacteria isolated from Manikaran thermal springs. Res J Biotechnol 10:33–42

Suzuki Y, Kishigami T, Inoue K, Mizoguchi Y, Eto N, Takagi M, Abe S (1983) *Bacillus thermoglucosidasius* sp-nov, a new species of obligately thermophilic *Bacilli*. Syst Appl Microbiol 4(4):487–495

Svendsen A (2000) Lipase protein engineering. BBA-Protein Struct 1543(2):223–238

Tambekar DH, Kalikar MV, Shinde RS, Vanjari LB, Pawar RG (2009) Isolation and characterization of multilple enzyme producer *Bacillus* species from saline belt of Purna river. J Appl Sci Res 5:1064–1066

Tang J, Liang YM, Jiang D, Li LH, Luo YF, Shah MR, Daroch M (2018) Temperature-controlled thermophilic bacterial communities in hot springs of Western Sichuan, China. BMC Microbiol 18:134

Traore MK, Buschle-Diller G (2000) Environmentally friendly scouring processes. Text Chem Color Am 32(12):40–43

Turner P, Labes A, Fridjónsson ÓH, Hreggvidson GO, Schönheit P, Kristjánsson JK et al (2005) Two novel cyclodextrin-degrading enzymes isolated from thermophilic bacteria have similar domain structures but differ in oligomeric state and activity profile. J Biosci Bioeng 100:380–390

Validation List no. 132 (2010) List of new names and new combinations previously effectively, but not validly, published. Int J Syst Evol Microbiol 60:469–472

Van WJ, Mogale A, Seseng T (2001) Bioconversion of wastepaper to sugars by cellulase from *Aspergillus niger*, *Trichoderma viride* and *Penicillium funiculosum*. J Solid Waste Technol Manage 27:82–86

Vieille C, Zeikus GJ (2001) Hyperthermophilic enzymes: sources, uses, and molecular mechanisms for thermostability. Microbiol Mol Biol R 65(1):1

Wang J, Bai Y, Yang P, Shi P, Luo H, Meng K et al (2010) A new xylanase from thermoalkaline *Anoxybacillus* sp. E2 with high activity and stability over a broad pH range. World J Microbiol Biotechnol 26:917–924

Wang S, Hou W, Dong H, Jiang H, Huang L (2014) Control of temperature on microbial community structure in hot springs of the Tibetan Plateau (vol 8, e62901, 2013). PLoS One 9 (6):e62901

Whitaker RJ, Grogan DW, Taylor JW (2003) Geographic barriers isolate endemic populations of hyperthermophilic archaea. Science 301(5635):976–978

Wiegand S, Rabausch U, Chow J, Daniel R, Streit WR, Liesegang H (2013) Complete genome sequence of *Geobacillus* sp. Strain ghh01, a thermophilic lipase-secreting bacterium. Genome Announc 1(2):e0009213

Wilson P, Remigio Z (2012) Production and characterisation of protease enzyme produced by a novel moderate thermophilic bacterium (ep1001) isolated from an alkaline hot spring, Zimbabwe. Afr J Microbiol Res 6(27):5542–5551

Wu XY, Jaaskelainen S, Linko YY (1996) An investigation of crude lipases for hydrolysis, esterification, and transesterification. Enzym Microb Technol 19(3):226–231

Xian W-D, Rao MPN, Zhou E-M, Liu L, Xiao M, Li W-J (2018) Diversity of thermophiles in terrestrial hot springs of Yunnan and Tibet, China worldwide: diversity and biotechnological applications. In: Microorganisms for sustainability: extremophiles in Eurasian ecosystems: Ecology, diversity, and applications. Springer, Singapore, pp 57–79

Yadav AN (2015) Bacterial diversity of cold deserts and mining of genes for low temperature tolerance. PhD thesis, BIT Mesra/IARI, New Delhi, pp 234

Yadav AN, Sachan SG, Verma P, Saxena AK (2015) Prospecting cold deserts of North Western Himalayas for microbial diversity and plant growth promoting attributes. J Biosci Bioeng 119:683–693

Yao N, Ren Y, Wang W (2013) Genome sequence of a thermophilic *Bacillus*, *Geobacillus thermodenitrificans* dsm465. Genome Announc 1:6

Zaigham NA, Nayyar ZA, Hisamuddin N (2009) Review of geothermal energy resources in Pakistan. Renew Sust Energ Rev 13(1):223–232

Zarilla KA, Perry JJ (1987) *Bacillus thermoleovorans*, sp-nov, a species of obligately thermophilic hydrocarbon utilizing endospore-forming bacteria. Syst Appl Microbiol 9(3):258–264

Zhou L, Yeung K, Yuen C (2001) Combined cellulase and wrinkle free treatment on cotton fabric. J Dong Hua Univ 18:11–15

Chapter 4
Study of Bacterial Diversity from Saline Environments (Salt Mines) of Pakistan and their Applications at Regional Level

Inam Ullah Khan, Muhammad Saqib, Neeli Habib, Min Xiao, Shakeeb Ullah, Shah Irum, Iftikhar Ahmed, and Wen-Jun Li

Abstract Pakistan has diverse extreme environments such as salt mines in the Khyber Pakhtunkhwa (Karak/Bahadur khel salt mines) and in the Punjab (Khewra salt mines), which provide diverse habitats for halophilic microorganisms have been studied for microbial communities by using culture-dependent and culture-independent techniques. Most of the isolated bacterial strains by laboratory culture method were belonging to the genus *Bacillus*. Two novel bacterial species were isolated and characterized by polyphasic taxonomy approach. Some of the isolated

---

I. U. Khan
State Key Laboratory of Biocontrol and Guangdong Key Laboratory of Plant Resources, School of Life Sciences, Sun Yat-Sen University, Guangzhou, People's Republic of China

Faculty of Veterinary and Animal Sciences, Gomal University, Dera Ismail Khan, Khyber Pakhtunkhwa, Pakistan

M. Saqib
Government Post Graduate College No.1 Bannu, Bannu, Khyber Pakhtunkhwa, Pakistan

N. Habib
Department of Microbiology, Shaheed Benazir Bhutto Women University, Peshawar, Khyber Pakhtunkhwa, Pakistan

M. Xiao · W.-J. Li (✉)
State Key Laboratory of Biocontrol and Guangdong Key Laboratory of Plant Resources, School of Life Sciences, Sun Yat-Sen University, Guangzhou, People's Republic of China
e-mail: liwenjun3@mail.sysu.edu.cn

S. Ullah
Faculty of Veterinary and Animal Sciences, Gomal University, Dera Ismail Khan, Khyber Pakhtunkhwa, Pakistan

S. Irum
Institute of Chemical Sciences, Gomal University, Dera Ismail Khan, Khyber Pakhtunkhwa, Pakistan

I. Ahmed
National Culture Collection of Pakistan (NCCP), Bioresources Conservation Institute (BCI), National Agricultural Research Centre (NARC), Islamabad, Pakistan

D. Egamberdieva et al. (eds.), *Microbial Communities and their Interactions in the Extreme Environment*, Microorganisms for Sustainability 32,
https://doi.org/10.1007/978-981-16-3731-5_4

bacterial strains were found to possess potential applications in industry, medicine, and agriculture fields.

Keywords Salt mines · Karak/Bahadur khel · Khewra · Culture-dependent and culture-independent techniques · Potential applications

4.1 Introduction

Extreme environments are those habitats in which organism's diversity is extremely restrained by different physicochemical factors (Ventosa 2006). For inhabitants of hypersaline environments, the main challenge and limiting factor is the high salt concentration. The other factors which increase the harshness of the hypersaline environments include: pH extremeness, hypoxic condition, and scarcity of nutrients, pressure, high temperatures and the presence of harmful compounds. Marine salterns, deep-sea brine pools, saline lakes, saline soils and sediments and salted foods are included in such environments (Ventosa et al. 2015).

Halophiles are belonging to all three main domains of life including Archaea, Bacteria, and Eucarya. They have been classified based on their salt (NaCl) requirements such as slight, moderate and extreme halophiles. Slight halophiles optimally grow at 1–5% of NaCl while moderate and extreme halophiles can optimally grow at 5–20%, and 20–30% of NaCl, respectively. Halotolerant organisms can be defined as the organisms which can survive in the presence or absence of high concentrations. On the other hand the non-halophilic organisms require 1% NaCl concentrations for their optimal growth (Singh et al. 2019). Microbial life has adapted to halophilic environments by applying various strategies (Galinski 1993, 1995; Oren 1999, 2000, 2008).

Halophiles have got great attention due to have many applications, such as in field of biotechnology used for producing extremozymes, carotene production, stabilizers, exopolysaccharides, and other valuable compounds (Quesada et al. 1982, 2004; Ventosa et al. 2008). The halophilic microorganisms particularly bacteria have many applications in the fields of medicine and agriculture (Amoozegar et al. 2003; Chakraborty et al. 2009; Ai et al. 2018; Ali et al. 2015; Shafiei et al. 2012; Zhang et al. 2012).

Pakistan possesses saline environments in the form of sea, rock and lake which are distributed throughout Pakistan. The deposition of Pakistan salt range on the Asian subcontinent represents the most ancient one. It shows a series of rock strata formations from the earlier pre-Cambrian era to the present geological periods (Cremo 2001; Jehangiri et al. 2015; Leena et al. 2018).

Pakistan saline environments have been investigated in order to probe into their microbial diversity particularly bacterial diversity by culture-dependent and culture-independent methods and also the industrial, agricultural and pharmaceutical values of their bacterial isolates have been explored (Akhtar et al. 2008; Roohi et al. 2012,

2014a, b; Jamil et al. 2013; Aftab et al. 2015; Bangash et al. 2015; Ali et al. 2016; Shah et al. 2017, 2018; Leena et al. 2018; Mukhtar et al. 2018a, b; Cycil et al. 2020).

The below sections highlight the phylogenetic analysis of halophiles, strategies of halophiles for adaptation to high salt environments, geographical distribution of salt mines in Pakistan, bacterial diversity of halophilic environments in Pakistan by culture-dependent and culture-independent approaches and potential applications of bacterial strains isolated from saline habitats of Pakistan.

4.2 Phylogenetic Analysis of Halophiles

Halophilic and halotolerant organisms are belonging to the three domains of life: Archaea, Bacteria, and Eucarya. The organisms which grow in the saline habitat having >100 g/l salinity have been identified in the phylogenetic analysis (tree of life) (Oren 2002). Among the halophiles, the halophilic aerobic Archaea of the family *Halobacteriaceae* are well known biomass in the hyper saline environments such as Dead Sea, saltern crystallizer ponds, and Lake Magadi. The only Eucaryotic microorganism, the green alga *Dunaliella* is among the important halophiles and found in the habitats like Dead Sea and lakes and ponds of high salt concentrations, otherwise very few members of the domain Eucarya are found to be halophiles.

High number of members of halophilic and halotolerant microorganisms are belonging to the domain Bacteria (Ventosa et al. 1998). Proteobacteria has different branches of their halophilic representatives and their close relatives are non-halophilic. The prokaryotes phylum Cyanophyta has halophiles (Oren 2000) and also halophiles are found in actinomycetes, *Flavobacterium–Cytophaga* branch and the spirochetes.

The order *Halanaerobiales* possessing two families (*Halobacteroidaceae* and *Halanaerobiaceae*) found to consists of anaerobic halophilic microorganisms (Oren 2001; Rainey et al. 1995) (Fig. 4.1).

4.3 Strategies of Halophiles for Adaptation to High Salt Environments

As we mentioned before, the big challenge for halophiles is high salt concentration. In order to tackle with this, the halophiles accumulate different types of solutes within their cytoplasm (Galinski 1993). Halophilic Archaea osmoregulate by pushing Na^+ ions (4M) from internal environment to outside of the cell, while taking K^+ ions (5M) from external environment and accumulate within the cell cytoplasm. Halotolerant algae and bacteria keep balance the osmotic pressure by accumulating organic molecule (glycerol) in the interior of their cells (Litchfield 1998; Oren 2008).

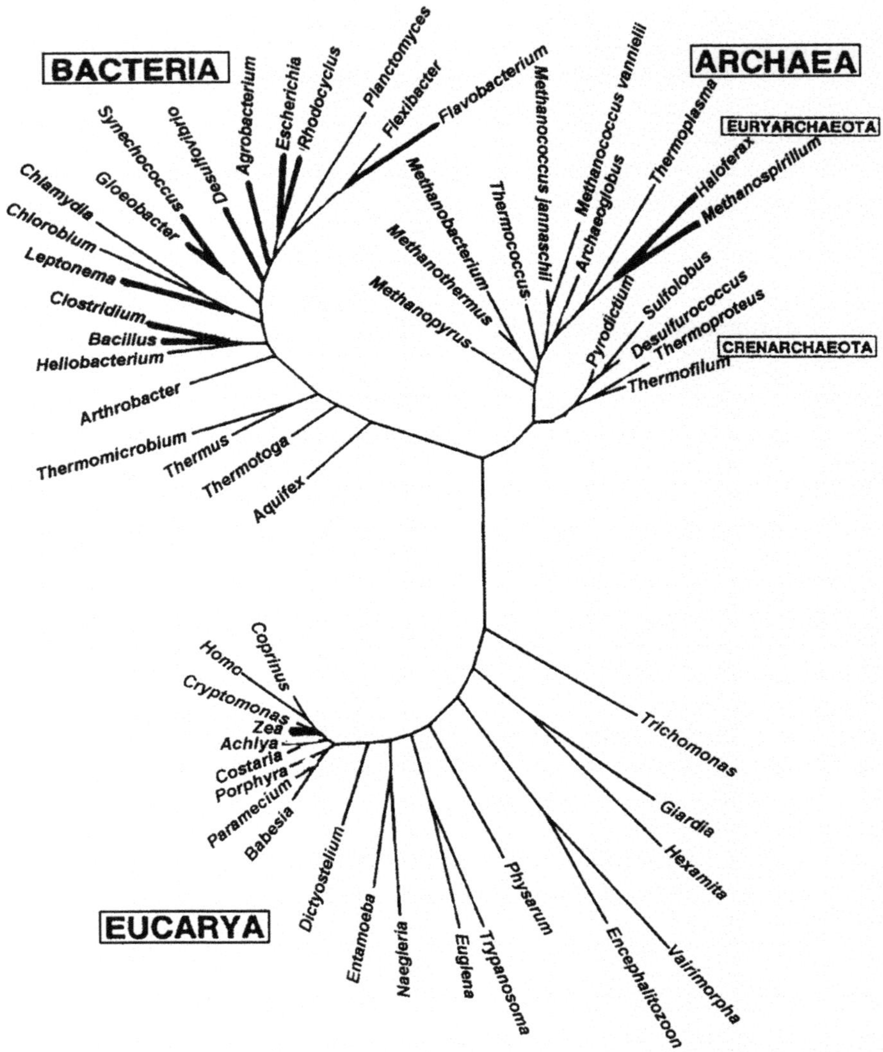

Fig. 4.1 The phylogenetic tree of life, analyzed based on small subunit rRNA gene sequences. Those microorganisms that grow well in the environment having >100 g/l salinity are shown in bold lines (Oren 2002)

Another strategy to cope with high salt concentration Archaea, Bacteria, and Eucarya concentrate compounds such as glycerol, amino acids, betaine, ectoine and glycine, 5-hydroxy derivative, and sugars such as trehalose and sucrose (Galinski 1995).

4.4 Geographical Distribution of Salt Mines in Pakistan

In Pakistan, the Salt mines are geographically distributed in different areas. Karak/Bahadur salt mines are located in the Khyber Pakhtunkhwa (KPK) province, while Khewra, Kalabagh, and Warcha salt mines are distributed in Punjab province (Fig. 4.2). Karak/Bahadur salt mines are rich source of salt reservoirs. A geological survey conducted in Pakistan reported that these salt mines are ancient as the early Eocene period (Roohi et al. 2014a; Khattak et al. 2016). So from these salt mines we can study the early microbial life of Eocene period (Fig. 4.1).

4.5 Bacterial Diversity of Halophilic Environments in Pakistan by Culture-Dependent and Culture-Independent Methods

4.5.1 Methodology

Microbial diversity of Pakistan salt mines has been studied by using culture-dependent and culture-independent techniques. In this culture-dependent method, we culture the organisms in the laboratory on different culture media at different culture conditions while in culture-independent approach we take a sample directly from an environment and extract DNA by Kit. After PCR amplification of 16S rRNA gene, we do next generation sequencing (Fig. 4.3).

4.6 Study of Bacterial Diversity by Culture-Dependent Method from Pakistan Salt Mines

4.6.1 Karak/Bahadur Khel Salt Mines

From halophilic environments of Pakistan, Karak/Bahadur khel salt mines are very important for microbial diversity. Earlier studies were conducted to explore their bacterial diversity. A culture-dependent method was used to investigate the bacterial diversity of karak salt mines and isolated twenty one halophilic and halotolerant bacterial strains. These strains were belonging to the genera *Brevibacterium*, *Bacillus*, *Oceanobacillus*, *Enterobacter*, *Halomonas*, *Halobacillus*, *Pseudomonas*, *Staphylococcus*, *Terribacillus*, *Thalassobacillus,* and *Virgibacillus* (Roohi et al. 2012) (Table 4.1). In another study, fifty seven (57) halotolerant and halophilic bacterial strains were isolated from karak salt mines. The 16S rRNA gene sequencing analyses revealed that the isolated strains were closely related to the genera *Bacillus*,

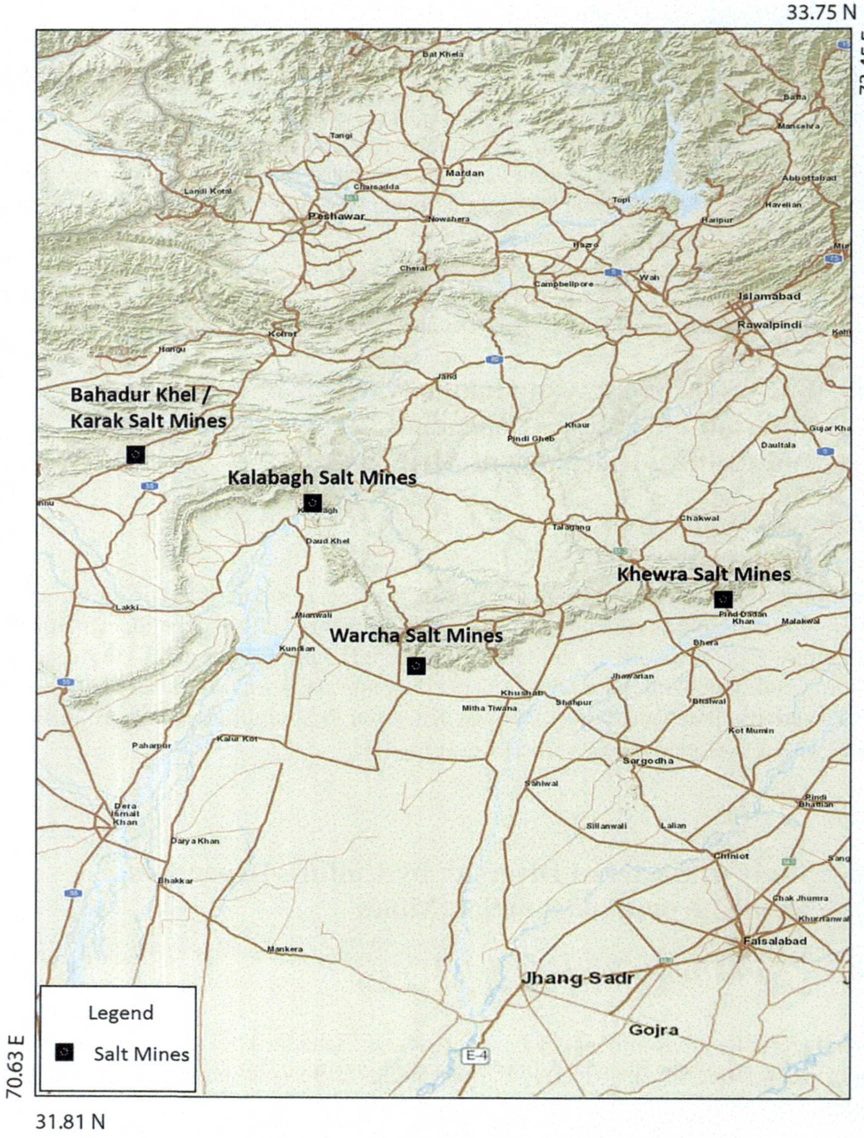

Fig. 4.2 Map showing the geographical locations of Salt Mines of Pakistan

Brevibacterium, *Gracilibacillus*, *Halomonas*, *Halobacillus*, *Kocuria*, *Jeotgalicoccus*, *Oceanobacillus*, *Planococcus*, *Salinicoccus*, *Staphylococcus,* and *Salinivibrio* (Roohi et al. 2014a) (Table 4.1).

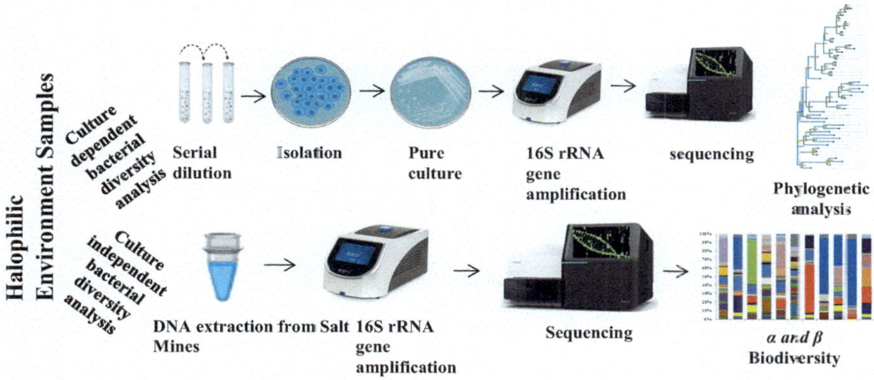

Fig. 4.3 Flow chart for the bacterial diversity analysis of Pakistan saline environments

Recently a study was conducted to probe into the microbial diversity of Karak/ Bahadur khel salt mines of Khyber Pakhtunkhwa, Pakistan. The isolated bacterial strains were belonging to the genera *Brevibacterium*, *Salicola*, *Chromohalobacter*, and *Oceanobacillus* (Cycil et al. 2020) (Table 4.1).

4.7 Validly Published Novel Bacterial Species Isolated from Karak Salt Mines, Pakistan

4.7.1 Bacillus pakistanensis *sp. nov and* Kushneria pakistanensis *sp. nov*

A Gram-staining positive, rod shaped, endospore forming, nonmotile and moderately halotolerant bacterial strain was isolated from salt mines in Karak, Khyber Pakhtunkhwa, Pakistan. This isolate was designated as NCCP-168^T. The taxonomic position of the strain NCCP-168^T was determined by polyphasic taxonomic approach. The phylogenetic analysis based on 16S rRNA gene sequences revealed that the organism is closely related to the genus *Bacillus*. The data obtained from the phenotypic and chemotaxonomic characteristics and phylogenetic analysis clearly distinguished the isolated organism from the closely related species of the genus *Bacillus*. So the isolate NCCP-168^T was novel species for which the name *Bacillus pakistanensis* sp. nov. was proposed (Roohi et al. 2014b). Another moderately halophilic bacterial strain NCCP-934^T was isolated from rhizosphere of *Saccharum spontaneum* obtained from salt mines of karak, Pakistan. The isolated strain NCCP-934^T was characterized using polyphasic taxonomy. The data obtained from differential characteristics distinguished the organism from closely related type strains and

Table 4.1 Study of bacterial diversity from salt mine areas of Pakistan

Organism	Strain ID	NCBI accession numbers	Source of isolation	References
Escherichia coli	BPT-2	AY430288	Salt mine (Khewra)	Akhtar et al. (2008)
Bacillus fumarioli	BPT-3	AY430289	Salt mine (Khewra)	Akhtar et al. (2008)
Bacillus sphaericus	BPT-4	AY430290	Salt mine (Khewra)	Akhtar et al. (2008)
Bacillus amyloliquefaciens	BPT-5	AY430292	Salt mine (Khewra)	Akhtar et al. (2008)
Bacillus cereus	BPT-6	AY430293	Salt mine (Khewra)	Akhtar et al. (2008)
Bacillus licheniformis	BPT-7	AY430294	Salt mine (Khewra)	Akhtar et al. (2008)
Staphylococcus arlettae	BPT-8	AY430291	Salt mine (Khewra)	Akhtar et al. (2008)
Bacillus licheniformis	BPT-11	AY430302	Salt mine (Khewra)	Akhtar et al. (2008)
Bacillus licheniformis	BPT-12	AY430295	Salt mine (Khewra)	Akhtar et al. (2008)
Staphylococcus gallinarum	BPT-15	AY430296	Salt mine (Khewra)	Akhtar et al. (2008)
Bacillus licheniformis	BPT-18	Y430297	Salt mine (Khewra)	Akhtar et al. (2008)
Bacillus licheniformis	BPT-19	AY430298	Salt mine (Khewra)	Akhtar et al. (2008)
Bacillus licheniformis	BPT-20	AY430299	Salt mine (Khewra)	Akhtar et al. (2008)
Bacillus licheniformis	BPT-23	AY430300	Salt mine (Khewra)	Akhtar et al. (2008)
Bacillus pumilus	BPT-25	AY430301	Salt mine (Khewra)	Akhtar et al. (2008)

(continued)

Table 4.1 (continued)

Organism	Strain ID	NCBI accession numbers	Source of isolation	References
Thalassobacillus sp.	NCCP-58	AB541110	Salt mine (Karak)	Roohi et al. (2012)
Thalassobacillus sp.	NCCP-64	AB698781	Salt mine (Karak)	Roohi et al. (2012)
Halomonas sp.	NCCP-67	AB698782	Salt mine (Karak)	Roohi et al. (2012)
Brevibacterium sp.	NCCP-68	AB698783	Salt mine (Karak)	Roohi et al. (2012)
Brevibacterium sp.	NCCP-69	AB698784	Salt mine (Karak)	Roohi et al. (2012)
Brevibacterium sp.	NCCP-70	AB698785	Salt mine (Karak)	Roohi et al. (2012)
Thalassobacillus sp.	NCCP-72	AB698786	Salt mine (Karak)	Roohi et al. (2012)
Oceanobacillus sp.	NCCP-76	AB698787	Salt mine (Karak)	Roohi et al. (2012)
Terribacillus sp.	NCCP-89	AB698791	Salt mine (Karak)	Roohi et al. (2012)
Terribacillus sp.	NCCP-90	AB698792	Salt mine (Karak)	Roohi et al. (2012)
Pseudomonas sp.	NCCP-164	AB698796	Salt mine (Karak)	Roohi et al. (2012)
Bacillus sp.	NCCP-165	AB698797	Salt mine (Karak)	Roohi et al. (2012)
Enterobacter sp.	NCCP-167	AB698799	Salt mine (Karak)	Roohi et al. (2012)
Oceanobacillus sp.	NCCP-169	AB698800	Salt mine (Karak)	Roohi et al. (2012)
Halobacillus sp.	NCCP-177	AB698808	Salt mine (Karak)	Roohi et al. (2012)

(continued)

Table 4.1 (continued)

Organism	Strain ID	NCBI accession numbers	Source of isolation	References
Halomonas sp.	NCCP-178	AB698809	Salt mine (Karak)	Roohi et al. (2012)
Halomonas sp.	NCCP-179	AB698810	Salt mine (Karak)	Roohi et al. (2012)
Staphylococcus sp.	NCCP-180	AB698811	Salt mine (Karak)	Roohi et al. (2012)
Halomonas sp.	NCCP-181	AB698812	Salt mine (Karak)	Roohi et al. (2012)
Virgibacillus sp.	NCCP-182	AB698813	Salt mine (Karak)	Roohi et al. (2012)
Halomonas sp.	NCCP-183	AB698814	Salt mine (Karak)	Roohi et al. (2012)
Bacillus sp.	NCCP-59	AB698777	Salt mine (Karak)	Roohi et al. (2014a)
Planococcus sp.	NCCP-60	AB698778	Salt mine (Karak)	Roohi et al. (2014a)
Bacillus sp.	NCCP-61	AB715332	Salt mine (Karak)	Roohi et al. (2014a)
Bacillus sp.	NCCP-62	AB698779	Salt mine (Karak)	Roohi et al. (2014a)
Bacillus sp.	NCCP-63	AB698780	Salt mine (Karak)	Roohi et al. (2014a)
Bacillus sp.	NCCP-65	AB715333	Salt mine (Karak)	Roohi et al. (2014a)
Jeotgalicoccus sp.	NCCP-66	AB735682	Salt mine (Karak)	Roohi et al. (2014a)
Bacillus sp.	NCCP-71	AB575949	Salt mine (Karak)	Roohi et al. (2014a)
Bacillus sp.	NCCP-73	AB715334	Salt mine (Karak)	Roohi et al. (2014a)

(continued)

Table 4.1 (continued)

Organism	Strain ID	NCBI accession numbers	Source of isolation	References
Bacillus sp.	NCCP-74	AB715335	Salt mine (Karak)	Roohi et al. (2014a)
Bacillus sp.	NCCP-75	AB715336	Salt mine (Karak)	Roohi et al. (2014a)
Bacillus sp.	NCCP-77	AB698788	Salt mine (Karak)	Roohi et al. (2014a)
Bacillus sp.	NCCP-79	AB698789	Salt mine (Karak)	Roohi et al. (2014a)
Bacillus sp.	NCCP-80	AB715337	Salt mine (Karak)	Roohi et al. (2014a)
Bacillus sp.	NCCP-81	AB715338	Salt mine (Karak)	Roohi et al. (2014a)
Bacillus sp.	NCCP-82	AB715339	Salt mine (Karak)	Roohi et al. (2014a)
Bacillus sp.	NCCP-83	AB715340	Salt mine (Karak)	Roohi et al. (2014a)
Staphylococcus sp.	NCCP-84	AB715341	Salt mine (Karak)	Roohi et al. (2014a)
Bacillus sp.	NCCP-85	AB715342	Salt mine (Karak)	Roohi et al. (2014a)
Thalasobacillus sp.	NCCP-86	AB698790	Salt mine (Karak)	Roohi et al. (2014a)
Halomonas sp.	NCCP-87	AB735683	Salt mine (Karak)	Roohi et al. (2014a)
Bacillus sp.	NCCP-88	AB715343	Salt mine (Karak)	Roohi et al. (2014a)
Bacillus sp.	NCCP-91	AB698793	Salt mine (Karak)	Roohi et al. (2014a)
Brevibacterium sp.	NCCP-92	AB698794	Salt mine (Karak)	Roohi et al. (2014a)

(continued)

Table 4.1 (continued)

Organism	Strain ID	NCBI accession numbers	Source of isolation	References
Bacillus sp.	NCCP-93	AB698795	Salt mine (Karak)	Roohi et al. (2014a)
Bacillus sp.	NCCP-166	AB698798	Salt mine (Karak)	Roohi et al. (2014a)
Bacillus sp.	NCCP-168	AB618147	Salt mine (Karak)	Roohi et al. (2014a)
Gracilibacillus sp.	NCCP-170	AB698801	Salt mine (Karak)	Roohi et al. (2014a)
Halomonas sp.	NCCP-171	AB698802	Salt mine (Karak)	Roohi et al. (2014a)
Bacillus sp.	NCCP-172	AB698803	Salt mine (Karak)	Roohi et al. (2014a)
Bacillus sp.	NCCP-173	AB698804	Salt mine (Karak)	Roohi et al. (2014a)
Bacillus sp.	NCCP-174	AB698805	Salt mine (Karak)	Roohi et al. (2014a)
Kocuria sp.	NCCP-175	AB698806	Salt mine (Karak)	Roohi et al. (2014a)
Halobacillus sp.	NCCP-176	AB698807	Salt mine (Karak)	Roohi et al. (2014a)
Staphylococcus sp.	NCCP-184	AB735684	Salt mine (Karak)	Roohi et al. (2014a)
Staphylococcus sp.	NCCP-204	AB698815	Salt mine (Karak)	Roohi et al. (2014a)
Salinivibrio sp.	NCCP-701	AB715344	Salt mine (Karak)	Roohi et al. (2014a)
Staphylococcus sp.	NCCP-704	AB715345	Salt mine (Karak)	Roohi et al. (2014a)
Salinicoccus sp.	NCCP-705	AB715346	Salt mine (Karak)	Roohi et al. (2014a)

(continued)

Table 4.1 (continued)

Organism	Strain ID	NCBI accession numbers	Source of isolation	References
Bacillus sp.	NCCP-706	AB715347	Salt mine (Karak)	Roohi et al. (2014a)
Salinivibrio sp.	NCCP-708	AB735685	Salt mine (Karak)	Roohi et al. (2014a)
Halomonas sp.	NCCP-713	AB735686	Salt mine (Karak)	Roohi et al. (2014a)
Bacillus sp.	NCCP-716	AB735687	Salt mine (Karak)	Roohi et al. (2014a)
Jeotgalicoccus sp.	NCCP-717	AB735688	Salt mine (Karak)	Roohi et al. (2014a)
Kocuria sp.	NCCP-718	AB735689	Salt mine (Karak)	Roohi et al. (2014a)
Oceanobacillus sp.	NCCP-721	AB735690	Salt mine (Karak)	Roohi et al. (2014a)
Staphylococcus sp.	NCCP-722	AB735691	Salt mine (Karak)	Roohi et al. (2014a)
Oceanobacillus sp.	NCCP-724	AB735692	Salt mine (Karak)	Roohi et al. (2014a)
Bacillus sp.	NCCP-726	AB735693	Salt mine (Karak)	Roohi et al. (2014a)
Bacillus sp.	NCCP-727	AB735694	Salt mine (Karak)	Roohi et al. (2014a)
Bacillus sp.	NCCP-729	AB735695	Salt mine (Karak)	Roohi et al. (2014a)
Staphylococcus sp.	NCCP-730	AB735696	Salt mine (Karak)	Roohi et al. (2014a)
Bacillus sp.	NCCP-731	AB735697	Salt mine (Karak)	Roohi et al. (2014a)
Bacillus sp.	NCCP-732	AB735698	Salt mine (Karak)	Roohi et al. (2014a)

(continued)

Table 4.1 (continued)

Organism	Strain ID	NCBI accession numbers	Source of isolation	References
Bacillus sp.	NCCP-734	AB735699	Salt mine (Karak)	Roohi et al. (2014a)
Halomonas sp.	NCCP-735	AB735700	Salt mine (Karak)	Roohi et al. (2014a)
Streptomyces sp.	KML-2	KJ009562	Salt mine (Khewra)	Aftab et al. (2015)
Bacillus subtilis	BLK-1.5	KX385201	Salt mine (Karak)	Ali et al. (2016)
Bacillus subtilis	SBA-5	NR116188	Salt mine (Karak)	Shah et al. (2017)
Oceanobacillus onchorhynci subsp. *incaldanensis*	HSL1	KP866216	Salt mine (Khewra)	Leena et al. (2018)
Staphylococcus lentus	HSL4	KP866217	Salt mine (Khewra)	Leena et al. (2018)
Bacillus endophyticus	HSL6	KP866218	Salt mine (Khewra)	Leena et al. (2018)
Bacillus aquimaris	HSL7	KP866219	Salt mine (Khewra)	Leena et al. (2018)
Exiguobacterium mexicanum DSM 6208	PGRS1	MH489029	Rhizosphere of *Dichanthium annulatum* collected from Salt mine (Khewra)	Mukhtar et al. (2018a)
Bacillus pseudofirmus ATCC 700159	PGRS2	MH489030	Rhizosphere of *Dichanthium annulatum* collected from Salt mine (Khewra)	Mukhtar et al. (2018a)
Exiguobacterium mexicanum DSM 16483	PGRS3	MH489031	Rhizosphere of *Dichanthium annulatum* collected from Salt mine (Khewra)	Mukhtar et al. (2018a)
Kocuria rosea ATCC 186	PGRS5	MH489032	Rhizosphere of *Dichanthium annulatum* collected from Salt mine (Khewra)	Mukhtar et al. (2018a)
Oceanobacillus oncorhynchi DSM 16557	PGRS6	MH489033	Rhizosphere of *Dichanthium annulatum* collected from Salt mine (Khewra)	Mukhtar et al. (2018a)
Bacillus cohnii DSM 6307	PGRS7	MH489034	Rhizosphere of *Dichanthium annulatum* collected from Salt mine (Khewra)	Mukhtar et al. (2018a)
Bacillus alcalophilus JCM 5262	PGRS9	MH489035	Rhizosphere of *Dichanthium annulatum* collected from Salt mine (Khewra)	Mukhtar et al. (2018a)

(continued)

Table 4.1 (continued)

Organism	Strain ID	NCBI accession numbers	Source of isolation	References
Bacillus polygoni NCIMB 14282	PGRS10	MH489036	Rhizosphere of *Dichanthium annulatum* collected from Salt mine (Khewra)	Mukhtar et al. (2018a)
Enterococcus durans ATCC 19432	PGRS11	MH489037	Rhizosphere of *Dichanthium annulatum* collected from Salt mine (Khewra)	Mukhtar et al. (2018a)
Virgibacillus halodenitrificans DSM	PGRS12	MH489038	Rhizosphere of *Dichanthium annulatum* collected from Salt mine (Khewra)	Mukhtar et al. (2018a)
Citricoccus alkalitolerans KCTC 19012	PGRP2	MH489039	Rhizosphere of *Dichanthium annulatum* collected from Salt mine (Khewra)	Mukhtar et al. (2018a)
Bacillus alcalophilus ATCC 27647	PGRP3	MH489040	Rhizosphere of *Dichanthium annulatum* collected from Salt mine (Khewra)	Mukhtar et al. (2018a)
Kocuria polaris CIP 107764	PGRP4	MH489041	Rhizosphere of *Dichanthium annulatum* collected from Salt mine (Khewra)	Mukhtar et al. (2018a)
Bacillus halodurans NRRL B-3881	PGRP6	MH489042	Rhizosphere of *Dichanthium annulatum* collected from Salt mine (Khewra)	Mukhtar et al. (2018a)
Bacillus alkalinitrilicus DSM 22532	PGRP7	MH489043	Rhizosphere of *Dichanthium annulatum* collected from Salt mine (Khewra)	Mukhtar et al. (2018a)
Exiguobacterium mexicanum DSM 16483	PGHP1	MH489044	Rhizosphere of *Dichanthium annulatum* collected from Salt mine (Khewra)	Mukhtar et al. (2018a)
Bacillus clarkii DSM 8720	PGHP2	MH489045	Rhizosphere of *Dichanthium annulatum* collected from Salt mine (Khewra)	Mukhtar et al. (2018a)
Citricoccus alkalitolerans DSM 15665	PGHP4	MH489046	Rhizosphere of *Dichanthium annulatum* collected from Salt mine (Khewra)	Mukhtar et al. (2018a)
Staphylococcus equorum ATCC 43958	PGHP5	MH489047	Rhizosphere of *Dichanthium annulatum* collected from Salt mine (Khewra)	Mukhtar et al. (2018a)
Micrococcus luteus CCM 169	PGHP6	MH489048	Rhizosphere of *Dichanthium annulatum* collected from Salt mine (Khewra)	Mukhtar et al. (2018a)
Bacillus pseudofirmus DSM 8715	PGHP8	MH489049	Rhizosphere of *Dichanthium annulatum* collected from Salt mine (Khewra)	Mukhtar et al. (2018a)

(continued)

Table 4.1 (continued)

Organism	Strain ID	NCBI accession numbers	Source of isolation	References
Kocuria rosea DSM 11630	PGHP9	MH489050	Rhizosphere of *Dichanthium annulatum* collected from Salt mine (Khewra)	Mukhtar et al. (2018a)
Oceanobacillus sp.	KPS3A	MT406254	Salt mine (Karak)	Cycil et al. (2020)
Oceanobacillus oncorhynchi subsp. *incaldanensis*	KPS8A	MT406253	Salt mine (Karak)	Cycil et al. (2020)
Brevibacterium linens/aureum	SWM2A	MT406257	Salt mine (Karak)	Cycil et al. (2020)
Salicola marasensis	KK1A	MT406258	Salt mine (Karak)	Cycil et al. (2020)
Chromohalobacter salexigens	KK2A	MT406255	Salt mine (Karak)	Cycil et al. (2020)

was declared a novel species for which the name *Kushneria pakistanensis* sp. nov. was proposed (Bangash et al. 2015) (Table 4.2).

4.7.2 Khewra Salt Mines

Khewra salt mines, which are distributed in Punjab Province, Pakistan, have rich sources of microbial diversity. A study was conducted by Leena et al. (2018) to isolate the potential halophilic bacteria from salt mines in Khewra, Pakistan. Total eight (8) bacterial strains were isolated in which four (4) were characterized by using polyphasic taxonomic approach. The isolated organisms were able grow in a medium containing 16% of NaCl. The 16S rRNA gene sequences comparison identified the strains to belonging to *Oceanobacillus*, *Staphylococcus*, and *Bacillus* genera (Table 4.1).

Mukhtar et al. (2018a) conducted a study; to explore the bacterial diversity from underground soil and root parts of *Dichanthium annulatum* collected from salt mine in Khewra, Pakistan. Total forty one (41) bacterial strains were isolated in which twenty two (22) were characterized based on 16S rRNA gene sequences. *Bacillus*, *Citricoccus*, *Enterococcus*, *Exiguobacterium*, *Micrococcus*, *Oceanobacillus*, *Staphylococcus*, *Virgibacillus*, and *Kocuria* were found to be the dominant genera (Table 4.1).

Table 4.2 Bacterial novel species isolated from the Karak salt mines, Pakistan

Novel bacterial strains	Source of isolation	GenBank accession number	Type strain/deposition number	Growth requirements			References
				Temperature range ($T_{opt.}$ °C)	pH range (pH_{opt})	NaCl range (optimum) (w/v)	
Bacillus pakistanensis	Karak salt mine, Khyber Pakhtunkhwa, Pakistan	AB618147	NCCP-168^T (=KCTC 13786^T = DSM 24834^T = JCM 18975^T)	10–40 (30–35)	5.0–9.0 (8)	0–17 (2–3)	Roohi et al. (2014b)
Kushneria pakistanensis	Rhizosphere of *Saccharum spontaneum* salt mines of Karak, Pakistan	AB970675	NCCP-934^T (=LMG 28525^T = KCTC 42082^T = JCM 18802^T)	10–40 (30–33)	6–10.5 (7–9)	1–30 (3–9)	Bangash et al. (2015)

4.8 Study of Bacterial Diversity by Culture-Independent Method from Pakistan Salt Mines

4.8.1 Karak/Bahadur Khel Salt Mines (16S rRNA Illumina Amplicon Sequencing)

Very recently, a metagenomic technique was used to probe into the microbial diversity of the salt mine in Karak, Pakistan. The 16S rRNA Illumina amplicon sequencing data showed that the dominant phyla were *Bacteroidetes* and *Proteobacteria* (Cycil et al. 2020).

4.8.2 Khewra Salt Mines (High-Throughput Sequencing of the 16S rRNA Gene)

Culture-independent study was conducted to compare the microbial community among the saline (Khewra salt mines) and non-saline (rhizosphere of *Triticum*) and moderately saline (rhizosphere of *Kochia* and *Urochloa*) environments. Pyrosequencing of 16S rRNA gene declared that the dominant group was Actinobacteria from saline soil samples while non-saline soil samples were predominated by Proteobacteria. *Acidobacteria*, *Bacteriodetes*, *Firmicutes*, and *Thaumarchaeota* were predominant phyla in saline and non-saline soils. The dominant halophilic bacterial strains were identified belonging to *Agrococcus*, *Haloferula*, *Armatimonadetes* gp4, and *Halobacterium* (Mukhtar et al. 2018b).

4.9 Potential Applications of Bacterial Strains Isolated from Saline Habitats of Pakistan

From worldwide saline environments, various prokaryotes particularly the bacterial strains have been isolated and have proven to possess potential applications in medicine, agriculture, and biotechnology fields. From Pakistan saline environments various bacterial strains have been isolated and proven their potential applications in industrial, agricultural and pharmaceutical fields.

Industrially important bacterial isolates belonging to the genera *Bacillus*, *Escherichia* and *Staphylococcus* were isolated from Khewra salt mine. The isolated organisms were able to produce important enzymes, such as cellulase, carboxymethylcellulase, amylase, xylanase, and protease (Akhtar et al. 2008).

Extremozymes are very important for industrial processes. A study was conducted to explore the extremophilic bacteria diversity salt mines of Karak, Pakistan. Total fifty three (53) bacterial strains were isolated in forty three

(43) were found to be positive for amylase production (Shah et al. 2017). A halotolerant bacterial strain *Bacillus subtilis* strain (BLK-1.5) was isolated from Karak salt mines, Pakistan. The isolated strain was able to produce protease enzyme (Ali et al. 2016).

The eight halophilic bacterial strains isolated from Khewra salt mines were potential candidates. The enzyme assays result showed, that these strains were able to produce amylase, cellulose, DNase, gelatinase, lipase, protease, urease, and xylanase (Leena et al. 2018).

Industrially important bacterial strains were isolated from the root soil of *Dichanthium annulatum* collected from the Khewra salt mine, Pakistan. The isolated strains were found to produce amylase, cellulose, lipase, and protease activities (Mukhtar et al. 2018a).

A halophilic bacterium *Bacillus licheniformis* NCCP-59 was isolated from salt mines in karak, Pakistan. This bacterial strain was found to have the promising ability to protect the plants from the harmful effects of nickel (Ni) and maintain the plants growth in nickel (Ni) contaminated soil (Jamil et al. 2013).

An actinobacterial strain designated KML-2, belonging to the genus *Streptomyces* was isolated from the saline soil sample of Khewra salt mines, Pakistan. Pharmaceutically this strain was a potential candidate due to its cytotoxic and antitumor activities (Aftab et al. 2015).

A halophilic bacterial strain designated SAL-15 was isolated from alkali-saline soils of Pakistan and was found to have ACC-deaminase activity. This strain can improve growth of plants in saline conditions, so can be used as biofertilizer for saline areas and in water stress environment (Rajput et al. 2013). These results showed that haloalkaliphilic bacterial diversity identified in this study had great agricultural values.

4.10 Conclusions and Future Perspectives

The Bacterial diversity of halophilic environments of Pakistan has been investigated by culture-dependent and culture-independent methods. The dominant genera were belonging to *Bacillus* and *Staphylococcus*. Two novel species namely *Bacillus pakistanensis* NCCP-168^T and *Kushneria pakistanensis* NCCP-934^T have been isolated from saline environments of Pakistan and validly published (Roohi et al. 2014a, b; Bangash et al. 2015). Extremozymes have got great attention due to its application in industrial processes. From Karak and Khewra salt mines of Pakistan, several bacterial species have been isolated and proven to have the catalytic properties (amylase, carboxymethylcellulase, xylanase, cellulase and protease (Akhtar et al. 2008). Some bacterial strains were found to have the activities of protease, amylase, lipase, xylanase, urease, gelatinase, cellulose, and DNase (Leena et al. 2018). In another study conducted by Mukhtar et al. (2018a, b) found that the bacterial strains have protease, amylase, lipase and cellulase activities.

Due to the increasing rate of antimicrobial drug resistance, discovery of new drugs will be highly appreciated. So in order to cope with challenge, the halophilic environments (salt mines) of Pakistan will be a rich source of novel bacterial species with production of drugs which will have high efficacy against MDR bacterial strains.

The whole genome sequencing of the halophilic bacteria from saline environments will validate the taxonomic position (phylogenomic analysis based on core genome). Furthermore, from the genome mining we will be able, to probe into the adaptation strategies of halophiles to hypersaline environments and biosynthetic gene clusters (BGCs) for the production of novel bioactive compounds, which will have agricultural, industrial and pharmaceutical values.

Proteomics, transcriptomics, metatranscriptomics, and metabolomics studies are essential to reveal the ecological role of bacterial strains in Pakistan saline environments and will be also for the understanding of interaction of theses halophiles with other organisms.

A very recent study is conducted by Cycil et al. (2020) on microbial diversity of karak salt mines (Pakistan) proposed, that due the perchlorateresistant ability of the bacterial strains belonging to the genera *Marinobacter* and *Halomonas* make this environment and isolates the best model for astrobiological studies.

The microbial diversity of Kalabagh and Warcha salt mines in the Punjab province of Pakistan has not yet been explored. The biodiversity of these halophilic environments will be a rich source of microbial novel taxa which will produce novel enzymes and natural products.

In order to utilize in better way the Pakistan salt mines for microbial diversity, we will need a foreign collaboration for exchange of students and bilateral projects.

Acknowledgments This research was supported by National Natural Science Foundation of China (No. 32061143043) and Xinjiang Uygur Autonomous Region, regional coordinated Innovation Project and Shanghai Cooperation Organization Science and Technology Partnership Program (Nos. 2020E01047 and 2021E01018). Iftikhar Ahmed was also supported by the Chinese Academy of Sciences President's International Fellowship Initiative (Grant No. 2020VBA0020).

References

Aftab U, Zechel DL, Sajid I (2015) Antitumor compounds from *Streptomyces* sp. KML-2, isolated from Khewra salt mines, Pakistan. Biol Res 48:58

Ai L, Huang Y, Wang C (2018) Purification and characterization of halophilic lipase of Chromohalobacter sp. from ancient salt well. J Basic Microbiol 58:647–657

Akhtar N, Ghauri MA, Iqbal A, Anwar MA, Akhtar K (2008) Biodiversity and phylogenetic analysis of culturable bacteria indigenous to Khewra salt mine of Pakistan and their industrial importance. Braz J Microbiol 39:143–150

Ali I, Akbar A, Anwar M, Prasongsuk S, Lotrakul P, Punnapayak H (2015) Purification and characterization of a polyextremophilic a-amylase from an obligate halophilic *Aspergillus penicillioides* isolate and its potential for souse with detergents. Biomed Res Int 1:1–8

Ali N, Ullah N, Qasim M, Rahman H, Khan SN, Sadiq A, Adnan M (2016) Molecular characterization and growth optimization of halo-tolerant protease producing *Bacillus Subtilis* Strain BLK-1.5 isolated from salt mines of Karak, Pakistan. Extremophiles 20(4):395–402
Amoozegar MA, Malekzadeh F, Malik KA (2003) Production of amylase by newly isolated moderate halophile, Halobacillus sp. strain MA-2. J Microbiol Methods 52:353–359
Bangash A, Ahmed I, Abbas S, Kudo T, Shahzad A, Fujiwara T, Ohkuma M (2015) *Kushneria pakistanensis* sp. nov., a novel moderately halophilic bacterium isolated from rhizosphere of a plant (*Saccharum spontaneum*) growing in salt mines of the Karak area in Pakistan. Antonie Van Leeuwenhoek 107:991–1000
Chakraborty S, Khopade A, Kokare C, Mahadik K, Chopade B (2009) Isolation and characterization of novel a-amylase from marine *Streptomyces* sp. J Mol 58:17–23
Cremo MA (2001) Paleobotanical anomalies bearing on the age of the salt range formation of Pakistan, a historical survey of an unresolved scientific controversy. In: Proceedings of the XXIst international congress of history of science, Mexico
Cycil LM, DasSarma S, Pecher W, McDonald R, AbdulSalam M, Hasan F (2020) Metagenomic insights into the diversity of halophilic microorganisms indigenous to the Karak Salt Mine, Pakistan. Front Microbiol 11:1567
Galinski EA (1993) Compatible solutes of halophilic eubacteria: molecular principles, water-solute interactions, stress protection. Experientia 49:487–496
Galinski EA (1995) Osmoadaptation in bacteria. Adv Microb Physiol 37:273–328
Jamil M, Zeb S, Anees M, Roohi A, Ahmed I, Rehman S, Shik RE (2013) Role of *Bacillus licheniformis* in phytoremediation of nickel contaminated soil cultivated with rice. Int J Phytoremediation 16(6):554–571
Jehangiri M, Hanif M, Arif M, Jan IF, Ahmad S (2015) The early Cambrian Khewra sandstone, salt range, Pakistan, endorsing southern Indian provenance. Arab J Geosci 8:6169–6187
Khattak NU, Khan AM, Ali N, Ahmed F, Tahir Shah M (2016) Recognition and characterization of a tectonically active Karak Thrust using radon measurement technique in the Southern Kohat Plateau, Pakistan. J Himal Earth Sci 49:40–49
Leena MC, Aamer AS, Abdul H, Fariha H (2018) Physiological, biochemical and phylogenetic characterization of extremely halophilic bacteria isolated from Khewra mine, Pakistan. Appl Ecol Environ Res 16:1243–1256
Litchfield CD (1998) Surrvival strategies for microorganisms. Hypersaline environments and their relevance to life on early mars. Meteorit Planet Sci 33:813–819
Mukhtar S, Kauser Abdulla M, Samina M (2018a) Isolation and characterization of haloalkaliphilic bacteria isolated from the rhizosphere of *Dichanthium annulatum*. J Adv Res Biotechnol 3 (1):1–9
Mukhtar S, Mirza BS, Mehnaz S, Mirza MS, Mclean J, Malik KA (2018b) Impact of soil salinity on the microbial structure of halophyte rhizosphere microbiome. World J Microbiol Biotechnol 34:136
Oren A (1999) Bioenergetic aspects of halophilism. Microbiol Mol Biol Rev 63:334–348
Oren A (2000) Salts and brines. In: Whitton BA, Potts M (eds) Ecology of cyanobacteria: their diversity in time and space. Kluwer Academic Publishers, Dordrecht, pp 281–306
Oren A (2001) The order haloanaerobiales. In: Dworkin M, Falkow S, Rosenberg E, Schleifer KH, Stackebrandt E (eds) The prokaryotes. A handbook on the biology of bacteria: ecophysiology, isolation, identification, applications, 3rd edn. Springer, New York
Oren A (2002) Diversity of halophilic microorganisms: environments, phylogeny, physiology, and applications. J Ind Microbiol Biotechnol 28:56–63
Oren A (2008) Microbial life at high salt concentrations: phylogenetic and metabolic diversity. Saline Syst 4:1–13
Quesada E, Ventosa A, RodriguezValera F, RamosCormenzana A (1982) Types and properties of some bacteria isolated from hypersaline soils. J Appl Microbiol 53:155–161
Quesada E, Béjar V, Ferrer M, Calvo C, Llamas I, MartínezCheca F, Arias S, RuizGarcía C, Páez R, MartinezCánovas M, del Moral A (2004) Moderately halophilic, exopolysaccharide-producing bacteria. Springer, Berlin, Berlin

Rainey FA, Zhilina TN, Boulygina ES, Stackebrandt E, Tourova TP, Zavarzin GA (1995) The taxonomic status of the fermentative halophilic anaerobic bacteria: description of Halobacteriales ord. nov., Halobacteroidaceae fam. nov., Orenia gen. nov. and further taxonomic rearrangements at the genus and species level. Anaerobe 1:185–199

Rajput R, Imran A, Mubeen F, Hafeez FY (2013) Salt-tolerant PGPR strain *Planococcus rifietoensis* promotes the growth and yield of wheat (*Triticum aestivum*) cultivated in saline soil. Pak J Bot 45(6):1955–1962

Roohi A, Ahmed I, Iqbal M, Jamil M (2012) Preliminary isolation and characterization of halotolerant and halophilic bacteria from salt mines of karak. Pak J Bot 44:365–370

Roohi A, Ahmed I, Khalid N, Iqbal M, Jamil M (2014a) Isolation and phylogenetic identification of halotolerant/halophilic bacteria from the salt mines of Karak, Pakistan. Int J Agric Biol 16:564–570

Roohi A, Ahmed I, Paek J, Sin Y, Abbas S, Jamil M, Chang YH (2014b) *Bacillus pakistanensis* sp. nov., a halotolerant bacterium isolated from salt mines of the Karak Area in Pakistan. Antonie Leeuwenhoek 105:1163–1172

Shafiei M, Ziaee AA, Amoozegar MA (2012) Purification and characterization of a halophilic a-amylase with increased activity in the presence of organic solvents from the moderately halophilic *Nesterenkonia* sp. strain F. Extremophiles 16:627–635

Shah W, Nawab A, Shahbaz A, Muhammad Q, Hazir R, Sami U, Farman U, Khattak B, Khan MD (2017) Molecular characterization and growth optimization of halo-tolerant amylase producing bacteria isolated from salt mines of Karak. Pure Appl Biol 6:385–393

Shah M, Hasan F, Hashmi SS, Muhammad W, Zubair M, Asad Ullah M (2018) Diversity of halophiles in Karak salt mine, KP, Pakistan and their ability to produce enzyme of industrial importance. Int J Biosci 13:257–266

Singh P, Jain K, Desai C, Tiwari O, Madamwar D (2019) Microbial community dynamics of extremophiles/extreme environment. In: Das S, Dash HR (eds) Microbial diversity in the genomic era. Academic, San Diego, pp 323–332

Ventosa A (2006) Unusual micro-organisms from unusual habitats: hypersaline environments. In: Logan N, Lappin-Scott H, Oyston P (eds) Prokaryotic diversity: mechanism and significance. Cambridge University Press, Cambridge, pp 223–253

Ventosa A, Nieto JJ, Oren A (1998) Biology of aerobic moderately halophilic bacteria. Microbiol Mol Biol Rev 62:504–544

Ventosa A, Mellado E, Sánchez Porro C, Marquez M (2008) Halophilic and halotolerant microorganisms from soils. Springer, Berlin

Ventosa A, de la Haba RR, Sánchez-Porro C, Papke RT (2015) Microbial diversity of hypersaline environments: a metagenomic approach. Curr Opin Microbiol 25:80–87

Zhang G, Li S, Xue Y, Mao L, Ma Y (2012) Effects of salts on activity of halophilic cellulase with glucomannanase activity isolated from alkaliphilic and halophilic *Bacillus* sp. BG-CS10. Extremophiles 16:35–43

Chapter 5
Taxonomic Characteristics of Dominant Microbial Communities in Hot Spring Sediments in Western Georgia

Natia Geliashvili, Ekaterine Jaiani, Marina Tediashvili, and Nils-Kåre Birkeland

Abstract Hot springs are one of the extreme environments where only specific populations of microbes, thermophiles can thrive. Microbial diversity of thermal springs in different geographical regions of the world is relatively under-explored. Studying extremophiles in various environmental niches and their ecological roles in their environment also gives hand to identification of new enzymes which are remarkable for industrial reasons. Owing to its geological location, Georgia has considerable resources of natural thermal waters and has long tradition of their exploitation. Over 250 natural thermal springs and artificial wells as well as spring clusters are known with water temperatures ranging from 30 °C to 108 °C. Despite increased worldwide interest in extremophiles in thermal waters, only few reports have been published on microbial diversity of geothermal springs in Georgia. In this chapter taxonomic profiles of microorganisms from two high-temperature geothermal spring sediments in Western Georgia will be presented and discussed.

Keywords Hot spring · Microbial community · Extremophiles

N. Geliashvili
G. Eliava Institute of Bacteriophages, Microbiology and Virology, Tbilisi, Georgia

Department of Biological Sciences, University of Bergen, Bergen, Norway
e-mail: nageli2014@agruni.edu.ge

E. Jaiani (✉) · M. Tediashvili
G. Eliava Institute of Bacteriophages, Microbiology and Virology, Tbilisi, Georgia
e-mail: m.tediashvili@pha.ge

N.-K. Birkeland
Department of Biological Sciences, University of Bergen, Bergen, Norway
e-mail: nils.birkeland@uib.no

D. Egamberdieva et al. (eds.), *Microbial Communities and their Interactions in the Extreme Environment*, Microorganisms for Sustainability 32,
https://doi.org/10.1007/978-981-16-3731-5_5

5.1 Introduction

Microbial communities in terrestrial hydrothermal systems and hot springs, due to its unique features and evolutionary research interests, have been extensively studied worldwide (Guo et al. 2020). Microbes are the major biological component of the environment and they can survive well in unfavorable, extreme, and harsh environments. Thus, they can occupy all possible ecological niches where the environment offers minimum conditions for survival. Hot springs are one of the extreme environments where only specific populations of microbes, thermophiles can thrive. Geological and physicochemical features such as depth of bore hole, chemical composition, temperature, and pH may vary largely between hot springs. Due to highly extreme conditions hot spring environments are suitable mainly for archaea and bacteria (Rawat and Joshi 2019). The temperature and pH are reported as two of the most influential factors affecting microbial community distribution in hot spring ecosystems. When the spring water temperature exceeds 75 °C, the thermophilic or hyperthermophilic bacteria, including *Aquificae*, *Deinococcus-Thermus*, *Thermodesulfobacteria*, or *Thermotogae* are enriched (Purcell et al. 2007; Wang et al. 2013). When temperature is suitable for photosynthesis (<75 °C), moderately thermophilic and mesophilic *Bacteria* are important members in terrestrial thermal springs, such as Cyanobacteria, Chloroflexi, and Proteobacteria. Comparative studies on hot springs with different temperatures give insights to better understand the distribution and prevalence of various microbial groups in different geographical regions. Even in the same sampling area, different springs could host different communities, because local physicochemical conditions are important factors in shaping community structure. Therefore, it is reasonable to observe spatial and temporal differences in bacterial community composition (Wang et al. 2013)

Studying extremophiles in various environmental niches and their ecological roles in their environment also gives hand to identify new enzymes which are remarkable for industrial reasons. Extremophile-derived enzymes, or extremozymes, are able to catalyze chemical reactions under harsh conditions that often are highly suitable for industrial processes. Due to their optimal activity and stability under extreme conditions, extremozymes offer new catalytic alternatives for many current industrial applications. There is wide range of commercially available extremozymes such as amylases, glucose oxidases, lipases, proteases, xylanases, celullases, and others that are widely used by the industry (Sarmiento et al. 2015).

Among the extreme environments, microbial diversity of hot springs is comparatively less explored. Culture-independent approaches are often preferred over cultivation-based methods for diversity assessment of hyperthermophiles, because few are capable of cultivation under laboratory conditions (Mori and Kamagata 2014; Blank et al. 2002). To overcome this limitation, metagenomic studies have been applied as one of the best molecular approaches for microbial ecology studies. More recent developments of metagenomic approaches have considerably increased the information related to microbial diversity and function (Najar et al. 2018).

Owing to its geological location between the Greater Caucasus in the north and the Lesser Caucasus range in the South, Georgia has considerable resources of natural thermal waters and has long tradition of their exploitation. Over 250 natural thermal springs and artificial wells as well as spring clusters are known with water temperatures ranging from 30 to 108 °C. West Georgia and in particular, Zugdidi region is rich in high-temperature springs where 25 wells have been drilled producing thermal waters with temperature ranging from 82 to 102 °C (Tsertsvadze et al. 1998). The drilled wells are mainly used in balneology and as a thermal source for greenhouses (Melikadze et al. 2010).

Despite increased worldwide interest in microbial assemblages of the thermal waters, only few reports have been published on microbial diversity of geothermal springs in Georgia. Here we provide data on microbial communities of two high-temperature hot springs located in Western Georgia, Zugdidi region using a culture-independent, metagenomic approach.

5.2 Zugdidi-Tsaishi Geothermal Field

Among the thermal waters described in Georgia, the Zugdidi-Tsaishi deposit is of special importance due to its high temperature, availability, and exploitation. The thermal water-containing horizon is represented by a Lower Cretaceous limestone complex composed of layered and massive dolomitized fractured and karstic limestone by which the Urtian Brachyanticline is build. At the west end of Tsaishi village, the anticline is broken by submeridional fault, along which the two blocks are shifted by 1000 m. The explorations allowed to establish that the above-mentioned thermal aquifer favorable within the deposit as the Senaki-Tsaishi tectonic fracture has been ruptured by a 1500 m amplitude and that there is no favorable connection between its up-thrown and downthrown sides (Tsertsvadze et al. 1998; Melikadze et al. 2010; Kapanadze et al. 2010).

5.3 Physicochemical Parameters of Geothermal Waters in the Zugdidi-Tsaishi Region

The geothermal springs of Zugdidi-Tsaishi area are characterized by diverse physicochemical parameters. The wells drilled in the up-thrown side are relatively shallow (1272–2661 m) and consequently, waters tapped here are characterized by low mineralization (0.87 g/L–1.6 g/L), and temperature (82 °C–90 °C), while in the downthrown side the water is of relatively high mineralization (2.4 g/L) and temperature (95 °C–98 °C) (Kapanadze et al. 2010).

The thermal waters in the region have mixed-cation and mixed-anion ratios mainly composed of hydrocarbonate, chloride, sulfate, sodium, potassium,

Fig. 5.1 Map of Georgia, sampling site—village Tsaishi

Table 5.1 Location and physicochemical properties of studied thermal waters

Sample	Year of drilling	Depth (M)	Temperature °C in situ	pH	Chemical formula and mineralization (g/L)
Tsaishi 4-K	1977	800	78	6.0	$M_{1,7}\frac{SO_4 51Cl38HCO_3 10}{Ca46(Na+K)39Mg14}$
Tsaishi 1	1973	3728	91	6.0	$M_{1,5}\frac{SO_4 64Cl26HCO_3 10}{Ca58Na19Mg19K4}$

magnesium, and calcium ions (Tsertsvadze et al. 1998). Two Geothermal hot springs described in this chapter in more detail are situated in village Tsaishi (Fig. 5.1; 42°27′0″N, 41°49′10″E) approximately 600 m apart. Both springs are used for heating of local greenhouses. Despite a difference in borehole depths both springs have similar mineralization and belong to hydrocarbonate-chloride-sulfate-sodium-potassium-magnesium-calcium sources (Tsertsvadze et al. 1998).

Our *in situ* measurements showed that the water temperature of 4-K spring was 78 °C while Tsaishi N1 water was characterized by higher temperature –91 °C. Water was slightly acidic in both thermal springs. The physicochemical parameters of the hot springs are summarized in Table 5.1. A white-grayish biofilm is formed at the bottom of the 4-K spring, while Tsaishi N1 water forms yellowish-white sediment (Fig. 5.2).

a b

Fig. 5.2 Close-up photographs of studied geothermal springs: (**a**) Tsaishi 4-K and (**b**) Tsaishi 1

5.4 Taxonomic Profiles of the Thermal Spring Sediments

Shotgun metagenomic study using Illumina HiSeq technology revealed that the studied hot springs harbor high diversity of microbial assemblages. According to our data, the prokaryotic communities of the 4-K sediment sample appeared to be more diverse compared to Tsaishi 1 sample (Fig. 5.3). The major phyla in 4-K were Firmicutes (35%), Proteobacteria (32%), and Aquificae (19%). At class level Bacilli (Firmicutes), Gammaproteobacteria, and Aquificae dominated the sample (Yadav et al. 2018).

Thermophilic bacilli were highly represented in the 4-K sediment, notably *Anoxybacillus* was determined as a dominant genus. Interestingly, *Anoxybacillus* members were shown to secrete a variety of heat-stable lignocellulolytic enzymes with biotechnological potential such as cellulase, xylanase, and xylosidases important in biomass degradation (Yadav et al. 2018).

The high temperature of the Tsaishi 4-K spring appeared to be a major factor determining high representation of Aquificae genera, such as *Sulfurihydrogenebium, Hydrogenobacter,* and *Thermocrinis* (Fig. 5.3). Hyperthomophilic, anaerobic, sulfur-oxidizing bacteria of genus *Sulfurihydrogenibium* were also described in various other thermal springs (Flores et al. 2008). Members of the genus *Sulfurihydrogenibium* were detected first by culture-independent molecular analyses from terrestrial hot environments (Hugenholtz et al. 1998; Yamaguchi et al. 2004) and three new strains were isolated from a subsurface hot aquifer, terrestrial hot springs and Calcite Springs in Yellowstone National Park, USA (Takai et al. 2003; Aguiar et al. 2004; Nakagawa et al. 2005). Members of the genus *Sulfurihydrogenibium* play a significant role in sulfur-cycling (Skirnisdottir et al. 2000) and iron mineralization (Reysenbach et al. 1999). These bacteria are capable to produce thermostable enzymes such as carbonic anhydrases with biotechnological applications (Del Prete et al. 2019). The other hyperthermophile within the class *Aquificae,* found in our sample, *Hydrogenobacter* includes obligately

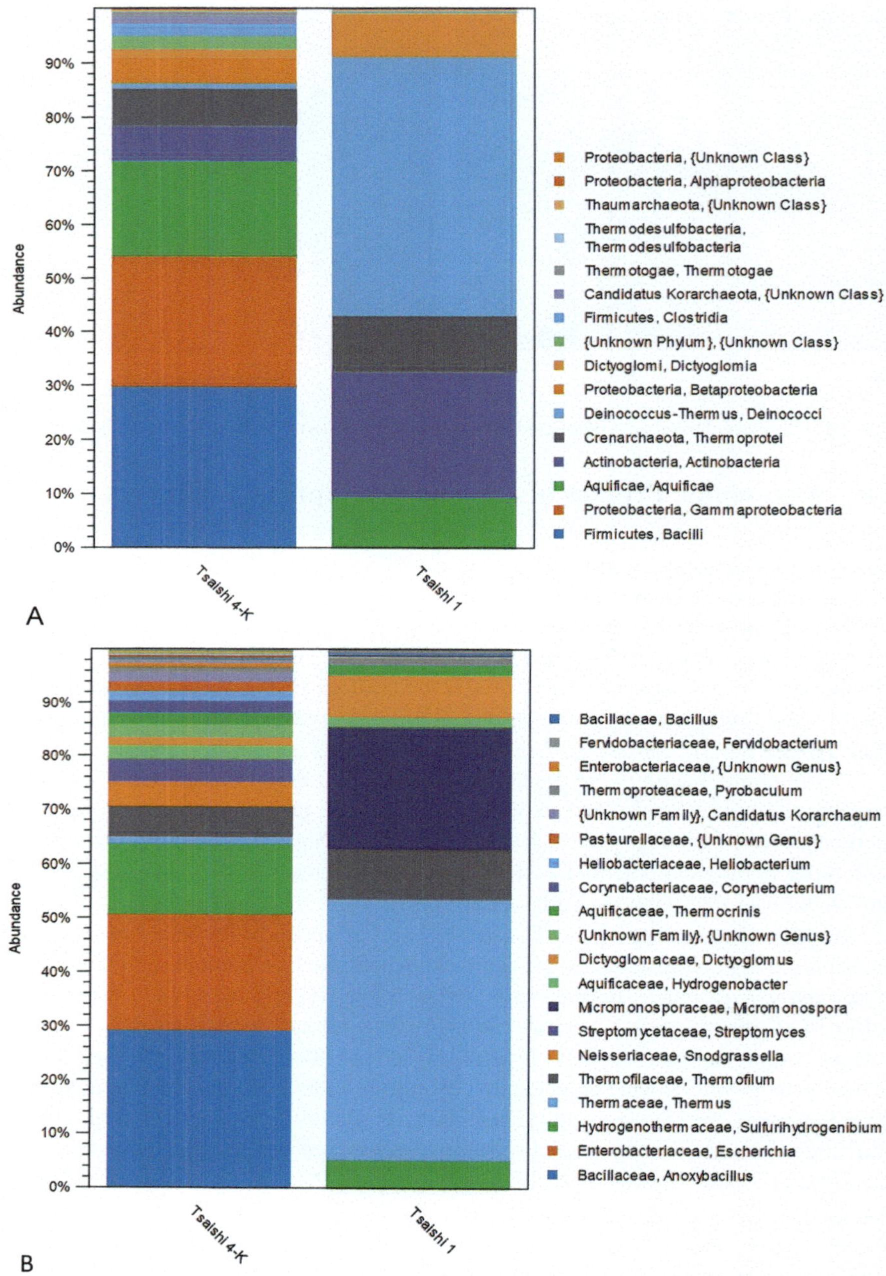

Fig. 5.3 Distribution of different microbial phyla in Tsaishi 4-K and Tsaishi 1 hot springs based on shotgun metagenomic sequencing. (**a**) Phylum and class level grouping of sequences and (**b**) Family and genus level grouping of sequences

chemolithoautotrophic and hydrogen-oxidizing bacteria that were found in various other geographical areas such as Japan, Italy and Iceland (Zeytun et al. 2011) indicating wide distribution of these bacteria in such extreme habitats. The genus *Hydrogenobacter* have attracted appreciable interest due to the thermostable enzymes they produce. Enzymes and proteins involved in the detoxification of oxygen, e.g., peroxidase, superoxide dismutase, and cytochromes, are of increased interest (Manelius et al. 1997). Another dominant genus, *Thermocrinis* (class Aquificae) detected in the Tsaishi 4-K sediment was described in a spring in the Yellowstone National Park as well (Jahnke et al. 2001). These bacteria were shown to grow on hydrogen, elemental sulfur, and thiosulfate and also utilize thioarsenate, common arsenic species in sulfidic geothermal waters (Härtig et al. 2014). Besides thermophiles, bacteria belonging to family Enterobacteriaceae were also found in the Tsaishi 4-K sediment that can be considered as contaminant bacteria as the spring is intensively used by the local population for washing of animal products.

As a minor phylum, Thermotogae bacteria and genus *Fervidobacterium* were also identified in the 4-K sediment sample. This group of bacteria were shown to degrade keratin, a process that could aid in complete degradation of feathers that are currently only partially utilized and most of the essential amino acids they could provide (serine, cysteine, and proline) are wasted (Papadopoulos 1989). Based on environmental considerations the use of keratinolytic enzymes in the production of amino acids and peptides is becoming attractive for biotechnological applications (Friedricht and Antranikian 1996). Interestingly, *Fervidobacterium* spp., capable of keratin degradation was isolated from the geothermal spring located in central Georgia (Geliashvili et al., unpublished data) suggesting that Georgian thermal waters might harbor Fervidobacteria producing enzymes of high industrial potential.

As a harsh, high-temperature environment, Tsaishi 4-K sediment was enriched with Archaea as well. Genus *Thermofilum* belonging to heterotrophic, hyperthermophilic Crenarchaeota was detected in the 4-K sediment sample. These extremophilic archaea were found to hydrolyze oligo- and polysaccharides by producing heat-stable cellulases (Zayulina et al. 2020).

Microbial assemblages of the second studied spring, Tsaishi 1 sediment, were dominated by *Deinococcus-Thermus* (54%), *Actinobacteria* (26%), and *Aquificae* (10%) (Fig. 5.3).

The major class in Taishi 1 sediment sample, Deinococci was mainly represented by the genus *Thermus*, the bacteria commonly found in high temperature (55–100 °C) and weak acid to alkaline pH (5–9) habitats (Henne et al. 2004). The genus *Thermus* have held biochemical and industrial attention already for several decades. The discovery of *Thermus aquaticus* (Brock and Freeze 1969) and its *Taq* DNA polymerase has revolutionized the field of extremophile research. Enzymes produced by *Thermus*, such as thermostable hydrolases, proteases, phosphohydrolases, catalases are of considerable biotechnological interest. The purification of proteins from *Thermus thermophilus* or the cloning of their genes and expression in mesophilic microorganisms showed that they have potential to be used in a variety of biotechnological processes (Pantazaki et al. 2002).

The second major genus in the Tsaishi 1 sediment, *Micromonospora* within the class *Actinobacteria,* are widely distributed in a variety of habitats, notably in soils rich in humus, but have been described in extreme habitats such as hot water soils as well (Thawai et al. 2019). The members of the genus *Micromonospora* are considered as important microbial sources for the pharmaceutical industry (Boumehira et al. 2016). In general, *Actinobacteria* are recognized for their ability to produce bioactive metabolites, notably antibiotics, thus characterization of novel Actinobacteria from thermal waters can provide clues to new antimicrobial discoveries (Hussein et al. 2018).

Interestingly, in the Tsaishi 1 sediment, Archaea and Bacteria within the class Aquificae were dominated by the genera found in 4-K sediment as well, such as *Thermofilum, Sulfurihydrogenibium, Thermocrinis* and *Hydrogenobacter* that can be explained by certain similarities in physicochemical properties of the geothermal waters. Both springs have high temperature, slightly acidic pH and contain salt ions like Na^+, K^+, and Cl^-, conditions that has previously been described as favorable for Aquificae (Xian et al. 2018).

Minor phyla, such as *Thermodesulfobacteria* a sulfate-reducing obligate anaerobic thermophilic bacterium and anaerobic (Bhatnagar et al. 2015), hyperthermophilic members of the *Dictyoglomus* genus were also found in both sediment samples. The latter was isolated from paper-pulp factory effluent as well (Brumm et al. 2016).

5.5 Conclusion

The culture-independent studies showed that high-temperature hot springs located in Western Georgia are diverse in terms of microbial composition. The metagenomic approach revealed number of microbial groups with ability to produce bioactive metabolites and enzymes of industrial applications.

Despite increased worldwide interest in extremophilic microbes and their products, the vast majority of geothermal springs in Georgia still remain unexplored. Further studies are needed to investigate the microbial assemblages and their biotechnological potential in thermal environments of Georgia using culture-dependent and culture-independent approaches.

References

Aguiar P, Beveridge TJ, Reysenbach AL (2004) Sulfurihydrogenibium Azorense, sp. Nov., a thermophilic hydrogen-oxidizing microaerophile from terrestial hot springs in the azores. Int J Syst Evol Microbiol. https://doi.org/10.1099/ijs.0.02790-0

Bhatnagar S, Badger JH, Madupu R, Khouri HM, O'Connor EM, Robb FT, Ward NL, Eisen JA (2015) Genome sequence of a sulfate-reducing thermophilic bacterium,

thermodesulfobacterium commune DSM 2178T (phylum thermodesulfobacteria). Genome Announc. https://doi.org/10.1128/genomeA.01490-14
Blank CE, Cady SL, Pace NR (2002) Microbial composition of near-boiling silica-depositing thermal springs throughout yellowstone national park. Appl Environ Microbiol. https://doi.org/10.1128/AEM.68.10.5123-5135.2002
Boumehira AZ, El-Enshasy HA, Hacène H, Elsayed EA, Aziz R, Park EY (2016) Recent progress on the development of antibiotics from the Genus Micromonospora. Biotechnol Bioprocess Eng. https://doi.org/10.1007/s12257-015-0574-2
Brock TD, Freeze H (1969) Thermus aquaticus Gen. n. and Sp. n., a nonsporulating extreme thermophile. J Bacteriol. https://doi.org/10.1128/jb.98.1.289-297.1969
Brumm PJ, Gowda K, Robb FT, Mead DA (2016) The complete genome sequence of hyperthermophile dictyoglomus turgidum DSM 6724™ reveals a specialized carbohydrate fermentor. Front Microbiol. https://doi.org/10.3389/fmicb.2016.01979
Del Prete S, Merlo R, Valenti A, Mattossovich R, Rossi M, Carginale V, Supuran CT, Perugino G, Capasso C (2019) Thermostability enhancement of the α-carbonic anhydrase from sulfurihydrogenibium yellowstonense by using the anchoring-and-self-labelling-protein-tag system (ASLtag). J Enzyme Inhib Med Chem. https://doi.org/10.1080/14756366.2019.1605991
Flores GE, Liu Y, Ferrera I, Beveridge TJ, Reysenbach AL (2008) Sulfurihydrogenibium Kristjanssonii Sp. Nov., a hydrogen- and sulfur-oxidizing thermophile isolated from a terrestrial icelandic hot spring. Int J Syst Evol Microbiol. https://doi.org/10.1099/ijs.0.65570-0
Friedricht AB, Antranikian G (1996) Keratin degradation by feividobacterium pennavorans, a novel thermophilic anaerobic species of the order thermotogales. Appl Environ Microbiol. https://doi.org/10.1128/aem.62.8.2875-2882.1996
Guo L, Wang G, Sheng Y, Sun X, Shi Z, Xu Q, Wenqing M (2020) Temperature governs the distribution of hot spring microbial community in three hydrothermal fields, Eastern Tibetan Plateau Geothermal Belt, Western China. Sci Total Environ. https://doi.org/10.1016/j.scitotenv.2020.137574
Härtig C, Lohmayer R, Kolb S, Horn MA, Inskeep WP, Planer-Friedrich B (2014) Chemolithotrophic growth of the aerobic hyperthermophilic bacterium thermocrinis ruber OC 14/7/2 on monothioarsenate and arsenite. FEMS Microbiol Ecol. https://doi.org/10.1111/1574-6941.12431
Henne A, Brüggemann H, Raasch C, Wiezer A, Hartsch T, Liesegang H, Johann A et al (2004) The genome sequence of the extreme thermophile thermus thermophilus. Nat Biotechnol. https://doi.org/10.1038/nbt956
Hugenholtz P, Pitulle C, Hershberger KL, Pace NR (1998) Novel division level bacterial diversity in a yellowstone hot spring. J Bacteriol. https://doi.org/10.1128/jb.180.2.366-376.1998
Hussein EI, Jacob JH, Shakhatreh MAK, Abd Al-Razaq MA, Juhmani ASF, Cornelison CT (2018) Detection of antibiotic-producing actinobacteria in the sediment and water of Ma'in thermal springs (Jordan). GERMS. https://doi.org/10.18683/germs.2018.1146
Jahnke LL, Eder W, Huber R, Hope JM, Hinrichs KU, Hayes JM, Des DJ, Cady SL, Summons RE (2001) Signature lipids and stable carbon isotope analyses of octopus spring hyperthermophilic communities compared with those of aquificales representatives. Appl Environ Microbiol. https://doi.org/10.1128/aem.67.11.5179-5189.2001
Kapanadze N, Melikadze GI, Janelidze P (2010) Estimation of Zugdidi and Tbilisi thermal water deposits. In: Weller A, Melikadze G, Kapanadze N (eds) Exploration and exploitation of groundwater and thermal water systems in Georgia, Tbilisi
Manelius Å, Holst O, Hreggvidsson GO, Kristjansson JK, Raven N, Sharp R (1997) Continuous cultivation of hydrogenobacter thermophilus at low partial pressures of H2. Biotechnol Tech. https://doi.org/10.1023/a:1018405614530
Melikadze G, Tsertsvadze L, Tsertsvadze N, Vardigoreli O, Barabadze T (2010) Country update from Georgia. In: Proceedings of world geothermal congress. Bali, Indonesia, pp 25–29
Mori K, Kamagata Y (2014) The challenges of studying the anaerobic microbial world. Microbes Environ. https://doi.org/10.1264/jsme2.ME2904rh

Najar IN, Sherpa MT, Das S, Das S, Thakur N (2018) Microbial ecology of two hot springs of sikkim: predominate population and geochemistry. Sci Total Environ. https://doi.org/10.1016/j.scitotenv.2018.05.037

Nakagawa S, Shataih Z, Banta A, Beveridge TJ, Sako Y, Reysenbach AL (2005) Sulfurihydrogenibium yellowstonense sp. nov., an extremely thermophilic, facultatively heterotrophic, sulfur-oxidizing bacterium from yellowstone national park, and emended descriptions of the genus sulfurihydrogenibium, sulfurihydrogenibium subterraneum. Int J Syst Evol Microbiol. https://doi.org/10.1099/ijs.0.63708-0

Pantazaki AA, Pritsa AA, Kyriakidis DA (2002) Biotechnologically relevant enzymes from thermus thermophilus. Appl Microbiol Biotechnol. https://doi.org/10.1007/s00253-001-0843-1

Papadopoulos MC (1989) Effect of processing on high-protein feedstuffs: a review. Biol Wastes. https://doi.org/10.1016/0269-7483(89)90092-X

Purcell D, Sompong U, Yim LC, Barraclough TG, Peerapornpisal Y, Pointing SB (2007) The effects of temperature, PH and sulphide on the community structure of hyperthermophilic streamers in hot springs of northern Thailand. FEMS Microbiol Ecol. https://doi.org/10.1111/j.1574-6941.2007.00302.x

Rawat N, Joshi GK (2019) Bacterial community structure analysis of a hot spring soil by next generation sequencing of ribosomal RNA. Genomics. https://doi.org/10.1016/j.ygeno.2018.06.008

Reysenbach AL, Seitzinger S, Kirshtein J, Mclaughlin E (1999) Molecular constraints on a high-temperature evolution of early life. Biol Bull. https://doi.org/10.2307/1542972

Sarmiento F, Peralta R, Blamey JM (2015) Cold and hot extremozymes: industrial relevance and current trends. Front Bioeng Biotechnol. https://doi.org/10.3389/fbioe.2015.00148

Skirnisdottir S, Hreggvidsson GO, Hjörleifsdottir S, Marteinsson VT, Petursdottir SK, Holst O, Kristjansson JK (2000) Influence of sulfide and temperature on species composition and community structure of hot spring microbial mats. Appl Environ Microbiol. https://doi.org/10.1128/AEM.66.7.2835-2841.2000

Takai K, Kobayashi H, Nealson KH, Horikoshi K (2003) Sulfurihydrogenibium Subterraneum Gen. Nov., Sp. Nov., from a subsurface hot aquifer. Int J Syst Evol Microbiol. https://doi.org/10.1099/ijs.0.02506-0

Thawai C, Tanasupawat S, Kudo T (2019) Micromonospora Caldifontis Sp. Nov., isolated from hot spring soil. Int J Syst Evol Microbiol. https://doi.org/10.1099/ijsem.0.003321

Tsertsvadze N, Buachidze G, Vardigoreli O, Vashakidze B, Inaishvili T, Kotrikadze N, Tsertsvadze L (1998) Thermal water of Georgia. Georgian Geothermal Association, Tbilisi

Wang S, Hou W, Dong H, Jiang H, Huang L, Wu G, Zhang C et al (2013) Control of temperature on microbial community structure in hot springs of the Tibetan plateau. PLoS ONE. https://doi.org/10.1371/journal.pone.0062901

Xian W-D, Rao MPN, Zhou E-M, Liu L, Xiao M, Li W-J (2018) Diversity of thermophiles in terrestrial hot springs of Yunnan and Tibet, China. Springer, Singapore. https://doi.org/10.1007/978-981-13-0329-6_3

Yadav P, Maharjan J, Korpole S, Prasad GS, Sahni G, Bhattarai T, Sreerama L (2018) Production, purification, and characterization of thermostable alkaline xylanase from Anoxybacillus Kamchatkensis NASTPD13. Front Bioeng Biotechnol. https://doi.org/10.3389/fbioe.2018.00065

Yamaguchi H, Furuhata K, Fukushima T, Yamamoto H, Sekiguchi J (2004) Characterization of a new Bacillus Subtilis peptidoglycan hydrolase gene, YvcE (Named CwlO), and the enzymatic properties of its encoded protein. J Biosci Bioeng. https://doi.org/10.1016/S1389-1723(04)00262-2

Zayulina KS, Kochetkova TV, Piunova UE, Ziganshin RH, Podosokorskaya OA, Kublanov IV (2020) Novel hyperthermophilic crenarchaeon thermofilum Adornatum Sp. Nov. Uses GH1, GH3, and two novel glycosidases for cellulose hydrolysis. Front Microbiol. https://doi.org/10.3389/fmicb.2019.02972

Zeytun A, Sikorski J, Nolan M, Lapidus A, Lucas S, Han J, Tice H et al (2011) Complete genome sequence of hydrogenobacter thermophilus type strain (TK-6 T). Stand Genomic Sci. https://doi.org/10.4056/sigs.1463589

Chapter 6
Analysis and Characteristics of Thermal Springs in Kazakhstan

Akzhigit Mashzhan, Rubén Javier-López, Aida Kistaubayeva, Irina Savitskaya, and Nils-Kåre Birkeland

Abstract Hot springs are extensively distributed across the world, and the territory of Kazakhstan covered with folded and mountainous areas is no exception. They harbor plenty of microorganisms that can be the source of complex bioactive compounds. Although several microbiological studies on geothermal hot springs from different areas of the planet are available, there is limited information regarding the microbial diversity of similar ecosystems from Kazakhstan. This chapter provides information about thermal springs located in the territory of Kazakhstan as well as microbial diversity analyses of Zharkent hot spring. In Kazakhstan, thermal waters are widespread, which is due to the presence of large artesian basins with the immersion of water-bearing rocks to great depths, as well as the development of folded areas experiencing the impact of the latest tectogenesis. A survey of the available literature has revealed a considerable amount of information on the location of the springs, flow rates, temperature, chemical composition of the water and how the thermal spring has been used. All of these data have been tabulated by geographic areas of Kazakhstan and this tabular information for each geographic zone includes a summarized description of the geology and a map showing the situation of the thermal springs.

Keywords Hot springs · Artesian basin · Thermophiles · Kazakhstan

---

A. Mashzhan (✉)
Department of Biotechnology, Al-Farabi Kazakh National University, Almaty, Kazakhstan

Department of Biological Sciences, University of Bergen, Bergen, Norway

R. Javier-López · N.-K. Birkeland
Department of Biological Sciences, University of Bergen, Bergen, Norway
e-mail: Ruben.Javier-Lopez@uib.no; Nils.Birkeland@uib.no

A. Kistaubayeva · I. Savitskaya
Department of Biotechnology, Al-Farabi Kazakh National University, Almaty, Kazakhstan
e-mail: irina.savickaya@kaznu.kz

D. Egamberdieva et al. (eds.), *Microbial Communities and their Interactions in the Extreme Environment*, Microorganisms for Sustainability 32,
https://doi.org/10.1007/978-981-16-3731-5_6

6.1 Introduction

The term "geothermal springs" refers to natural groundwater heated above 20 °C (Prokhorov 1969). The depth of occurrence of the 20 °C isotherm in the earth's crust depends on climatic zoning: in areas of permafrost development, this depth of occurrence is set between −1500 and −2000 m; up to 100 m in the subtropical areas, and on the surface in the tropics. A low thermal regime is observed mainly in the territory of ancient shields and ancient folded systems. Whereas the moderate thermal regime can be seen within the ancient artesian platform areas, and the increased one in artesian areas of the Epipaleozoic plates and also in associated intermountain depressions and troughs, as well as in hydrogeological regions of alpine folding, where systems of tectonic faults. The high thermal regime is associated with areas of artesian basins (discharge of thermal waters from the deep parts of the basins) and hydrogeological folded areas that have experienced intense neotectonic impact (Ivanov and Nevraev 1964; Mavriczkij 1971). These extreme habitats have attracted wide interest so far as they are analogs for primal land (Stan-Lotter and Fendrihan 2012). Geothermal areas are considered selective habitats for thermophiles due to their unique properties, well adapted to these extreme environments (Raddadi et al. 2015; DeCastro et al. 2016). Thermophilic microorganisms need to live at high temperatures in hot springs, hot hydrothermal vents, and black smokers; these microorganisms not only survive but even thrive in these harsh conditions (Deepika and Satyanarayana 2013). The most studied and well described areas with thermal springs are in the United States (Boomer et al. 2009), Turkey (Adiguzel et al. 2009), Iceland (Krebs et al. 2014), New Zealand (Hetzer et al. 2007), Italy (Maugeri et al. 2001), Indonesia (Xu et al. 2013), and India (Saxena et al. 2016). The appealing characteristic of hot springs is the ecology, with its variety of thermophilic microorganisms with a unique ability. Due to their outstanding resistance to high temperatures and their metabolic flexibility, thermophilic microorganisms are as interesting for biotechnological research as they are for industrial applications (Raddadi et al. 2015; Huber and Stetter 2001). Thermozymes adapted to higher temperatures bring privileges to industrial processes, contribute to speed up chemical reactions while decreasing the risk for contamination of the system, increase the solubility of the substrate, and also decrease the solution viscosity and improve the miscibility of the solvent (Liszka et al. 2012). Currently more than 500 products are produced using enzymes and about 150 industrial processes benefit from the use of enzymes or catalysts from microorganisms. Moreover, more than 3000 enzymes are known; among them, approximately 65% are hydrolases used in the detergent, textile, pulp, paper and starch industries, and almost 25% of these are used for food processing (Adrio and Demain 2014; Niehaus et al. 1999).

Geothermal springs located in Kazakhstan remain still unexplored and can be an important source of undescribed thermophilic microorganisms. The first large geothermal research in our country was carried out during the USSR period in the oil regions of Kazakhstan. This is because geothermal research was mainly carried out along the way in connection with the search for deposits of oil, gas, coal and other minerals, and the development of their deposits (Zhevago 1963). More than forty hot springs have been discovered during the last decades in Kazakhstan territory, due to the presence of large artesian basins (Zhevago 1963). Most of the sources are located in mountainous regions in the fracturing zone of the crystalline basement, in deep structures of loose-detrital deposits of artesian basins, and in zones of tectonic faults. In spite of a broad spread of thermal springs throughout Kazakhstan with hints of intrinsic scientific interest, restricted attention has been paid toward the microbiological investigation of these areas. In this scene, the microbial communities in Kazakhstan hot springs need to be explored and studied in detail. The initial purpose of this chapter is to review the geothermal springs in Kazakhstan and the results of the first investigation of bacterial diversity in Zharkent hot spring using cultural dependent methods. The results of this study gave an assessment of the microbiological potential of the Zharkent thermal spring and it will be a basis for comparison with other geothermal springs in the world.

6.2 Geographical Distribution and Formation of Geothermal Springs in the Territory of the Republic of Kazakhstan

In Kazakhstan, thermal springs are widespread, due to the presence of large artesian basins with immersion of water-bearing rocks to great depths. As well as the development of folded areas experiencing the impact of the latest tectogenesis. Thermal springs are exposed in mountainous regions in the zone of fracturing of the crystalline basement at great depths, in zones of tectonic faults, and also in deep structures of loose-detrital deposits of artesian basins.

According to the results of more than 40 years of studies on geothermal resources in Kazakhstan, about a hundred prospecting wells were drilled. These wells opened thermal waters with conditioned characteristics in terms of flow rates, temperature and salinity, gas and chemical composition, identifying promising geothermal sources for research (Smolyar et al. 2002; Sydykova 1977, 1981; Zhevago 1972, 1976; Abdulina 1999; Mukhamedzhanova 1990).

Thermal waters have local development and belong to the fissure-vein type; Epipaleozoic platforms, foredeeps, and intermountain depressions filled with Mesozoic and Cenozoic deposits with areal distribution of reservoir-porous and reservoir-fissured waters with salinity not exceeding 35 g/l.

Thermal waters with a temperature of more than 40 °C are common in the Ili, Syrdarya, Shu-Sarysu, Mangistau-Ustyurt, and southern Caspian artesian basins.

Medium thermal (75–100 °C) groundwater is widespread in the Ili, Syrdarya, and Mangistau-Ustyurt artesian basins.

Highly thermal (more than 100 °C) groundwater is recorded in the Zharkent and Almaty depressions, the Mangyshlak-Ustyurt artesian basin. The most promising for

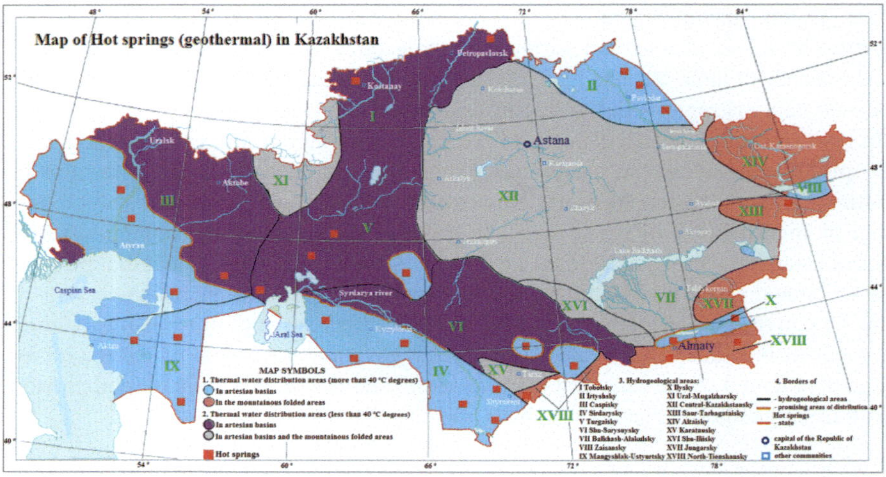

Fig. 6.1 Thermal springs map of Kazakhstan. The percentage of thermal springs in Kazakhstan: western Kazakhstan—75.9 (78.2%); south Kazakhstan—15.6 (16%); central Kazakhstan—5.3 (5.5%); north Kazakhstan—0.3 (0.3%); east Kazakhstan—0.003

our research are thermal springs with temperatures ranging from 40 to 100 °C, and above.

The main places of distribution for hydrogeothermal waters are confined to the Kazakhstani parts of the Turan plate (56.3%) and the Caspian basin (36%) (Mavriczkij 1971) (Fig. 6.1).

The following areas of geothermal waters in Kazakhstan fall under this criterion; the Mangyshlak-Ustyurt system, the Syrdarya artesian basin, and the Ili artesian basin.

6.2.1 Mangyshlak-Ustyurt System of Thermal Springs

The Mangyshlak-Ustyurt system of thermal springs is confined to the Aral-Caspian watershed and occupies the western part of the Turan plate. Geothermal resources are associated with Cretaceous and Jurassic formations. The composition of thermal waters includes industrially significant concentrations of iodine, boron, bromine and other microcomponents (Tyumenev 2008).

The Cretaceous thermal water-bearing complex is almost ubiquitous and lies at a depth of up to 2000 m and more in the troughs of Mangyshlak and Ustyurt. Piezometric levels are set from 160–250 m below to the first tens of meters above the earth's surface. Well flow rates vary within 140–3500 m^3/day. Salinity of water ranges from 1–10 g/dm^3 in the area of uplifts to 6–35 g/dm^3 in the Zhetybai-Uzek zone and up to 50–100 g/dm^3 in the troughs of Mangyshlak and Ustyurt, with a predominant sodium chloride composition. The reservoir temperature of groundwater varies from 50–65 °C in the Zhetybai-Uzek zone to 100–120 °C in the North Ustyurt trough and up to 120–150 °C in the depressions of South Mangyshlak and South Ustyurt. The water temperature at the outlet of the springs is 40–60 °C.

The Jurassic thermal water-bearing complex is also widely developed and is exposed at a depth of 1650–3200 m and more. The levels are set at a depth of 10–60 to 240–290 m. Well flow rates vary within 8–260 m^3/day. Brine waters (100–195 g/dm^3) with sodium chloride composition. The reservoir temperature of the water reaches 130–175 °C in the most submerged parts, and at the wellhead the water temperature ranges from 40–60 to 80–110 °C (Vol'vovskiy et al. 1966; Zhevago 1972; Kononov 1965).

6.2.2 *Syrdarya Artesian System of Thermal Springs*

Syrdarya artesian basin is located within the South Kazakhstan and Kyzylorda regions. In its section, thermal waters are confined to Cretaceous thermal water-bearing complexes. The depth of thermal springs reaches up to 2000 m, and their salinity is not higher than 3 g/dm^3. Well flow rates up to 2000 m^3/day. On the territory of the Syrdarya artesian basin, two deposits of geothermal springs have been identified: Shaulderskoe and Arys (Abdulina 1999) (Fig. 6.2). Based on the results of the survey on the territories of the Shaulderskoe and Arysskoe thermal springs, the following wells can be distinguished, as indicated in Table 6.1 (Geotherm Manufacturing Company 2020).

The Shaulder thermal spring is located 149 km northwest of the city of Shymkent. The explored site is located on the territory of the Shaulder regional center. The

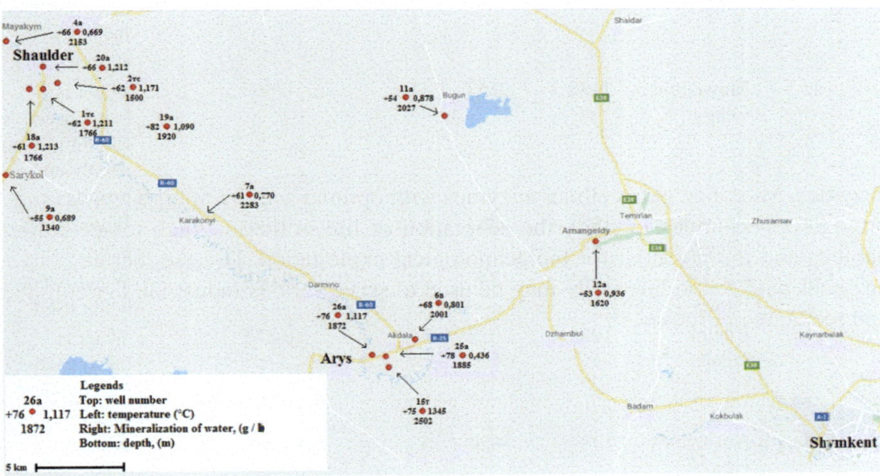

Fig. 6.2 Shaulderskoe and Arysskoe system of hot springs (adopted from Google maps 2020)

Table 6.1 Wells located on the territory of Shaulderskoe and Arysskoe thermal springs

No.	Well No.	Location of the wells	Drilling year	Well depth (m)	Water temperature (°C)	Mineralization of water (g/l)	Usage
1	15т	Arys city	1962	2502	75	1.345	Water is being used for the bath
2	26a	Arys city	1990	1872	76	1.117	Water is not being used
3	25a	Arys city	1990	1885	78	0.436	Water is being used for hot water supply and heating
4	6в	Akdala village	1962	2001	68	0.801	Water is not being used
5	18a	Kokmardan village	1991	1766	61	1.213	Water is being used for hot water supply and heating
6	1тс	Kokmardan village	1969	1800	62	1.211	Water is being used for hot water supply
7	20a	Shaulder village	1991	–	66	1.212	Water is being used for the bath
8	2тс	Shaulder village	1969	1500	62	1.171	Water is not being used
9	19a	Shaulder village	1989	1920	82	1.090	Water is being used for hot water supply
10	4a	Mayakum village	1988	2153	66	0.669	Water is being used for hot water supply and heating
11	7a	Karakonyr village	1987	2283	61	0.770	Water is not being used
12	9a	Sarykol village	1985	1340	55	0.689	Water is not being used
13	11a	Bugun village	1986	2027	54	0.878	Water is being used for hot water supply
14	12a	Amangeldy village	1984	1620	53	0.936	Water is being used for hot water supply

underground waters of the deposit are characterized as highly thermal with a water temperature at the wellhead of 50–70 °C (Absametov et al. 2014).

The Arys thermal spring is confined to the city of Arys, which is a regional center and one of the largest railway stations in the south of Kazakhstan. Thermal waters are characterized as thermal since the base of the Cenomanian aquifer the temperature is 90 °C, and at the wellheads, it is 75 °C. The most abundant component is hydrocarbon-chloride-sulfate sodium. Groundwater contains very small amounts of dissolved gases, of which nitrogen predominates. The corrosive aggressiveness of thermal waters is assessed as slightly aggressive (Absametov et al. 2014).

6.2.3 Ili Artesian System of Thermal Springs

The Ili artesian basin is confined to the latitudinal depression of the same name, located between the Tien Shan and Zhetysu Alatau ridges. There are two geothermal hot springs: Almaty and Zharkent (Mukhamedzhanov et al. 1965; Yanshin 1965).

The Almaty thermal spring occupies the western part of the depression (Fig. 6.3). In its section, the Neogene and Paleogene thermal water-bearing complexes were discovered, the depths of which in the axial part, respectively, are up to 650 and 1500–2600 m. Dozens of deep wells have been drilled within the Almaty thermal spring (Table 6.2). Wells are usually self-flowing with a productivity from 10–500 to 800–2200 m^3/day and mineralization of water from <3 to 10–15 and more g/dm^3 with sulfate chloride and sodium chloride composition. Water temperature at a depth of 700–800 m is up to 40 °C, and at a depth of up to 2600–3000 m is at 75–84 °C.

The Zharkent thermal spring is confined to the depression of the same name in the eastern part of the Ili depression (Fig. 6.4). Thermal groundwater here is associated with formations from Cretaceous to Triassic age.

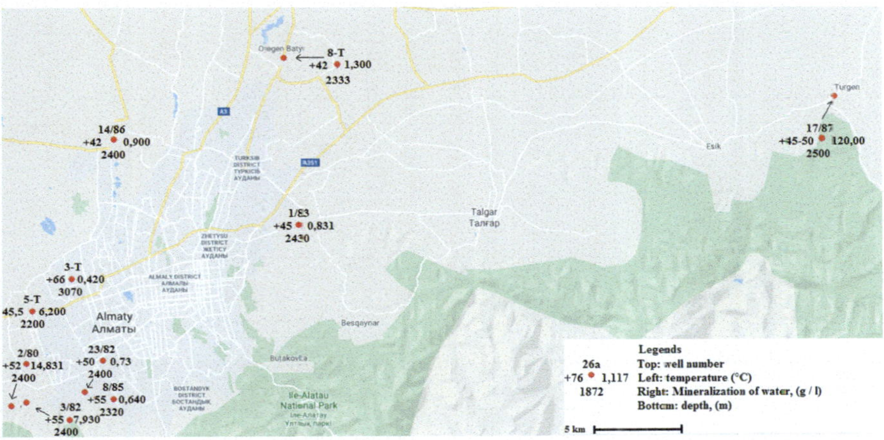

Fig. 6.3 Location of Almaty thermal spring wells on the map (adopted from Google maps 2020)

Table 6.2 Description of Almaty thermal spring wells

No.	Well No.	Drilling year	Well depth (m)	Water temperature (°C)	Mineralization of water (g/l)	Usage
1	8-T	1980	2333	42	1.300	Water is being used for hot water supply
2	8/85	–	2320	55	0.640	Water is being used for hot water supply
3	5-T	–	2200	45.5	6.200	Water is being used for hot water supply
4	3-T	1958	3070	66	0.420	Water is being used for hot water supply
5	3/82	–	2400	55	7.930	Water is being used for hot water supply
6	23/82	–	2400	50	0.673	Water is being used for hot water supply
7	2/80	1961	2400	52	14.831	Water is being used for hot water supply
8	17/87	–	2500	45–50	120.00	Water is not being used
9	14/86	1963	2400	42	0.900	Water is being used for hot water supply
10	1/83	1962	2430	45	0.831	Water is being used for hot water supply

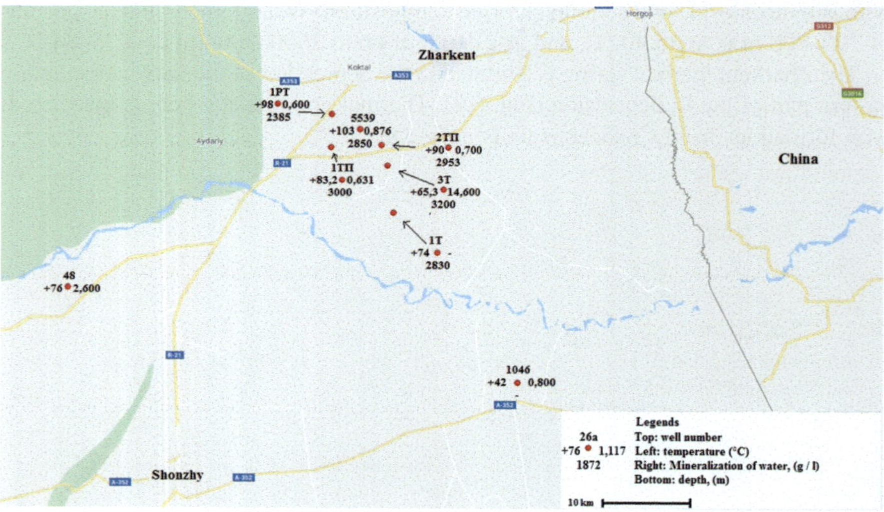

Fig. 6.4 Location of Zharkent thermal spring wells on the map (adopted from Google maps 2020)

Table 6.3 Wells located on the territory of Zharkent thermal spring

No.	Well No.	Drilling year	Well depth, (m)	Water temperature (°C)	Mineralization of water (g/l)	Usage
1	5539	2016	2850	103	0.876	Water is being using for farming
2	1-PT	–	2385	98	0.600	Water is being used for fish farming
3	1-ТП	–	3000	83	0.631	Water is being used for farming
4	2-ТП	–	2953	90	0.700	Water is being used for farming
5	3T	–	3200	65	14.600	Water is being used for farming
6	1T	–	2830	74	–	Water is being used for farming
7	1046	–	–	42	0.800	Water is being used for farming
8	48	–	–	76	2.600	Water is not being used

Plenty of deep wells have been drilled within the Zharkent thermal spring (Table 6.3). The depth of occurrence increases from the foothills to the axial part of the depression from 20–150 to 3300 m and more. On the foothill plain of the Ketmen ridge (the Karadala tract), thermal waters occur at a depth of 300–600 m. The levels are set at 20–70 m above the ground. The productivity of self-flowing wells is 900–12000 m^3/day. The waters are usually fresh (up to 1 g/dm^3), and their chemical composition varies from bicarbonate calcium to mixed three-anionic sodium and sodium-calcium. In the central part of the artesian basin, the thermal water-bearing complex was tested at a depth of 1400–2900 m. The waters are high-pressure, piezometric levels are set 70–240 m above the earth's surface, and the flow rate of wells on self-pouring is 1900–5200 m^3/day. Mineralization of water is less than 1 g/dm^3 with hydrocarbonate-sulfate and chloride-hydrocarbonate sodium composition. The water temperature at the wellhead is from 47 to 96 °C. In the most submerged parts of the depression, the water temperature is expected to be 100–125 °C (Mashzhan et al. 2021; Mukhamedzhanov 1971; Zektser and Everett 2004).

Triassic and Jurassic thermal water-bearing complexes have been tested in the southern half of the Zharkent basin. The depth of their occurrence varies from 250–400 m in the foothills to 4000–4500 m in the central part. The water abundance of the complex is quite variable, the flow rates of wells on self-flow vary from 110 to 4700 m^3/day. The mineralization of water ranges from less than 1 to 3 g/dm^3, and the chemical composition—from bicarbonate calcium and chloride-bicarbonate calcium-sodium to chloride sodium. The water temperature of the Triassic and Jurassic thermal water-bearing complexes at the outflow is 38–78 °C. According to calculations, the temperature at the bottom of thermal water-bearing complexes, depending on the depth of occurrence, varies from 40–75 to 155–165 °C (Lyubimova 1959; Makarenko and Dvorov 1967; Mukhamedzhanov 1971).

6.3 Microbiological Analysis

Molecular and cultivation-dependent approaches have been used to analyze spore-forming bacteria from Zharkent hot spring. During this investigation, five different bacteria have been isolated. The isolates were given codes viz., 4Ak1, 4Ak2, 4Ak3, 4Ak4, 4Ak5. A water sample was used for enrichment in nutrient broth (HiMedia, Mumbai) at 65 °C for 3 days, and then the enriched culture was streaked on nutrient agar to obtain pure colonies. The isolates were considered as pure after detecting under the optical microscope a single cell per culture. Unto inoculation, water samples were incubated at 80 °C for 10 min so that only spore-forming bacteria were obtained (Panosyan and Birkeland 2014).

To determine the effect of temperature on the growth of the isolates, microorganisms were grown for 24 h at different incubation temperatures in the range of 35–95 °C with regular 5 °C increment. Besides, the effect of pH on the growth of the isolated microorganisms was studied by growing the organism for 24 hours at optimal temperature in nutrient broth medium adjusted to different pH, ranging from 4.0 to 9.0. The growth of the organisms was assessed by determining the optical density at 620 nm on a spectrophotometer (PD-303 spectrophotometer, Apel, Japan). Carbohydrate assimilation and fermentation of 49 compounds plus one control was determined on API 50 CH strips (bioMerieux, France). API was carried out by following the manufacturer's instructions, using API 50 CHB/E medium. Additional enzymatic activities were determined using the API ZYM strips (bioMerieux, France) (Table 6.4).

It was found that all the isolates grew well at alkaline pH, as the pH of the hot geothermal spring is naturally alkaline. The effect of pH on the growth profile of all the strain was between 6 and 9. The bacterial isolates were screened to check their thermotolerance in a range of temperatures starting from 35 °C to 95 °C. Isolates 4Ak1, 4Ak2, 4Ak5 showed an optimum temperature of 75°C, and maximum and minimum temperature were 80 °C and 40 °C, respectively. With isolates 4Ak3, the optimum temperature was 70 °C, with maximum 75 °C and minimum 40°C. For isolates 4Ak4 the optimal temperature was 65°C and maximum and minimum temperature were 70 °C and 38 °C, respectively. The isolates 4Ak1, 4Ak2, 4Ak5 showed the highest growth temperature among the genus *Anoxybacillus*.

The screening and identification of thermophilic bacteria producing extracellular hydrolases was performed using various carbon sources such as carboxymethyl cellulose (CMC), potato starch, tween 80 and skim milk (Table 6.5). Casein peptonization was tested on plates containing (w/v) skimmed milk powder (20%), glucose (0.5%) and agarose (1.5%); clear zones of hydrolysis around colonies indicated positive results (Wehr and Frank 2004). Decomposition of carboxymethyl cellulose was tested on a medium containing (w/v) CMC (1%), NaCl (0.01%),

Table 6.4 Different characteristics of isolates

Characteristics	Isolates				
	4Ak1	4Ak2	4Ak3	4Ak4	4Ak5
Optimal temperature for growth	75 °C	75 °C	70 °C	65 °C	75 °C
Optimal pH for growth	6–8	6–8	6–7.5	6–9	6–8
Fermentation patterns of (API 50 CH)					
Glycerol	–	+	–	–	+
Erythritol	–	+	–	–	–
D-Arabinose	–	+	–	–	–
L-Arabinose	–	+	–	–	–
Ribose	+	–	+	+	+
D-xylose	–	–	–	–	–
L-xylose	–	–	–	–	–
Adonitol	–	–	–	–	–
Methyl-D-xylopyranoside	–	–	–	–	–
Galactose	–	–	–	–	–
D-Glucose	+	–	+	–	+
D-Fructose	+	±	+	+	+
D-Mannose	–	–	–	–	+
L-Sorbose	+	±	+	±	–
Rhamnose	–	–	–	–	±
Dulcitol	–	–	–	–	–
Inositol	–	–	–	–	–
D-Mannitol	+	+	+	–	+
D-Sorbitol	–	–	–	–	–
Methyl-D-mannopyranoside	–	–	–	–	–
Methyl-D-glucopyranoside	+	–	+	–	–
N-Acetylglucosamine	+	–	–	–	–
Amygdalin	+	–	+	–	+
Arbutin	–	–	–	–	–
Esculin	+	+	+	+	+
Salicin	–	–	–	–	+
D-Cellobiose	+	–	+	–	+
D-Maltose	+	–	+	–	+
D-Lactose	+	–	–	–	–
D-Melibiose	–	–	–	–	–
D-Sucrose	±	–	+	–	+
D-Trehalose	+	–	+	+	+
Inulin	–	–	–	–	+
D-Melezitose	+	–	+	–	±
D-Raffinose	+	–	+	–	+
Starch	+	–	+	+	+
Glycogen	–	–	–	–	±
Xylitol	–	–	–	–	–

(continued)

Table 6.4 (continued)

Characteristics	Isolates				
	4Ak1	4Ak2	4Ak3	4Ak4	4Ak5
Gentiobiose	–	–	–	–	–
D-Turanose	+	–	+	+	+
D-Lyxose	+	+	–	–	±
D-Tagatose	+	+	+	–	+
D-Fucose	–	–	–	–	–
L-Fucose	–	–	–	–	–
D-Arabitol	+	–	–	–	–
L-Arabitol	+	–	–	–	–
Gluconate	–	–	–	–	–
2-ketogluconate	–	–	–	–	–
5-ketogluconate	–	–	–	–	–
Presence of (API ZYM)					
Alkaline phosphatase	+	+	+	+	+
Esterase (C 4)	+	+	+	+	+
Esterase Lipase (C 8)	+	+	+	+	+
Lipase (C 14)	±	–	+	–	+
Leucine arylamidase	–	–	+	–	+
Valine arylamidase	–	–	+	–	+
Cystine arylamidase	–	–	+	–	+
Trypsin	–	–	+	+	+
α-chymotrypsin	–	–	+	–	+
Acid phosphatase	+	±	+	±	+
Naphthol-AS-BI-phosphohydrolase	+	+	+	+	+
α-galactosidase	+	–	±	–	+
β-galactosidase	–	–	+	–	+
β-glucuronidase	–	–	–	±	–
α-glucosidase	±	+	+	+	+
β-glucosidase	–	±	+	±	+
N-acetyl-β-glucosaminidase	–	–	–	–	–
α-mannosidase	–	–	–	–	–
α-fucosidase	–	–	–	–	–

+ positive; ± weakly positive; – negative. All of the isolates were positive to D-Fructose, esculin, but do not assimilated D-xylose, L-xylose, adonitol, Methyl-D-xylopyranoside, Galactose, Dulcitol, Inositol, D-Sorbitol, Methyl-D-mannopyranoside, Arbutin, D-Melibiose, Xylitol, Gentiobiose, D-Fucose, L-Fucose, Gluconate, 2-ketogluconate, 5-ketogluconate. Isolate 3A2KZ was only one that assimilated Erythritol, D-Arabinose, L-Arabinose, but do not assimilated Ribose, D-Trehalose, Starch, D-Turanose. Isolate assimilated D-Mannose, Rhamnose, Salicin, Inulin, Glycogen, but did not assimilate L-Sorbose. All of the isolates in API Zym tests were negative for N-acetyl-β-glucosaminidase, α-mannosidase, α-fucosidase, but all of them positive for Alkaline phosphatase, Esterase (C 4), Esterase Lipase (C 8), Acid phosphatase, α-glucosidase. None of the isolates showed equal characteristics during these tests

Table 6.5 Hydrolytic enzyme production by obtained bacterial isolates

	Enzyme activity[a]											
	Protease			Cellulase			Amylase			Lipase		
Isolates	55 °C	65 °C	75 °C	55 °C	65 °C	75 °C	55 °C	65 °C	75 °C	55 °C	65 °C	75 °C
4Ak1	–	–	–	+	+	++	–	–	–	++	++	++
4Ak2	–	–	–	+	+	+	–	+	+	++	++	+++
4Ak3	–	–	–	+	++	++	+	++	++	+	++	+
4Ak4	–	–	–	+	++	–	++	++	–	+	+	–
4Ak5	–	–	–	–	–	–	+	+	+	–	+	–

[a]Enzyme activity was expressed by diameter of clear zone (in case of, amylase and cellulose) and precipitation (in case of lipase and protease) around colonies: (<4 mm; +) (4–8 mm: ++) (>8 mm; +++)

$NaNO_3$ (0.02%), K_2HPO_4 (0.01%), $MgSO_4$ (0.003%), KCl (0.003%), peptone (0.01%) and agar (1.5%) (Shaikh et al. 2013; Shokatayeva et al. 2019). Amylase activity was tested by determining starch hydrolysis on plates containing (w/v) soluble starch (2%), tryptone (1%), yeast extract (0.1%) and agarose (1.5%). A transparent halo around the colony after staining with Lugol's solution (0.5% I_2 and 1.0% KI (w/v) in distilled water) indicates the presence of enzymatic activity (Springham et al. 1999; Kasana et al. 2008). The production of lipolytic enzymes was assessed in Spirit blue agar medium (Sigma, USA) supplemented with 1% (w/v) Tween 80 (Sigma, USA). The presence of lipase activity was indicated by visible precipitation of calcium salts of fatty acids (Wehr and Frank 2004).

Isolate 4Ak1 produced cellulase and lipase while isolate 4Ak5 was capable to produce amylase and lipase. In addition, three isolates (4Ak2, 4Ak3, 4Ak4) produced a combination of three extracellular hydrolytic enzymes.

The phylogenetic analysis of the strain began with DNA extraction. The genomic DNA was isolated from the pelleted bacterial cells with a NA2100 SIGMA GenElute™ Bacterial Genomic DNA extraction kit, by Sigma-Aldrich, following the manufacturer's protocol. The DNA quality and quantity were determined using a NanoDrop™ One/OneC Microvolume UV-Vis Spectrophotometer, by Thermofisher Scientific. Universal primer pairs 27f (5′-GAG TTT GAT CCT GGC TCA -3′) and 1525r (5′-GAA AGG AGG AGA TCC AGC C-3′) (*Escherichia coli* numbering) were used to amplify 16S rRNA genes. GenElute™ PCR Cleanup Kit (Sigma) was applied to purify PCR products. Then, the PCR products were used to perform a cycle sequencing reaction according to the Big Dye v3.1 protocol (Sequencing facility), and then sequenced in the UiB (Bergen university, Bergen, Norway) Sequencing facility. The obtained sequences were corrected using MEGA X software (Kumar et al. 2018). Then, they were merged and a consensus sequence obtained using the EMBOSS software package (Rice et al. 2000). This consensus sequence was then compared in BLAST through the Blastn suite, using the default options for megablast query (https://blast.ncbi.nlm.nih.gov/Blast.cgi). The isolate 4Ak1 showed a close relationship to *Anoxybacillus salavatliensis* strain DSM 22626 (99.53% identity) and at the same time with *Anoxybacillus gonensis* strain G2

(99.53% identity). The isolate 4Ak2 showed 99.61% identity relationship with *Anoxybacillus kamchatkensis* strain G10, *Anoxybacillus salavatliensis* strain DSM 22626, *Anoxybacillus gonensis* strain G2 and *Anoxybacillus flavithermus* clone LK4. The isolate 4Ak3 showed 100% a close relationship to *Anoxybacillus gonensis* strain G2 (99.59% identity). The isolate 4Ak4 showed close relationship to Uncultured bacterium clone bac50 (99.53% similarity) and to *Anoxybacillus kamchatkensis* strain G10 with 99.46% identity, while the isolate 4Ak5 showed a relationship to *Anoxybacillus kamchatkensis* strain TS13 (98.98% similarity). A phylogenetic tree based on the 16S rRNA gene sequence was reconstructed for the members of genus *Anoxybacillus,* and is shown in Fig. 6.5. Phylogenies were inferred by the GGDC web server (Meier-Kolthoff et al. 2013) using the DSMZ phylogenomics pipeline (Meier-Kolthoff et al. 2014) adapted to single genes (http://ggdc.dsmz.de/). A multiple sequence alignment was created with MUSCLE (Edgar 2004). Maximum likelihood (ML) and maximum parsimony (MP) trees were inferred from the alignment with RAxML (Stamatakis 2014) and TNT (Goloboff et al. 2008), respectively. For ML, rapid bootstrapping in conjunction with the autoMRE bootstopping criterion (Pattengale et al. 2010) and subsequent search for the best tree was used; for MP, 1000 bootstrapping replicates were used in conjunction with tree-bisection-and-reconnection branch swapping and ten random sequence addition replicates. The sequences were checked for a compositional bias using the X^2 test as implemented in PAUP (Swofford 2002).

The input nucleotide matrix comprised 30 operational taxonomic units and 1577 characters, 243 of which were variable and 120 of which were parsimony-informative. The base-frequency check indicated no compositional bias ($p = 1.00$, $\alpha = 0.05$). ML analysis under the GTR+GAMMA model yielded a highest log likelihood of -4737.61, whereas the estimated alpha parameter was 0.02. The ML bootstrapping did not converge; hence, 1000 replicates were conducted; the average support was 52.41%. MP analysis yielded a best score of 423 (consistency index 0.68, retention index 0.76) and 70 best trees. The MP bootstrapping average support was 62.19%. The phylogenetic tree showed that the relationship between isolate 4Ak2 clustered together with the strain most related to *Anoxybacillus bogrovensis* NBIMCC 8427 (NR_115021.1) and for isolate, 4Ak3 it was *Anoxybacillus gonensis* G2 (NR_025667.1), while the isolates 4Ak1, 4Ak4 and 4Ak5 formed their own clusters.

6.4 Conclusion

Microbial communities in thermal springs living at harsh conditions have been extensively studied worldwide. In this sense, the Kazakhstan folded highlands are valuable ecosystems for providing both the extreme "cold" and "hot" sites for exploring microbial diversity. A deep study of all issues related to the distribution of geothermal waters in Kazakhstan, and the conditions of their formation is not only of practical interest in geological research, but also has an important role in biology.

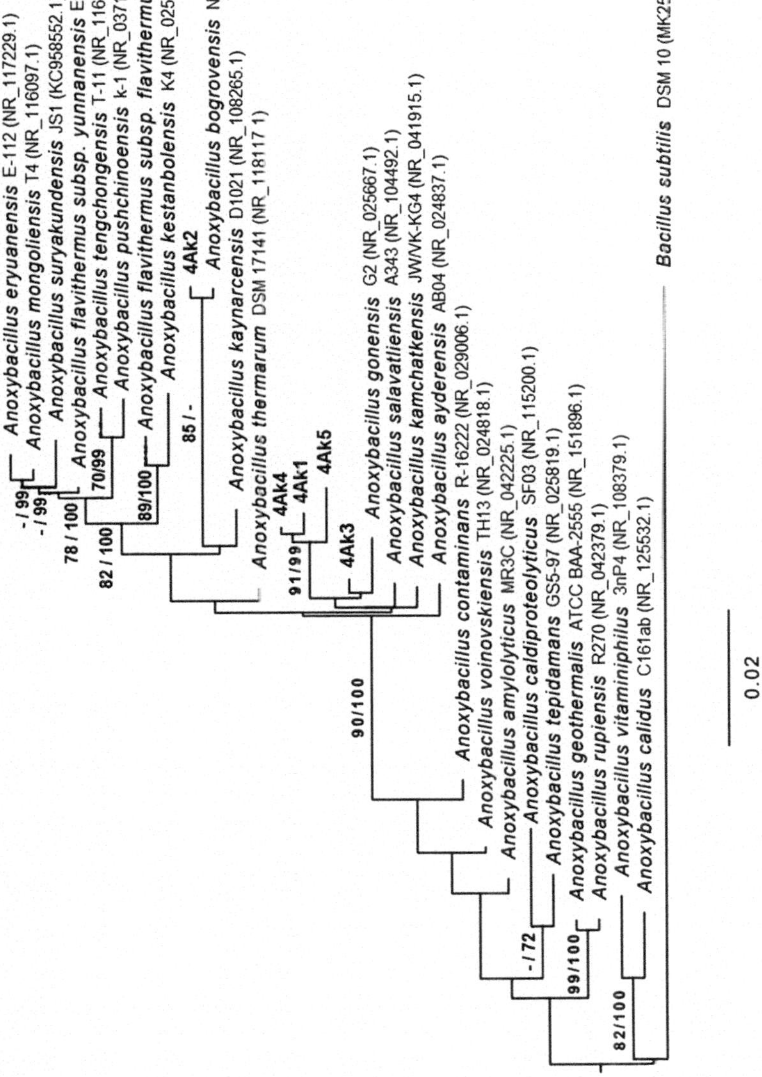

Fig. 6.5 ML tree inferred under the GTR+GAMMA model and rooted by midpoint-rooting. The branches are scaled in terms of the expected number of substitutions per site. The numbers over the branches are support values when higher than 60% from ML (left) and MP (right) bootstrapping. *Bacillus subtilis* DSM 10 used as an outgroup

Bacteria living in thermal springs and their extremely thermostable enzymes could be used in microbial biotechnology. Moreover, the ancient origin and the position in the phylogenetic system allow the possibility of obtaining new significant results in the field of fundamental microbiology and also evolution. Present study on different aspects of these springs reveals important findings on geological setting, hydrological character, chemical nature of thermal waters and their location. The presence of these minerals and chemicals along with higher temperature of these hot springs in Kazakhstan provide of good conditions to enhance the microorganisms diversity. The average temperature of the studied hot springs composed 70 °C. They often include significant concentrations of iodine, boron, bromine, and other micro components. The speed of discharge of thermal water from the springs varies from 8 to 12,000 m^3/day and the overall discharge speed is stable over the year. The high-temperature springs are described by gas and vapor ebullitions and are noticeable with rings of waves on the water surface. Moreover, most of them have a distinctive sulfurous smell. Thermal springs located on the territory of the Ili artesian system are of greatest interest as they are less susceptible to anthropogenic factors. The highest temperature in this region was 103 °C, this is the hottest thermal spring found in the Kazakhstan region. This is the first investigation of spore-forming bacteria from Zharkent hot spring. Other powerful factors forming the microbiota of the studied Zharkent geothermal spring it appears that high temperature, mineralization, pressure, and pH. Geological and biogeography history should not be disregarded in microbial ecology investigation, because all non-living chemical components collectively endow to the dynamics of the microbial communities. Few thermophilic spore-forming bacteria belonging to the *Anoxybacillus* genera have been purified, identified, and assessed taking into account their enzymatic activity. The present study, therefore, gives new information about the thermophilic bacterial diversity and enzymatic potential of thermal spring in the Zharkent and extends information regarding the geothermal springs in Kazakhstan.

References

Abdulina A (1999) Mineral'nye lechebnye i termal'nye (teploenergeticheskie) podzemnye vody (Mineral medicinal and thermal (heat and power) underground waters). Nauka, Almaty

Absametov M, Murtazin E, Kasymbekov D (2014) Prespektivy osvoeniya gidrogeotermal'nykh i gidrogeomeneral'nykh resursov Kazakhstana (Prospects for the development of hydro-geothermal and hydro-mineral resources in Kazakhstan). Iz Tom Pol Univ 1:110–118

Adiguzel A, Ozkan H, Baris O, Inan K, Gulluce M, Sahin F (2009) Identification and characterization of thermophilic bacteria isolated from hot springs in Turkey. J Microbiol Methods 79 (3):321–328

Adrio JL, Demain AL (2014) Microbial enzymes: tools for biotechnological processes. Biomol Ther 4(1):117–139

Boomer SM, Noll KL, Geesey GG, Dutton BE (2009) Formation of multilayered photosynthetic biofilms in an alkaline thermal spring in Yellowstone National Park, Wyoming. Appl Environ Microbiol 75(8):2464–2475

DeCastro ME, Rodriguez-Belmonte E, Gonzalez-Siso MI (2016) Metagenomics of thermophiles with a focus on discovery of novel thermozymes. Front Microbiol 7:1521

Deepika M, Satyanarayana T (2013) Diversity of hot environments and thermophilic microbes. In: Satyanarayana T, Littlechild J, Kawarabayasi Y (eds) Thermophilic microbes in environmental and industrial biotechnology biotechnology of thermophiles. Springer, London, pp 3–60

Edgar RC (2004) MUSCLE: multiple sequence alignment with high accuracy and high throughput. Nucleic Acids Res 32(5):1792–1797

Geotherm Manufacturing Company (2020). http://www.geotherm.kz. Accessed 20 Sept 2020

Goloboff PA, Farris JS, Nixon KC (2008) TNT, a free program for phylogenetic analysis. Cladistics 24(5):774–786

Hetzer A, Morgan HW, McDonald IR, Daughney CJ (2007) Microbial life in champagne pool, a geothermal spring in Waiotapu, New Zealand. Extremophiles 11(4):605–614

Huber R, Stetter KO (2001) Discovery of hyperthermophilic microorganisms. Methods Enzymol 330:11–24

Ivanov V, Nevraev G (1964) Klassifikatsiya podzemnykh mineral'nykh vod (Classification of underground mineral waters). Nauka, Moskva

Kasana RC, Salwan R, Dhar H, Dutt S, Gulati A (2008) A rapid and easy method for the detection of microbial cellulases on agar plates using gram's iodine. Curr Microbiol 57(5):503–507

Kononov V (1965) Vliyanie estestvennykh i iskusstvennykh ochagov tepla na formirovanie khimicheskogo sostava podzemnykh vod (Influence of natural and artificial heat centers on the formation of the chemical composition of groundwater). Nauka, Moskva

Krebs JE, Vaishampayan P, Probst AJ, Tom LM, Marteinsson VT, Andersen GL, Kasthuri V (2014) Microbial community structures of novel Icelandic hot spring systems revealed by PhyloChip G3 analysis. Astrobiology 14(3):229–240

Kumar S, Stecher G, Li M, Knyaz C, Tamura K (2018) MEGA X: molecular evolutionary genetics analysis across computing platforms. Mol Biol Evol 35(6):1547–1549

Liszka MJ, Clark ME, Schneider E, Clark DS (2012) Nature versus nurture: developing enzymes that function under extreme conditions. Annu Rev Chem Biomol Eng 3:77–102

Lyubimova E (1959) O temperaturnom gradiente v verkhnikh sloyakh Zemli i vozmozhnosti ob'yasneniya sloya ponizhennykh skorostey (On the temperature gradient in the upper layers of the Earth and the possibility of explaining the layer of reduced velocities). AN SSSR, Moskva

Makarenko F, Dvorov I (1967) Regional'naya geotermiya i rasprostranenie termal'nykh vod v SSSR (Regional geothermy and the distribution of thermal waters in the USSR). Nauka, Moskva

Mashzhan A, Javier-López R, Kistaubayeva A, Savitskaya I, Birkeland NK (2021) Metagenomics and culture-based diversity analysis of the bacterial community in the zharkent geothermal spring in Kazakhstan. Curr Microbiol 1–9

Maugeri TL, Gugliandolo C, Caccamo D, Stackebrandt E (2001) A polyphasic taxonomic study of thermophilic bacilli from shallow, marine vents. Syst Appl Microbiol 24(4):572–587

Mavriczkij B (1971) Termal'nye vody skladchatykh i platformennykh oblastey SSSR (Thermal waters of folded and platform regions of the USSR). Nauka, Moskva

Meier-Kolthoff JP, Auch AF, Klenk HP, Goker M (2013) Genome sequence-based species delimitation with confidence intervals and improved distance functions. BMC Bioinf 14:60

Meier-Kolthoff JP, Hahnke RL, Petersen J, Scheuner C, Michael V, Fiebig A, Rohde C, Rohde M, Fartmann B, Goodwin LA, Chertkov O, Reddy T, Pati A, Ivanova NN, Markowitz V, Kyrpides NC, Woyke T, Göker M, Klenk HP (2014) Complete genome sequence of DSM 30083(T), the type strain (U5/41(T)) of *Escherichia coli*, and a proposal for delineating subspecies in microbial taxonomy. Stand Genomic Sci 9(1):8

Mukhamedzhanov S (1971) Gidrogeologiya severo-vostochnoy chasti Kazakhstana (Hydrogeology of the north-eastern part of Kazakhstan). Nauka, Alma-Ata

Mukhamedzhanov S, Isabaev T, Kabiev F (1965) Podzemnye vody khrebta Tarbagatay i ego ravninnykh predgoriy (Underground waters of the Tarbagatai ridge and its flat foothills). Nauka, Alma-Ata

Mukhamedzhanova SM (1990) Podzemnye termal'nye vody Kazakhstana. Nauka (Underground thermal waters of Kazakhstan), Alma-Ata

Niehaus F, Bertoldo C, Kahler M, Antranikian G (1999) Extremophiles as a source of novel enzymes for industrial application. Appl Microbiol Biotechnol 51(6):711–729

Panosyan H, Birkeland NK (2014) Microbial diversity in an Armenian geothermal spring assessed by molecular and culture-based methods. J Basic Microbiol 54(11):1240–1250
Pattengale ND, Alipour M, Bininda-Emonds ORP, Moret BME, Stamatakis A (2010) How many bootstrap replicates are necessary? J Comput Biol 17(3):337–354
Prokhorov AM (1969) Termal'nye vody (thermal waters). Nauka, Moskva
Raddadi N, Cherif A, Daffonchio D, Neifar M, Fava F (2015) Biotechnological applications of extremophiles, extremozymes and extremolytes. Appl Microbiol Biotechnol 99(19):7907–7913
Rice P, Longden I, Bleasby A (2000) EMBOSS: The European molecular biology open software suite. Trends Genet 16(6):276–277
Saxena R, Dhakan DB, Mittal P, Waiker P, Chowdhury A, Ghatak A, Sharma VK (2016) Metagenomic analysis of hot springs in central india reveals hydrocarbon degrading thermophiles and pathways essential for survival in extreme environments. Front Microbiol 7:2123
Shaikh NM, Patel A, Mehta S, Patel N (2013) Isolation and screening of cellulolytic bacteria inhabiting different environment and optimization of cellulase production. Univers J Environ Res Technol 3(1):25–34
Shokatayeva D, Ignatova L, Savitskaya I, Kistaubaeva A, Talipova A, Asylbekova A, Abdulzhanova M, Mashzhan A (2019) Bacterial cellulose and pullulan from simple and low cost production media. Eurasian Chem-Technol J 21(3):247–258
Smolyar V, Burov B, Veselov V, Makhmutov T, Kasymbekov D (2002) Vodnye resursy Kazakhstana (Poverkhnostnye i podzemnye vody, sovremennoe sostoyanie) (Water resources of Kazakhstan (Surface and ground waters, current state)). NICz Fylym, Almaty
Springham D, Moses V, Cape R (1999) Biotechnology - the science and the business. CRC Press, London
Stamatakis A (2014) RAxML version 8: a tool for phylogenetic analysis and post-analysis of large phylogenies. Bioinformatics 30(9):1312–1313
Stan-Lotter H, Fendrihan S (2012) Adaption of microbial life to environmental extremes: novel research results and application. Springer, New York
Swofford D (2002) PAUP*. Phylogenetic analysis using parsimony (* and other methods), version 4.0 b10. Sinauer Associates, Sunderland
Sydykova ZS (1977) Geotermicheskie usloviya Aralo - Kaspiyskogo neftenosnogo regiona (Geothermal conditions of the Aral-Caspian oil-bearing region). Nauka KazSSR, Alma-Ata
Sydykova ZS (1981) Termoanomalii podzemnykh vod Kazakhstana (Thermal anomalies of groundwater in Kazakhstan). Nauka KazSSR, Alma-Ata
Tyumenev S (2008) Vodnye resursy i vodoobespechennost' territorii Kazakhstana (Water resources and water supply in Kazakhstan). Nauchno-tekhnicheskij izdatel`skij czentr KazNTU, Almata
Vol'vovskiy IS, Garetskiy RG, Shlezinger AE, Shraybman EG (1966) Tektonika turanskoy plity (Tectonics of the Turan plate). Nauka, Moskva
Wehr HM, Frank JF (2004) Standard methods for the examination of dairy products, 17th edn. American Public Health Association, Washington, DC
Xu SY, He PQ, Dewi SZ, Zhang XL, Ekowati C, Liu TJ, Huang XH (2013) Hydrogen-producing microflora and Fe-Fe hydrogenase diversities in seaweed bed associated with marine hot springs of Kalianda, Indonesia. Curr Microbiol 66(5):499–506
Yanshin A (1965) Tektonicheskoe stroenie Evrazii (Tectonic structure of Eurasia). GIN AN SSSR, Moskva
Zektser I, Everett L (2004) Groundwater resources of the world and their use. UNESCO, Saint-Denis
Zhevago V (1963) Termal'nye vody SSSR i voprosy ikh teploenergeticheskogo ispol'zoaniya (Thermal waters of the USSR and issues of their heat and power use). AN SSSR, Moskva
Zhevago V (1972) Geotermiya i termal'nye vody Kazakhstana (Geothermy and thermal waters of Kazakhstan). Nauka KazSSR, Alma-Ata
Zhevago V (1976) Teplovoy rezhim i geotermal'naya energiya nedr Yuzhnogo Kazakhstana (Thermal regime and geothermal energy of the bowels of South Kazakhstan). Nauka KazSSR, Alma-Ata

Chapter 7
Purple Photosynthetic Bacteria: A Brief Research Overview on Distribution in Armenia and Biotechnological Application

Vigen Goginyan, Baghish Harutyunyan, Ruzanna Hovhannisyan, Naira Andreasyan, and Narine Kalantaryan

Abstract The chapter summarizes the results of long-term studies of purple photosynthetic bacteria in Armenia, namely in the Institute of Microbiology NAS RA (National Academy of Sciences of the Republic of Armenia) and Scientific and Production Center "Armbiotechnology" NAS RA. Their biodiversity, ecology, taxonomy, morphological, physiological, and biochemical characteristics, as well as the prospects for obtaining biologically active compounds, such as enzymes, carotenoids, vitamins, organic acids, amino acids, polysaccharides, etc. are described. The results of studies conducted over the years show that Armenia, having a great variety of geographical zoning and climatic conditions, is endowed with a very rich and diverse microflora, the study of which is a topical issue both theoretically and practically. It is very important to study and preserve the vast biodiversity of phototrophic bacteria in Armenia, as these microorganisms have enormous genetic potential, rich metabolic pathways, different growth conditions, which greatly contribute to their application in photobiotechnology. On the other hand, the ability to assimilate different organic wastes can be used to improve the ecological condition of the human environment, at least to some extent, as the residues from the assimilation of waste can be used as biological fertilizers.

The publication is dedicated to the memory of Araksya Khosrov Paronyan (1943–2010), Head of the Laboratory of Photosynthetic Microorganisms (1989–2009) of the Institute of Microbiology of the National Academy of Sciences of the Republic of Armenia, an enthusiastic scientist who greatly contributed to the study of microbiological biodiversity in Armenia.

Keywords Purple photosynthetic bacteria · Armenian biodiversity · Distribution · Biologically active substances · Biotechnology · Application

---

V. Goginyan (✉) · B. Harutyunyan · R. Hovhannisyan · N. Andreasyan · N. Kalantaryan
Scientific and Production Center "Armbiotechnology", National Academy of Sciences, Republic of Armenia, Yerevan, Armenia

D. Egamberdieva et al. (eds.), *Microbial Communities and their Interactions in the Extreme Environment*, Microorganisms for Sustainability 32,
https://doi.org/10.1007/978-981-16-3731-5_7

7.1 Introduction

Photosynthetic bacteria are the most primitive of the forms of photosynthetic organisms currently living in nature. The photosynthetic bacteria can be found in fresh, salt, acidic and basic waters, and also in various wastewaters. This is a widespread group of microorganisms that plays a significant role in the formation of the biosphere and the preservation of the vital functions of ecological systems (Kondrat'eva et al. 1989).

The research areas of photosynthetic bacteria are quite extensive, and the ability to convert light energy, neutralize toxic gases and utilize heavy metals, lability, and versatile metabolism, as well as assimilation of organic compounds that are inaccessible to other groups of microorganisms determine their enormous ecological significance. The study of photosynthetic bacteria is also important due to the problem of fixing molecular nitrogen, using sulfur compounds by microorganisms and their cycle in nature (Kobayashi and Haque 1971; Pfennig 1975; Kobayashi 1982; Willems et al. 1992).

The production of bacterial biomass using photosynthetic bacteria can replenish the lack of assimilable protein while reducing the level of environmental pollution and the cost of the product (Tsygankov and Gogotov 1990; Malatyan and Paronyan 1997). At the same time, the biomass of photosynthetic bacteria is characterized by a high content of essential amino acids and various biologically active compounds, which makes it a promising and valuable source of food for farm animals, birds, and fish. The biomass of photosynthetic bacteria is used as a raw material for producing enzymes, vitamins, organic acids, polysaccharides, carotenoid pigments, food colors, etc. Photosynthetic bacteria are also capable of accumulating large amounts of poly-β-hydroxyalkanoates in cells, which are most often considered as an alternative source for the production of biodegradable polymers (Jensen et al. 1961; Sasikala and Ramana 1995; Butler 2005; Sasaki et al. 2005).

This paper for the first time summarizes a several results of long-term studies of purple non-sulfur and sulfur photosynthetic bacteria of Armenia, carried out in the Laboratory of Photosynthetic Microorganisms (now the Laboratory of Energy Alternative Sources) of the Institute of Microbiology of the NAS RA and the Scientific and Production Center "Armbiotechnology" NAS RA.

Biodiversity, ecology, taxonomy, morphological, physiological, and biochemical characteristics of bacteria are described, as well as the prospects for obtaining biologically active compounds: vitamins, organic acids, enzymes, amino acids, polysaccharides, and protein-vitamin concentrates.

7.2 Ecology and Biodiversity of Photosynthetic Bacteria of Different Geographical Zones of Armenia

Sharp differences in the ecological-geographic and climatic conditions of the Caucasus, in particular, on the territory of Armenia, have always attracted attention of scientists in terms of identifying the biodiversity of endemic species of animals, plants and microorganisms, as well as the patterns of their distribution in nature.

At the end of the nineteenth century, Academician V.V. Dokuchaev according to the results of the scientific expedition to the Caucasus published two works in which he pointed out the main regularities of the genesis and geographical location of soils, noted a special position of the soil in nature, which is determined by the participation of both mineral and organic compounds in its composition. He proved that an indispensable part of the soil constituted living organisms, including root systems of plants, insects, and animals living in the soil as well as microorganisms.

The patterns of distribution of various organisms in the soils of the Caucasus depending on the environmental and climatic life conditions have also been studied by other scientists. In particular, Academician E.M. Mishustin based on the long-term research, taking into account a huge amount of factual material has revealed the patterns of the vertical zoning of various groups of soil microorganisms (Mishustin 1950; Mishustin and Pertsovskaya 1954). Nevertheless, the ecology and biodiversity of photosynthetic bacteria in Armenia have not been sufficiently studied.

Later on, our laboratory studied in detail the distribution, morphological features and biological properties of photosynthetic bacteria living in mineral springs, large freshwater reservoirs and saline soils of Armenia depending on the influence of environmental factors, taking into account the seasonal and annual dynamics of the species contamination of natural habitats.

7.2.1 Photosynthetic Bacteria of Mineral Springs of Armenia

Mineral springs of Armenia are mainly hydrocarbonate. However, their physicochemical parameters vary to some extent, which is caused by the geographical location and the diversity of geological and structural conditions of their formation (Paronyan 2010). It is the physicochemical characteristics of individual ecological niches that are considered as the main factors determining the spread of photosynthetic bacteria (Fig. 7.1).

As can be seen from the data of Table 7.1, the biodiversity of microorganisms in mineral springs is presented by non-sulfur and purple sulfur photosynthetic bacteria (Malatyan et al. 1982; Paronyan 2002; Paronyan and Gasparyan 2009). Nevertheless, their quantitative and species compositions differ greatly.

In the mineral springs, located in the southern regions of Armenia, mainly purple non-sulfur bacteria predominate, while in the northern regions purple sulfur bacteria predominate.

Fig. 7.1 Distribution of photosynthesizing bacteria in mineral waters of Armenia ((**a**)—Hankavan, (**b**)—Arzakan, (**c**)—Arzni, (**d**)—Jermuk)

The intensive spread of photo- and heterotrophic microorganisms is facilitated by a relatively high average annual temperature, on the one hand, and by an increased content of organic substances, on the other hand.

Thus, in the mineral springs of the Ararat group, the genera of non-sulfur (*Rhodobacter*, *Rhodopseudomonas*, *Rhodospirillum*) and sulfur (*Allochromatium*) bacteria prevail. As the dominant species appear *Rhodobacter (Rba.) sphaeriodes*, *Rhodopseudomonas (Rps.) palustris* and *Allochromatium vinosum*, and, in limited quantities, the green sulfur bacterium *Chlorobium limicola*. Photosynthetic bacteria from the Jermuk mineral springs have a relatively rich species composition with a relatively low titer of microorganisms.

Along with the species detected in the Ararat springs, *Rba. capsulatus*, *Rba. sulfidophilum*, *Rps. acidophila*, *Allochromatium minus*, *Thiospirillum jenense* were revealed there. Moreover, non-sulfur purple bacteria were isolated from the thermal Jermuk springs, the temperature of which ranged from 50 to 62^{0}C with the maximum growth temperature of bacteria—42–50 °C. At the same time, halotolerant *Rba. sphaeroides* strains that can withstand NaCl concentration up to 8% in the nutrient medium, were isolated from the bottom sediments of the main reservoir of the spring. It should be noted that in the springs of Jermuk no sharp quantitative changes

Table 7.1 Physicochemical indices of mineral waters of Armenia and distribution of dominant photosynthetic bacteria

Groups of mineral springs	Types of mineral springs	t (°C)	pH	Total mineralization (g/)l	CO_2 (g/l)	Dominant species	Contamination (water/sludge) (CFU/ml)
Ararat	Calcium-sodium	24–28.5	6.5–6.8	1.2–4.27	2.0–3.0	*Rhodobacter sphaeroides* *Rba. palustris* *Rhodospirillum rubrum*	860/1160 500/850 200/650
Arzni	Sodium-chloride	13.3–22.5	6.3–6.6	1.4–16.8	1.8–8.5	*Rba. sphaeroides* *Rba. capsulatus* *Rsp. rubrum* *Rhodocyclus gelatinosus*	500/700 40/70 400/600 40/280
Hankavan	Chloride-magnesium	11–12	7.2–7.4	4.8–7.5	2.0–2.5	*Rba. sphaeroides* *Rhodopseudomonas acidophila* *Rsp. rubrum*	500/800 10/35 22/40
Alaverdi	Ferric chloride	7–16	5.5–6.7	2.7–4.1	0.2–1.9	*Rba. capsulatus* *Rsp. acidophila*	150/300 110/225
Bjni-Arzakan	Sodium-chloride	14.2–44	6.7–8.0	2.4–5.95	0.72–8.8	*Rba. sphaeroides* *Rps. palustris* *Thiocapsa roseopersicina*	1100/1550 140/250 10/30
Jermuk	Chloride-sulfate	11–58	5.9–7.6	0.7–5.3	0.2–2.9	*Rba. sphaeroides* *Rba. capsulatus* *Thiospirillum jenense*	1150/1420 115/100 8/14
Dilijan	Calcium-magnesium	10–13.5	6.3–6.8	3.5–3.9	2.2–2.5	*Rba. capsulatus* *Rsp. acidophila*	50/65 11/18
Lori	Sulfate-calcium	8.6–12.5	6.3–6.4	2.3–5.2	0.39–2.7	*Rba. sphaeroides* *Rba. capsulatus* *Rsp. rubrum*	130/150 116/25 250/780

(continued)

Table 7.1 (continued)

Groups of mineral springs	Types of mineral springs	t (°C)	pH	Total mineralization (g/)l	CO_2 (g/l)	Dominant species	Contamination (water/sludge) (CFU/ml)
Sevan-Gavar	Sodium-magnesium	4–18	6.3–6.7	3.5–5.1	2.0–4.2	*Rba. capsulatus* *Rsp. acidophila*	112/225 10/519
Shirak	Sulfide	12–15	6.8–7.2	4.0–5.9	1.0–1.6	*Rba. sulfidophilum* *Thiocapsa roseopersicina* *Thiospirillum jenense* *Allochromatium vinosum*	170/380 120/350 60/220 100/300

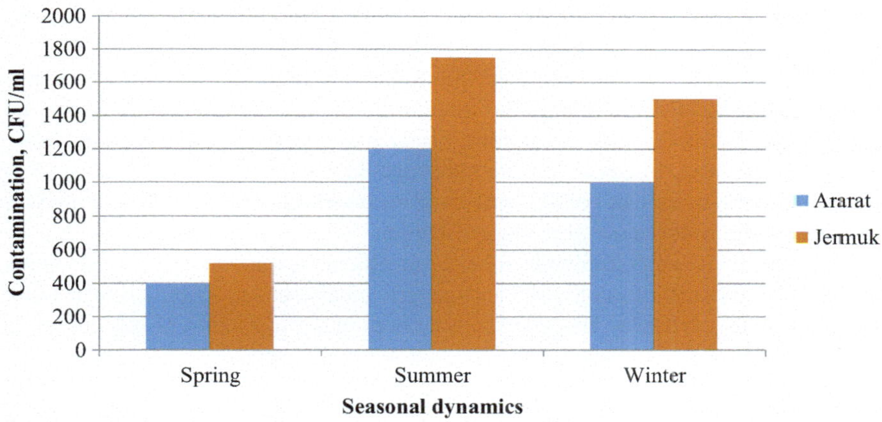

Fig. 7.2 Seasonal changes in the number of photosynthetic bacteria in the mineral springs of Ararat and Jermuk

in bacteria were detected as distinct from the Ararat mineral waters (Paronyan 2006a, b, 2007).

On the other hand, it is established that bacterial associations, rich in species diversity, are more resistant to fluctuations of climatic factors, and the seasonal dynamics of the total amount of photosynthetic bacteria in the mineral springs of Ararat and Jermuk are shown in Fig. 7.2 (Paronian et al. 1982).

The mineral springs of Bjni, Arzakan, and Hankavan are characterized by a wider and at the same time evenly distributed species composition of non-sulfur and sulfur photosynthetic bacteria (Paronyan 1997). The most common are the genera *Rhodobacter*, *Rhodopseudomonas*, *Rhodospirillum*, *Thiocapsa*, *Ectothiorhodospira*, *Allochromatium* and *Clorobium*, while the dominant species are *Rba. sphaeroides*, *Rps. palustris*, *Rps. acidophila*, *Rsp. rubrum*, *Thiocapsa roseopersicina*.

The interest in the Shirak group of mineral springs is stipulated exceptionally by the high content of hydrogen sulfide. Thus, for example, in the springs of Shirakavan-Noraber its concentration reaches 120–130 mg/l.

It is natural that purple sulfur bacteria of the genera *Thiocapsa*, *Allochromatium*, *Thiospirillum*, and *Lamprocystis* are widely represented here with the predominantly dominant species *Thiospirillum jenense* and *Allochromatium vinosum* in the rare associations with non-sulfur bacteria of the *Rba. sphaeroides* and *Rba. sulfidophilum* species.

The mineral springs of Arzni are characterized by a high content of non-sulfur purple bacteria of the genera *Rhodobacter*, *Rhodopseudomonas* and *Rhodospirillum*, the dominant species of which are *Rba. palustris*, *Rba. capsulatus*, *Rba. sphaeroides*, *Rsp. rubrum*, *Rhodocyclus gelatinosus*. Such a wide variety of non-sulfur bacteria is primarily due to the high content of organic compounds in the springs at the expense of the widespread intensive blooming of blue-green microalgae (cyanobacteria) (Eliazian et al. 1984).

In contrast to the afore-mentioned, the mineral springs of Alaverdi are distinguished by relatively low pH values (in the range of 5.5–6.8) which allow the development of moderately acidophilic bacteria of the *Rsp. acidophila* and *Rba. capsulatus* species. Moreover, the latter are capable of existing in the pH range from 5.5 to 10.5 with the increase of the hydrocarbon concentration. The same picture was observed in the acidic mineral springs of Dilijan.

Thus, the biodiversity of photosynthetic bacteria in the mineral waters of Armenia is due to a complex of ecological and geographical factors. Together with other microorganisms, they play an important role both in the formation of biocenoses in general, and in the composition of individual mineral waters.

7.2.2 Photosynthetic Bacteria of Natural Water Sources of Armenia

Natural water reservoirs in Armenia are not numerous. They are mainly represented by freshwater, hydrocarbonate, weakly acidic or neutral lakes with a low portion of mineralization (Table 7.2), fed by underground springs containing significant amounts of sulfates. The microbiological processes carried out by photosynthetic bacteria are practically identical. According to the studies of Gorlenko et al. (1977) and Overmann et al. (1996) three ecological niches are identified in water reservoirs: aerobic, microaerophilic, and anaerobic. Photosynthetic bacteria are concentrated in the microaerophilic zone characterized by low light penetration. However, due to the rich carotenoid composition, bacteria are able to use for their life short-term light penetrating to a depth of 5–10 m.

The high-mountainous Lake Sevan by the Shorzha underwater shaft-ridge, the Artanish peninsula and the protruding Noraduz cape is divided into two parts: Small Sevan (maximum depth 58.7 m) and Big Sevan (maximum depth 80 m). Due to the low average annual temperature, biological processes in the lake are rather slow. In summer months, the water temperature in the coastal zone is 18–19 °C with an average annual water temperature of +4 °C. In winter, the average air temperature near the lake ranges from −7 to −30 °C. In winter months, vertically moving streams of cold water saturated the lake with oxygen to a depth of 21 m, and after decrease in the water level—to a depth of 10–15 m. Artificial lowering of the water level caused disturbances in the heat and gas balances in the lake, which resulted in the formation of photosynthetic microflora of Sevan.

Since a small amount of hydrogen sulfide is generated in the lake, the non-sulfur purple bacteria *Rba. sphaeroides* and *Rps. palustris* are most common here, as well as *Rba. capsulatus* and *Rhodocyclus gelatinosus* in the areas of the reservoir mostly polluted with organic compounds. In the bottom sediments of the lake, the sulfur bacteria *Allochromatium* and the green photosynthetic bacteria *Chlorobium* are

Table 7.2 Physicochemical indices of the main water reservoirs and the distribution of dominant photosynthetic bacteria

Name of the lake	Maximum temperature (°C)	pH	Total mineralization (g/l)	Carbonates (%)	Soluble salts (%)	Dominant species	Contamination (water/sludge) (CFU/ml)
Lake Sevan	19	6.0–7.8	0.72	–	–	*Rhodobacter sphaeroides* *Rhodopseudomonus palustris* *Allochromatium vinosum*	120/240 120/314 14/10
Lake Aygr	26	7.0–7.2	0.52	–	–	*Rba. sphaeroides* *Rba. capsulatus* *Rhodocyclus gelatinosus* *A. vinosum*	580/1320 500/1100 400/1000 25/90
Lake Akna	18	7.1–7.2	0.1	–	–	*Rba. sphaeroides* *A. vinosum*	13/55 6/40
Lake Parz	12	6.9–7.0	0.1	–	–	*Rba. sphaeroides* *A. vinosum*	15/65 8/50
Lake Ahagel	28	6.8–6.9	5.9	–	–	*Rba. capsulatus* *Rsp. rubrum* *Thiospirillum jenense*	200/560 180/425 100/210

found, though rare. At that, the titer of bacteria in Lake Sevan in general is low. Their number increases in the summer period, when there is a massive seasonal bloom of algae and cyanobacteria at a depth of 2–40 m, accompanied by a significant increase in the content of organic compounds (Paronyan and Malatyan 1995). With the gradual degradation of the lake, its transition from an oligotrophic to a dystrophic state, not least due to vigorous human activity, the titer of non-sulfur bacteria has increased on average 5–10 times in recent years.

Lake Aygr (Metsamor) located at an altitude of 800 m above sea level, has a maximum depth of 9.4 m. The lake is fed by underground water sources and has a constant mineralization level of 500–520 mg/l. In some places, it is swampy due to which the level of light penetration is limited. This fact contributes to the uneven distribution of photosynthetic bacteria in the water mass. The purple sulfur bacteria of the *Allochromatium*, *Lamprocystis*, and *Thiospirillum* genera prevail here.

However, in summer months, with abundant bloom of algae and an excess of organic matter, the microflora balance shifts toward non-sulfur phototrophs represented by the *Rba. capsulatus*, *Rba. Sphaeroides*, and *Rhodocyclus gelatinosus* species. At the same time, in the bottom sediments of sludge, the green sulfur bacteria *Chlorobium limicola* and *C. limicola* ssp. *thiosulfatophilum* are found.

The freshwater Lakes Akna (height above sea level—3032 m, depth—up to 15 m) and Parz (Parzlich) (height above sea level—1350 m, depth—up to 10 m) belong to the water bodies of a dystrophic type. They are fed mainly by thawed or rain waters, therefore, the mineralization level in them is relatively low—100 mg/l. The lakes are characterized by a low content of non-sulfur and sulfur photosynthetic bacteria, but by their species diversity, they are comparable to Lake Sevan. Noteworthy is the fact that in winter it is practically impossible to find photosynthetic bacteria in these lakes.

Extremely interesting in terms of biodiversity is a small salt Lake Agagol. It is the salinity of the lake that determines the composition of the detected halophilic photosynthetic bacteria resistant to up to 8% NaCl concentration. The lake is saturated with hydrogen sulfide. Therefore, purple sulfur bacteria, mainly belonging to the *Allochromatium* and *Thiospirillum* genera, predominate. At the same time, in the aerobic-anaerobic zones of the water reservoir, in the presence of insignificant amounts of oxygen, mixed associations of sulfur and non-sulfur sulfate-reducing bacteria are found. It is established that all isolated non-sulfur bacteria (*Rba. capsulatus*, *Rba. sphaeroides*, *Rsp. rubrum*) are microaerophiles capable of utilizing small amounts of hydrogen sulfide.

Thus, the studies performed have shown that despite a number of ecological-geographical and climatic differences in the studied water reservoirs of Armenia, the species composition of photosynthetic bacteria is homogeneous. Nevertheless, being one of the most common groups of aquatic microorganisms, they play an important role in the transformation of organic and mineral compounds, thereby improving the ecological state of water bodies.

7.2.3 Photosynthetic Bacteria of the Soda Saline Soils of the Ararat Valley

The soda saline soils of the Ararat Valley occupy about 30,000 hectares. They are characterized by high alkalinity and low mineralization (Table 7.3). In summer months, under conditions of high air temperature (up to 42–45 °C), groundwater located close to the earth's surface intensively evaporate, leaving behind them sharply different in composition salts, including readily soluble ones (about 2% in a water extract). These soils are characterized by a poor humus layer, a significant content of carbonates (up to 15–18%) and exchangeable sodium (20–25 mg-eq).

During microbiological analysis of soil samples, we identified mainly extremophilic—alkalophilic forms of sulfur bacteria of the *Ectothiorhodospira*, *Thiospirillum*, *Thiocapsa*, *Allochromatium* and *Lamprocystis* genera, the most common species being *Ectothiorhodospira mobilis*, *Lamprocystis roseopersicina*, *Thiospirillum jenense*, and *Allochromatium vinosum.*

Most of these strains are characterized by mixotrophy, microaerophilic, halotolerance (from 6 to 8%), and resistance to pH values in the range of 9.0–10.5. The peak of development of photosynthetic bacteria in saline soils coincides with the late spring and early autumn periods. In summer, during dry weather, the number of bacteria is sharply reduced, and some species are practically not detected. Taking into account the morphological, physiological and biochemical peculiarities of species populations associated with the habitat, it is possible to judge the formation of new ecotypes created as a result of long-term exposure to similar regimes of ecological factors (Paronyan 2001, 2006a, b).

7.2.4 Taxonomy of Domestic Strains of Purple Photosynthetic Bacteria

The systematics of purple photosynthetic bacteria is constantly changing, undergoing significant changes due to detection of a great number of new species along with the discovery of new physiological and biochemical features.

In recent years, new methods of comparative analysis have been used to classify photosynthetic bacteria as well as other microorganisms: determination of the nucleotide composition of DNA, study of nucleotide sequences of 16S rDNA and sets of oligonucleotides of 5S rRNA, study of the composition of ribosomal proteins and determination of amino acid sequences in cytochromes type "C," as well as chemotaxonomic studies. Based on the results obtained it was concluded on the close evolutionary interconnection of photosynthetic purple bacteria, as well as of some aerobic chemotrophic bacteria and eukaryotic mitochondria.

In the Bergey's Manual purple photosynthetic bacteria are isolated in the *Rhodospirillales* order (with two suborders—*Rhodospirillineae* and *Chlorobiineae*), which consists of 39 species united in 18 genera.

Table 7.3 Physicochemical parameters of saline soils and distribution of dominant photosynthetic bacteria

Saline soils	Maximum temperature (°C)	pH	Total mineralization (g/l)	Carbonates (%)	Soluble salts (%)	Dominant species	Contamination (CFU/ml)
Soil	21	9.0–10.5	–	15–18	1.5–1.95	*E. mobilis* *L. roseopersicina* *T. jenense*	20 17 32
Ground waters	30	9.0–10.5	–	13.5–16	1.4–2.1	*L. roseopersicina* *A. vinosum* *E. mobilis* *Rba. sulfidophilus*	18 30 15 6

The division of purple bacteria into sulfur and non-sulfur bacteria was suggested by Molisch (1907) on the grounds that the former when oxidized with sulfide are capable of depositing elemental sulfur in the cells. As the data on the properties of these microorganisms accumulated, their characteristics on this feature changed. It was shown that representatives of *Ectothiorhodospira* genus, first described by Pelyp (1936) though they oxidized hydrogen sulfide and thiosulfate did not accumulate sulfur in the cells (Kondratyeva 1963; Raymond and Sistrom 1969). It has also been established that the ability to use sulfide and thiosulfate as electron donors in CO_2 photoassimilation is widespread in purple non-sulfur bacteria (Keppen and Gorlenko 1975; Hansen and Gemerden 1972; Pfennig 1969). Some of them, for example, *Rhodopseudomonas palustris*, *Rps. sulfoviridis*, *Rps. sulfidophilum* oxidize sulfide to sulfate without intermediate sulfur formation, others—*Rhodobacter sphaeroides*, *Rba. capsulatus*, *Rhodospirillum rubrum* only to sulfur but accumulate it outside the cells. Therefore, purple non-sulfur bacteria (the *Rhodospirillaceae* family) are characterized as microorganisms that can use hydrogen sulfide and thiosulfate as an electron donor to form either elemental sulfur (without accumulation in the cell) or sulfates but without intermediate formation of sulfur. Thus, it is emphasized that these microorganisms do not oxidize sulfur. When characterizing the purple sulfur bacteria (the *Chromatiaceae* family) it is noted that all species of this family can use sulfide or elemental sulfur as an electron donor, oxidizing them to sulfates.

In recent years purple bacteria that are close in a number of morphological and physiological properties to non-sulfur bacteria, but oxidize sulfide first to sulfur and then to sulfates, have been isolated (Kompantseva and Gorlenko 1984; Neutzling et al. 1984).

Purple non-sulfur bacteria have a pronounced tendency to a photo-heterotrophic life. They preferably use organic substances—acetate, pyruvate, lactate, and a number of others (Pfennig 1969). On the contrary, all purple sulfur bacteria can be considered as autotrophs, since they grow on purely mineral media containing carbon dioxide, or with the addition of vitamin B_{12}, but without requiring substrate amounts of organic compounds. Moreover, some of these microorganisms are referred to the category of the so-called obligate autotrophs since their abilities to assimilate other carbon compounds in addition to carbon dioxide are very limited.

Many species of *Rhodospirillaceae* can grow in the dark under aerobic and/or microaerobic conditions (Pfennig and Truper 1973, 1992). Purple sulfur bacteria are more sensitive to oxygen, but among them there are species that grow in the dark in the presence of oxygen. These include, for example, *Thiocapsa roseopersicina*, *Amoebobacter roseus* and some species of *Allochromatium* (Bogorov 1974; Gorlenko 1974; Kondratyeva et al. 1976).

The classification of purple sulfur bacteria is complicated by the fact that of the 27 known species of the *Chromatiaceae* family four species (*Thiosarcina rosea*, *Thiospirillum sanguineum*, *Thiospirillum rosenbergii*, *Thiopedia sevani*) have not yet been isolated as pure cultures (Pfennig and Truper 1992).

Thus, based on genetic analysis and considering the geographical features of the distribution of photosynthetic bacteria in Armenia, they can be divided into three

Table 7.4 The most widespread genera and species of purple photosynthetic bacteria in habitats of Armenia

Genera	Species
Purple non-sulfur photosynthetic bacteria	
Rhodobacter	*sphaeroides, capsulatus, sulfidophilus*
Rhodopseudomonas	*palustris, acidophila*
Rhodospirillum	*rubrum, fulvum*
Rhodocyclus	*Gelatinosus*
Purple sulfur photosynthetic bacteria	
Allochromatium	*vinosum, minus, okenii*
Ectothiorhodospira	*mobilis*
Thiocapsa	*roseopersicina*
Lamprocystis	*Roseopersicina*
Thiospirillum	*jenense, sanguineum*
Green sulfur bacteria	
Chlorobium	*limicola, limicola* ssp. *Thiosulfatophilum*

systematic groups (purple non-sulfur, purple sulfur and sulfur green) that sharply differ from each other in morphological, cultural and physiological characteristics, pigment composition and habitat. Most of the isolated cultures were identified on the basis of morphological, physiological, cultural, biochemical and genetic characteristics. Based on these data, the most common genera of non-sulfur photosynthetic bacteria in Armenia are *Rhodobacter*, *Rhodopseudomonas*, *Rhodospirillum*, and *Rhodocyclus* (Bokulich et al. 2015; Harutyunyan 2018a, b; Kalantaryan 2019).

Despite some similarity in morphological and physiological characteristics, individual representatives of these genera differ in cultural features (temperature, pH), assimilation of nitrogen and carbon sources, as well as ecological and geographical origin. The strains isolated from mineral springs, freshwater and salt water reservoirs, and saline soils differ in their need for NaCl; they are either moderate halophiles or obligate halophiles, or do not need salt at all. As distinct from many species of non-sulfur photosynthetic bacteria described in the literature, they are capable of assimilating thiosulfates.

Among purple sulfur photosynthetic bacteria, the most widespread are the genera *Allocromatium*, *Thiocapsa*, *Lamprocystis*, *Thiospirillum*, *Ectothiorhodospira* that are commonly found in the soda saline soils of the Ararat Valley. The alkalophilic strains of *Ectothiorhodospira mobilis*, *Lamprocystis roseopersicina*, *Thiocaps roseopersicina*, and *Thiospirillum jenense* isolated and also described as new ecotypes, are characterized by mixotrophic growth and halotolerance (Table 7.4).

At the same time, a narrow range of green sulfur bacteria represented by two species *Chlorobium limicola* and *C. limicola* ssp. *thiosulfatophilum* was identified.

Thus, the study of biodiversity of photosynthetic bacteria revealed their wide distribution both in various water and soil biocenoses of Armenia.

7.3 The Importance of Photosynthetic Bacteria in Biotechnology

Photosynthetic bacteria are of practical interest as producers of valuable biologically active compounds, such as enzymes, vitamins, amino acids, carotenoids, biohydrogen, etc. At the same time, it should be noted straight away that their cultivation does not require expensive and nutrient-rich nutrient media (Karapetian et al. 1980; Malatyan and Paronyan 1997; Paronian and Malatian 1996; Eliazyan and Paronyan 1997, 2003; Goginyan et al. 2019).

Here we present the results of our research with the use of cultures of purple photosynthetic bacteria isolated from various ecological-geographical zones of Armenia.

7.3.1 Enzymatic Activity of Photosynthetic Bacteria

7.3.1.1 Aspartase Activity

The enzyme aspartase (L-aspartate ammonium lyase) is widely used by plants and microorganisms (Plachy and Sikyta 1977). The enzyme catalyzes the reversible addition of ammonia to fumarate to form aspartate. Aspartate initiates the synthesis of pyrimidines and some amino acids: lysine, methionine, isoleucine, and asparagine. The conversion of fumarate to L-aspartic acid under the action of the enzyme aspartase has long been the basis for producing aspartic acid on an industrial scale (Chibata et al. 1960).

The most well-known producers of aspartic acid are some representatives of bacteria of the *Pseudomonas*, *Bacillus* and *Enterobacter* genera. Despite the fact that the number of producers with well pronounced aspartase activity is large, up to date there was a lack of information on such an activity in the photosynthetic bacteria. Table 7.5 presents data on the aspartase activity of non-sulfur and purple sulfur photosynthetic bacteria in a comparative aspect.

The obtained data prove that all the strains studied have aspartase activity, that is, they are capable of converting fumaric acid to aspartic acid under the anaerobic cultivation conditions. It is established that non-sulfur bacteria unlike sulfur ones exhibited the highest activity. At the same time, experiments have shown the absence of direct relationship between biomass accumulation and aspartase activity. The bacterial biomass reached its maximum growth on the 3rd day, and the aspartase activity—on the 2nd.

The factors stimulating conversion of fumarate to aspartic acid include the amount of fumarate introduced into the medium (15%), the cultivation temperature (37 °C), pH (6.0–9.0), the presence of metal-catalysts of the process (Mg^{2+} or Mn^{2+}) and the duration of the incubation period (48 h). In addition to *Rps. palustris* MDC

Table 7.5 Aspartase activity of photosynthetic bacteria

Strains	Yield of L-aspartic acid 100 mg of dry biomass ($\overline{X}$ ± m) (%)	Strains	Yield of L-aspartic acid 100 mg of dry biomass ($\overline{X}$ ± m) (%)
Rba. sphaeroides MDC[a] 6510	18.5 ± 0.8	*Rba. capsulatus* MDC 6508	28.0 ± 1.04
Rba. sphaeroides D-2[b]	19.7 ± 0.76	*Rps. palustris* MDC 6506	33.1 ± 1.12
Rba. sphaeroides D-3	27.5 ± 0.99	*Rsp. rubrum* MDC 6505	32.8 ± 0.98
Rba. sphaeroides D-5	16.6 ± 0.7	*Allochromatium* sp. MDC 6514	19.1 ± 0.59
Rba. sphaeroides D-8	30.7 ± 0.28	*Allochromatium vinosum* MDC 6515	12.6 ± 0.54
Rba. sphaeroides MDC 6509	31.5 ± 0.9	*Ectothiorhodospira mobilis* MDC 6517	22.1 ± 0.68
Rba. sphaeroides MDC 6507	29.4 ± 0.78		

[a]MDC—microbial depository center of the scientific and production center "armbiotechnology" NAS RA. Registered in the World Federation of Collections of Cultures (WFCC) under No 803
[b]Work numbers of the strains maintained in the culture collection of the laboratory of energy alternative sources of the scientific and production center "Armbiotechnology" NAS RA

6506, high aspartase activity was detected in the strains of *Rsp. rubrum* MDC 6505, *Rba. sphaeroides* D-8 and *Rba. sphaeroides* MDC 6509.

7.3.1.2 Aminoacylase Activity

The study of acylases of microbial origin is an urgent task due to the problem of separating chemically produced racemic mixtures of amino acids. Japanese scientists were the first who studied microbial acylases which subsequently served as a theoretical and practical basis to set up production of a number of amino acids, both in Japan and other countries of the world (Chibata et al. 1960). Studies of aminoacylases in various groups of microorganisms have shown that they differ in basic physicochemical properties (Afrikyan 1967; Rudakov et al. 1978).

Our studies have shown that the majority of strains of non-sulfur purple photosynthetic bacteria isolated from the springs of Arzni and Jermuk exhibits relatively high acylase activity in relation to a number of acyl derivatives of amino acids (Table 7.6).

Table 7.6 Acylase activity of non-sulfur purple photosynthetic bacteria (μmmol of amino acid/100 mg of dry biomass/h)

Strain	Acetyl-DL-Methionine	Alanine	Serine	Phenylalanine	Tryptophan
Rba. sphaeroides MDC 6510	55.2	60.7	21.5	99.8	12.5
Rba. sphaeroides D-2	44.6	56.6	16.1	86.7	8.4
Rba. sphaeroides D-3	68.7	35.5	6.9	57.8	3.1
Rba. sphaeroides D-8	59.5	42.0	12.2	56.5	7.9
Rba. sphaeroides MDC 6509	58.6	40.5	11.9	55.5	7.3
Rba. capsulatus MDC 6508	87.5	130.0	8.7	122.4	5.6
Rba. capsulatus A-5	120.7	110.2	53.5	130.2	2.9
Rba. sulfidophilus Sh-1	41.2	32.7	12.2	110.4	4.8
Rps. palustris MDC 6506	126.3	92.7	84.3	108.9	10.3
Rsp. rubrum MDC 6505	38.8	118.7	33.4	90.0	3.7

Table 7.7 The effect of nutrient medium constituents on the acylase activity of photosynthetic bacteria (μmmol of amino acid/100 mg of dry biomass/h)

Nutrient medium constituents (1.0%)	*Rba. capsulatus* MDC 6508		*Rba. palustris* MDC 6506	
	Growth (D_{660} HM)	Alanine	Growth (D_{660} HM)	Methionine
Peptone	0.51	134.6	0.47	129.9
Yeast extract	0.48	133.7	0.40	129.2
Corn extract	0.35	80.1	0.28	101.2
Control, constituents-free	0.30	130.1	0.23	126.2

Studies have shown that the strains differ in both the level of acylase activity and substrate specificity. The cultures have a wide range of activity, which affects the yield of *acyl* derivatives of methionine, alanine, serine, phenylalanine and tryptophan.

The maximum acylase activity was detected in 48-hour cultures. It was also found that the addition of 1% peptone or yeast extract to the nutrient medium positively affected the acylase activity with simultaneous yield of biomass (Table 7.7).

The highest activity of the enzyme was demonstrated in the presence of 1.0% peptone in a nutrient medium. An increase in the aspartase activity is also promoted by the addition of 0.1% yeast extract irrespective of the amount of peptone (Eliazyan and Paronyan 2001).

We have also found that the presence of certain metal ions in the nutrient medium, especially those of Co^{2+}, significantly stimulates the enzymatic activity of bacteria.

On the contrary, ions of other metals exert their inhibitory action of various degrees in the synthesis of enzyme (Table 7.8). As can be seen from the table, ions of Cu, Zn, and Ni unlike ions of Ca, Fe, and Mn significantly inhibit the acylase activity of strains.

Table 7.8 Influence of metal ions on the acylase activity of photosynthetic bacteria (μmmol of amino acid/100 mg of dry biomass/h)

	Rps. palustris MDC 6506	*Rba. capsulatus* MDC 6508
Metal ions (10^{-3} mol)	Methionine	Alanine
Cu	32.3	41.5
Mg	49.7	50.0
Ca	53.6	38.9
Zn	28.0	30.2
Mn	88.7	98.9
Fe	82.5	80.1
Co	130.8	145.3
Ni	30.0	38.0
Control without metal ions	100.8	120.3

Table 7.9 Asparaginase activity of non-sulfur purple photosynthetic bacteria (single induction)

Strains	Enzyme activity (mM)/g on dry weight of biomass ($\overline{X} \pm$ m)
Rba. sphaeroides MDC 6510	6.93 ± 0.68
Rba. sphaeroides D-2	5.10 ± 0.2
Rba. sphaeroides D-3	25.50 ± 0.99
Rba. sphaeroides D-8	15.00 ± 0.37
Rba. sphaeroides D-9	15.30 ± 0.18
Rba. sphaeroides MDC 6509	15.80 ± 0.24
Rba. capsulatus MDC 6508	17.78 ± 0.64
Rba. capsulatus A-5	14.2 ± 0.28
Rba. capsulatus D-5	12.05±0.52
Rps. palustris MDC 6508	7.73±0.15
Rps. palustris A-7	5.85±0.35
Rsp. rubrum MDC 6505	22.75 ± 0.3

7.3.1.3 Asparaginase Activity

L-asparaginase is a hydrolyzing enzyme participating in asparagine hydrolysis to generate aspartic acid. The enzyme, isolated for the first time from *Escherichia coli*, is efficient in the treatment of leukemia. When introduced into the blood flow, asparaginase reduces the entry into deceased cells of exogenous asparagine, which stimulates the rapid growth of tumor cells (Campbell et al. 1967; Heinemann and Howard 1969). At the same time, under the influence of asparaginase, the cells with low asparagine synthetase activity are damaged, which limits the clinical use of the enzyme in treatment of leukemia. L-asparaginase plays an important role in the metabolism of amino acids in microorganisms, although its synthesis in photosynthetic bacteria has been insufficiently studied (Tchan 1971).

Our studies have revealed high enzymatic activity in photosynthetic non-sulfur purple bacteria (Table 7.9).

Table 7.10 Influence of incubation duration on asparaginase activity of non-sulfur purple photosynthetic bacteria (single induction)

Strains	Enzyme activity (mM)/g on dry weight of biomass			
	24 ч	48 ч	72 ч	96 ч
Rba. sphaeroides D-3	25.50	40.10	19.15	2.05
Rba. capsulatus MDC 6508	17.80	33.75	17.59	1.95
Rsp. rubrum MDC 6505	22.75	39.57	18.75	1.98

We have explored the effect of a number of factors, such as pH 7.5–9.0, cultivation temperature and induction of enzyme synthesis in the substrate and the duration of incubation—48 h on average (Table 7.10) that facilitate an increase in the asparaginase activity of strains.

Thus, taking into account the data obtained, it can be concluded that while creating favorable conditions, the cultures of photosynthetic bacteria can exhibit a sufficiently high competing activity of asparaginase.

7.3.2 Synthesis of Carotenoid Pigments by Purple Photosynthetic Bacteria

In recent years due to the sharply increased environmental pollution, natural biologically active compounds with anti-inflammatory, immunomodulatory, and anticarcinogenic activities are widely used. In a series of these drugs, carotenoid pigments occupy a special place. The main source of natural carotenoids is single-celled green microalgae. Along with them, purple synthetic bacteria that synthesize acyclic carotenoids, β-carotene, lycopene, spheroidene, spheroidenone spiriloxanthin and others can also be considered as a natural source of these compounds (Vrati 1984).

Comparable characteristics of carotenoid pigments of sulfur photosynthetic bacteria *Allochromatium* sp. MDC 6514, *Ectothiorhodospira mobilis* MDC 6517, *Lamprocystis* sp. MDC 6518, *Thiospirillum jenense* MDC 6519, *Thiocapsa* sp. MDC 6520 showed that spiriloxanthin, lycopene, rhodopine, and okenone predominated in their biomass. Unlike sulfur purple photosynthetic bacteria, non-sulfur bacteria are characterized by a higher content of carotenoids of spheroidene-branch. The conditions for the biosynthesis of these pigments were studied on the strains isolated from the mineral springs of Jermuk. The most promising of them is *Rba. sphaeroides* MDC 6508, under microaerophilic conditions, synthesizes spheroidene (Rf 0.53) and hydroxyspheroidene (Rf 0.72). At that, it is established the intensity of illumination in the range of 750–1000 lux has a positive effect on the growth rate and accumulation of biomass by the culture. However, no direct relationship between the biomass accumulation and the synthesis of carotenoids was revealed, despite the fact that with an increase in the light intensity from 1500 to 3000 lux, there was a sharp decrease in the synthesis of pigments. It is necessary to note that as distinct from other carotenoids, the amount of detected spheroidene was about 60% of the total amount of synthesized pigments regardless of the light intensity.

At the same time, extremely interesting is the strain *Rba. sphaeroides* MDC 6511, isolated from the mineral springs of Arzni. Its distinctive features as compared to other non-sulfur bacteria are that the optimal cultivation conditions are as follows: growth temperature—32 °C, pH 6.8, illumination 2000 lux. The strain is characterized as one of the efficient producers of β-carotene, detected by the values of the absorption spectrum and melting temperature, as well as chromatographically. The yield of β-carotene per dry weight of biomass was 29.32 μg/g, and the total amount of carotenoid pigments was 497.0 μg/g (Paronyan 1997, 2004).

7.3.3 Synthesis of Vitamins

The ability of photosynthetic purple bacteria to synthesize vitamins on the nutrient media, which included mineral water from the Jermuk springs with addition of organic sludge hydrolyzate was studied (Paronyan and Eliazyan 2002; Paronyan and Markosyan 2008).

On the optimized nutrient media developed by us, the studied strains of the non-sulfur bacteria *Rba. sphaeroides* D1, D2, D3, and *Rba. capsulatus* D4 synthesized one or a group of vitamins in the following amounts (μg/g of dry biomass): thiamine—4.7–6.9; pyridoxine—0.6–4.0; biotin—0.03–0.5; pantothenic acid—290–500; nicotinic acid—29–500.

The important feature of these strains is that, in addition to the mentioned vitamins, they are also capable of synthesizing vitamin B_{12} (Table 7.11).

The results obtained indicate that the studied strains intracellularly synthesize vitamin B_{12} regardless of the nutrient media composition. The most active producers were the strains of *Rba. sphaeroides* MDC 6508, 6509 and *Rba. sphaeroides* D8, which synthesized vitamin in the range of 1.5–2.1 μg /g of dry biomass. To increase the yield of B_{12},—the effect of cobalt added to the nutrient medium was studied (Table 7.12).

Pursuant to the results obtained, the content of cobalt in the nutrient medium in the amount of 10 mg/l suppresses the growth of bacteria and inhibits synthesis of the vitamin. According to our data, the optimal amounts of cobalt at which the *R. sphaeroides* D8 strain synthesized 5.5 and 2.8 μg/g of the vitamin are from 0.5 to 1.0 mg/l, respectively.

Thus, the analysis of the data obtained shows that the cheap nutrient medium developed on the basis of the Jermuk mineral water completely suits for the biosynthesis of the vitamin and the growth of photosynthetic bacteria. According to our research results, this group of microorganisms can be used to produce both individual vitamins and vitamin complexes.

Table 7.11 Synthesis of vitamin B_{12} by non-sulfur purple photosynthetic bacteria on various nutrient media

Strains	Vitamin B_{12} (μg)/g of dry biomass		
	Ormerod nutrient medium	Ormerod medium with sludge hydrolyzate	Mineral water with sludge hydrolyzate
Rba. sphaeroides MDC 6510	0.70	0.87	1.0
Rba. sphaeroides D-2	1.10	1.30	1.30
Rba. sphaeroides D-3	0.40	0.51	0.63
Rba. sphaeroides D-8	1.90	1.90	2.10
Rba. sphaeroides MDC 6509	1.80	1.95	1.95
Rba. capsulatus MDC 6508	1.50	2.05	2.03
Rps. palustris MDC 6506	0.25	0.19	0.19
Rsp. rubrum MDC 6505	0.48	0.62	0.6

Table 7.12 Synthesis of vitamin B_{12} under the influence of various concentrations of cobalt

Concentration $C\hat{I}Cl_2 \times 6H_2O$ (mg/l)	*Rba. capsulatus* MDC 6508		*Rba. sphaeroides* D-8	
	Growth (D_{660nm})	Vitamin B_{12} (μg/g)	Growth (D_{660nm})	Vitamin B_{12} (μg/g)
0.5	0.65	5.00	0.67	5.50
1.0	1.68	5.10	0.69	2.80
10.0	0.20	–	0.25	–
Ormerod medium (control)	0.56	1.60	0.52	1.80

7.3.4 *Synthesis of 5-aminolevulinic Acid*

5-aminolevulinic acid (5-ALA) is an intermediate in the synthesis of different tetrapyrrole molecules in all living organisms, i.e., chlorophyll, hem or vitamin B_{12} (Kang et al. 2011). There are two different pathways in which 5-ALA can be produced: C4 pathway (Shemin pathway) which is present in purple bacteria, yeasts and mammalian cells and C5 pathway which is present in many plants and some microorganisms (Woodard and Dailey 1995). Today, 5-ALA is mostly produced using microbial fermentation, namely by photosynthetic bacteria because the chemical synthesis of 5-ALA has lower yields and is more complicated in comparison to microbial production (Liu et al. 2014). Production of 5-ALA has been reported using both wild strains of bacteria and their mutants. The application of mutant strains is far more suited for 5-ALA production. So far many different strains of photosynthetic bacteria together with their mutants have been tested for their 5-ALA production capacities (Sasaki et al. 1991; Harutyunyan 2018a, b). Both chemically defined and complex media can be used. 5-ALA can be used as an effective herbicide, or as a plant stress tolerance enhancer (Nunkaew et al. 2014). It is not harmful to crops, animals or humans and it is biodegradable which makes it interesting from ecological point of view. Much attention was also dedicated to its great potential in the field of medicine, namely tumor-localizing and photodynamic therapy.

The selection of potential strain-producers of 5-ALA was carried out taking into account available literature data indicating high possibility and prospects for obtaining the target product by microbiological synthesis and based on the presence of a vast collection of cultures of photosynthetic bacteria in the Laboratory of Energy Alternative Sources of the SPC "Armbiotechnology," NAS RA (Mikhalchuk et al. 2016; Minasyan et al. 2019).

Preliminary assessment of 5-ALA biosynthesis by purple non-sulfur photosynthetic bacteria of species *Rba. azotoformans* MDC 6523, *Rba. sphaeroides* MDC 6509, *Rsp. rubrum* MDC 6505, and *Rps. palustris* MDC 6506 was performed on the well-known Ormerod liquid nutrient medium under standard cultivation conditions. It is shown that when growing strains for 15 days, at a temperature of 28 °C and illumination of 2000 lux, *Rba. azotoformans* MDC 6523 synthesizes 2.5 mg/l, *Rba. sphaeroides* MDC 6509—13.0 mg/l, *Rsp. rubrum* MDC 6505—1.8 mg/l and *Rps. palustris* MDC 6506—13.3 mg/l of 5-ALA.

By induced chemical mutagenesis, a number of mutated strains of purple non-sulfur photosynthetic bacteria were obtained. Based on the results of comparative characteristics of 5-ALA biosynthesis, the most active mutant E10 of *Rba. azotoformans* MDC 6523 culture was selected, which synthesized 179 mg/l of 5-ALA in the nutrient medium, while the wild-type strain *Rba. azotoformans* under the same conditions of cultivation synthesized only 2.5 mg/l of 5-ALA.

Further studies aimed at increasing the yield of 5-ALA in the mutant strain *Rba. azotoformans* E10. In particular, the growth of the strain in microaerobic and aerobic conditions with 24-h illumination and in complete darkness was studied. It is shown that under microaerobic conditions of cultivation at 24 h illumination, for 28 h, strain *Rba. azotoformans* E10 synthesizes 179 mg/l of 5-ALA, while the amount of dry biomass formed was 2.6 g/l. At the same time, in the absence of illumination, the strain synthesized only 11 mg/l, and the amount of dry biomass formed was 1.0 g/l.

Meanwhile, under aerobic conditions at 24 h illumination, the *Rba. azotoformans* E10 strain synthesized 58.1 mg/l (weight of dry biomass—2.5 g/l), and in the absence of light—46.5 g/l (weight of dry biomass—1.4 g/l). Thus, microaerobic conditions and 24 h illumination are the most optimal for biomass accumulation and synthesis of 5-ALA by *Rba. azotoformans* strain.

The effect of illumination intensity, temperature and pH on the synthesis of 5-ALA in the mutant E10 of *Rba. azotoformans* MDC 6523 was studied. As a result of numerous experiments optimal values of light (2000 lux), temperature (30 °C), and pH (7.0) at which the highest yield of the target product was 182 mg/l, have been determined.

The effect of different concentrations of malate in the nutrient medium on the growth and synthesis of 5-ALA in the mutant E10 of *Rba. azotoformans* MDC 6523 was studied. It was found that with the initial malate concentration of 2.7 g/l the yield of 5-ALA is maximal and consists 242.8 mg/l. The concentrations of glycine, succinate, levulinic acid, and glutamate were optimized ensuring the best growth with producers and the synthesis of 5-ALA. It was established that at concentrations of glycine 40 mmol/l, succinate 30 mmol/l, levulinic acid 15 mmol/l, and glutamate 4 g/l, the synthesis of 5-ALA reached 385 mg/l. Optimized cultivation conditions were applied to the cultivation of mutant E10 of *Rba. azotoformans* in the bioreactor, where the maximum yield of 5-ALA was 579 mg/l (Novak et al. 2017, 2018; Harutyunyan et al. 2018).

7.4 Conclusion

The results of studies conducted over the years show that Armenia, having a great variety of geographical zoning and climatic conditions, is endowed with a very rich and diverse microflora, the study of which is a topical issue both theoretically and practically. It is very important to study and preserve the vast biodiversity of phototrophic bacteria in Armenia, as these microorganisms have enormous genetic potential, rich metabolic pathways, different growth conditions, which greatly contributes to their application in phototechnology. On the other hand, the ability to assimilate different organic wastes can be used to improve the ecological condition of the human environment, at least to some extent, as the residues from the assimilation of waste can be used as biological fertilizers.

Acknowledgments This work was partially supported by the RA MESCS Committee of Science, in the frames of the research projects №16A-1f37 "The study of features of 5-aminolevulinic acid biosynthesis by photosynthetic bacteria", № 18T-1F118 "Study of the systematics and taxonomic affiliation of purple photosynthetic bacteria" and the EC "Horizon-2020" (H2020-MSCA-RISE-2015) «Phoenix» (People for tHe eurOpean bio-ENergy mIX) Project (Agreement No690925).

References

Afrikyan EK (1967) Microbial acylases. Microbiol Synth 1:1–7

Bogorov LV (1974) On the properties of *Thiocapsa roseopersicina*, BBS strain isolated from the White Sea estuary. Microbiology 43(2):326–332

Bokulich NA, Aminarashvili L, Chitchyan K, Ghazanchyan N, Darbinyan K, Gagelidze N, Sadunishvili T, Goginyan V, Kvesitadze G, Torok T, Mills DA (2015) Microbial biogeography of the transnational fermented milk matsoni. J Food Microbiol 50:12–19

Butler M (2005) Animal cell cultures: recent achievements and perspectives in the production of biopharmaceuticals. Appl Microbiol Biotechnol 68(3):283–291

Campbell HA, Mashburn LT, Boyse EA, Old LJ (1967) Two L-asparaginases from *Escherichia coli* B. Their separation, purification and antitumor activity. Biochemistry 6:721–730

Chibata I, Ishikawa T, Tosa T (1960) Studies on aminoacids. Bull Agric Chem Soc 24(1):31–36
Eliazian AA, Paronian AK, Harutiunian SH, Malatian MN (1984) The study of the biomas of phototrophic bacteria, grown on the Arzni mineral water. Biol J Armenia 37(3):220–225
Eliazyan AA, Paronyan AK (1997) Glucose-isomerase activity of phototrophic bacteria. Biol J Armenia 50(3-4):226–228
Eliazyan AA, Paronyan AK (2001) On aminoacylase activity of photosynthetic bacteria. Biol J Armenia 53(1-2):46–50
Eliazyan AA, Paronyan AK (2003) Activity of some enzymes of phototrophic bacteria. Biol J Armenia 55(4):324–330
Goginyan V, Eusébio A, Neves A, Marques IP, Harutyunyan B, Hovhannisyan R, Andreasyan N (2019) Purple non-sulphur photosynthetic bacteria in brewery wastewater during anaerobic digestion. In: Proceedings, 5th international conference "WASTES: Solutions, Treatments and Opportunities", Guimarães, Portugal, pp 87–89
Gorlenko VM (1974) Oxidation of *Amoebobacter roseus* thiosulfate in the dark under microaerophilic conditions. Microbiology 43(4):729–732
Gorlenko VM, Dubinina GA, Kuznetsova SI (1977) The ecology of aquatic microorganisms. Nauka 1977:13–200
Hansen TA, Gemerden H (1972) Sulphide utilization by purple non-sulfur bacteria. Arch Microbiol 86(1):49–56
Harutyunyan B (2018a) Molecular-genetic verification of the taxonomic status of two strains of purple non-sulfur photosynthetic bacteria based on the analysis of nucleotide sequences of 16S rRNA genes. Electron J Nat Sci 31(2):57–60
Harutyunyan BA (2018b) Determination of 5-aminolevulinic acid producers among cultures of purple non-sulfur photosynthetic bacteria. Biol J Armenia 70(2):6–12
Harutyunyan B, Novak M, Pavlečić M, Goginyan V, Hovhannesyan R, Melkumyan I, Šantek B (2018) Influence of glycine, succinate, levulinic acid and glutamate concentrations on growth of purple non-sulfur photosynthetic bacteria and 5-aminolevulinic acid production. Int J Innovat Res 7(4):3839–3846
Heinemann B, Howard AJ (1969) Production of tumor inhibitory L-asparaginase by submerged growth of *Serratia marcescens*. Appl Microbiol 18(4):550–554
Jensen SL, Cohen-Bazire G, Stanier RY (1961) Biosynthesis of carotenoids in purple bacteria: a re-evaluation based on considerations of chemical structure. Nature 192:1168–1172
Kalantaryan NK (2019) Molecular genetic identification of unicellular green microalgae of the family *Chlorellaceae*. Biol J Armenia 71(3):6–13
Kang Z, Wang Y, Gu P, Wang Q, Qi Q (2011) Engineering *Escherichia coli* for efficient production of 5-aminolevulinic acid from glucose. Metab Eng 13(5):492–498
Karapetian SK, Balasanian HG, Malatian MN, Paronian AV (1980) Phototrophic bacterium biomass fodder value test and the influence of biomass on productivity and growth-development of chickens. Biol J Armenia 33(2):150–155
Keppen OI, Gorlenko VM (1975) Characteristics of a new species of budding purple bacteria containing bacteriochlorophyll. Microbiology 44(2):258–265
Kobayashi M (1982) The role of phototrophic bacteria in nature and their utilization. In: Advances in agricultural microbiology. Oxford and Publishing Co., Oxford, pp 643–661
Kobayashi M, Haque MZ (1971) Contribution to nitrogen fixation and soil fertility by photosynthetic bacteria. Plant Soil 35(1):443–456
Kompantseva EI, Gorlenko VM (1984) A new species of moderately halophilic purple bacteria *Rhodospirilum mediosalinum* sp.nov. Microbiology 53(6):954–961
Kondrat'eva EN, Maksimova IV, Samuilov VD (1989) Phototrophic microorganisms. Izd-vo MGU, Moscow, p 376
Kondratyeva EN (1963) Photosynthetic bacteria. M., Ed. USSR Academy of Sciences, p 280
Kondratyeva EN, Krasilnikova EN, Pedan LA (1976) Growth conditions in the dark in the presence of oxygen *Ectothiorhodospira shaposhnikovii*. Microbiology 42(1):172–174

Liu S, Zhang G, Li X, Zhang J (2014) Microbial production and applications of 5-aminolevulinic acid. Appl Microbiol Biotechnol 98(17):7349–7357

Malatyan MN, Paronyan AK (1997) Biomass production by phototrophic bacteria grown on whey. Appl Biochem Microbiol 33(2):208–210

Malatyan MN, Harutyunyan TG, Paronyan AK, Kepper OI, Alexandrushkina NI (1982) Purple non-sulfur bacteria isolated from mineral springs of Armenia. J Microbiol 51(3):517–519

Mikhalchuk AL, Rudak EV, Karman PV, Babitskaya SV, Kisel MA, Kalantaryan NK, Saghatelyan LO, Goginyan VB (2016) Synthesis of aliphatic esters of 5-aminolevulinic acid and their effect on the growth of phototrophic microalgae. Vesti NAS Belarus. Ser Chem Sci 2:64–68

Minasyan EV, Tsaturyan AH, Kalantaryan NK, Harutyunyan BA (2019) Determination of 5-aminolevulinic Acid in liquid cultures of purple non-sulfur photosynthesizing bacteria by high performance liquid chromatography. Chem J Armenia 72(3):243–248

Mishustin EN (1950) The doctrine of Dokuchaev-Kostychev-Williams about the soil and the question of the composition of microorganisms in plant formations. J Microbiol 19(1):1–18

Mishustin EN, Pertsovskaya MI (1954) Microorganisms and soil self-cleaning. Publishing house of the Academy of Sciences of the USSR, Moscow, pp 609–613

Molisch H (1907) The purple bacteria. Gustav Fischer, Jena, 95 pp

Neutzling O, Imhoff JF, Truper HG (1984) *Rhodopseudomonas adriatica* sp. nov., a new species of the Rhodospiriiiaceae, dependent on reduced sulfur compounds. Arch Microbiol 137 (3):256–262

Novak M, Pavlečić M, Harutyunyan B, Goginyan V, Horvat P, Šantek B (2017) Characteristics and selection of cultures of photosynthetic purple non-sulphur bacteria as a potential 5-aminolevulinic acid producers. Croat J Food Technol 12(3-4):113–119

Novak M, Harutyunyan B, Goginyan V, Pavlečić M, Horvat P, Šantek B (2018) Influence of initial malate concentration on 5-ALA synthesis during cultivation of *Rhodobacter capsulatus* B-6508. Electron J Nat Sci 31(2):3–7

Nunkaew T, Kantachote D, Kanzaki H, Nitoda T, Ritchie RJ (2014) Effects of 5-aminolevulinic acid (ALA)-containing supernatants from selected *Rhodopseudomonas palustris* strains on rice growth under NaCl stress, with mediating effects on chlorophyll, photosynthetic electron transport and antioxidative enzymes. Electron J Biotechnol 17:19–26

Overmann J, Beatty JT, Hall KJ (1996) Purple sulfur bacteria control the growth aerobic heterotrophic bacterioplankton in a meromictic salt lake. Appl Environ Microbiol 62(8):3251–3258

Paronyan AK, Malatyan MN (1995) Some properties of mixed growth of phototrophic and heterotrophic bacteria. Biol J Armenia 48(3-4):60–64

Paronian AK, Malatian MN (1996) Precipitation of phototrophic bacteria biomass by method of coagulation. Biol J Armenia 49(1-2):70–72

Paronian AK, Gabaeva TA, Malatyan MN (1982) *Rhodopseudomonas* sp. D-4-phototrophic bacterium separated from the mineral spring of Jermuk. Biol J Armenia 35(10):803–809

Paronyan AK (1997) Influence of illumination on the growth and carotinoids formation of *Rhodopseudomonas sphaeroides* D-10. Biol J Armenia 50(3-4):229–233

Paronyan AK (2001) Alcalophilic sulfur photosynthetic bacteria from soda. Biol J Armenia 53 (1-2):41–45

Paronyan AK (2002) Ecology and biodiversity of phototrophic bacteria of various ecosystems of Armenia. Biol J Armenia 54(1-2):91–98

Paronyan AK (2004) The influence of carbon compounds on the growth and carotenoid biosynthesis of *Rhodobacter sphaeroides* str. A-1. Biol J Armenia 56(3-4):133–137

Paronyan AK (2006a) Characteristic of alcalophilic sulfur purple bacterium isolated from soda salted soils of Ararat plain. Biol J Armenia 58(1-2):34–37

Paronyan AK (2006b) Ecological and physiological features of phototrophic bacteria *Rhodopseudomonas palustris* from mineral springs of Jermuk. Biol J Armenia 59(1-2):73–77

Paronyan AK (2007) Ecophysiological properties of phototrophic bacterium *Rhodopseudomonas palustris* from mineral springs of Jermuk. Biol J Armenia 59(1-2):73–77

Paronyan AK (2010) Biological peculiarities of distributed in Armenia phototrophic bacteria and their applications. Science Publishing House, New York, p 224
Paronyan AK, Eliazyan AA (2002) Vitamins requiring and vitamins producing properties of some phototrophic bacteria. Biol J Armenia 54(1-2):99–103
Paronyan AK, Gasparyan AV (2009) Production of biomass of photosynthetic bacteria on base of stock-breeding and poultry wastes. Biol J Armenia 61(1):5–10
Paronyan AK, Markosyan LS (2008) The influence of group B vitamins on the content of carotenoids of photosynthetic bacterium *Rhodobacter sphaeroides*. Biol J Armenia 60 (1-2):22–27
Pelyp AD (1936) Hydrobiology of Karabugaz. Proc Sapropel Lab USSR Acad Sci 5:48–80
Pfennig N (1969) *Rhodopseudomonas acidophila* sp. nov., a new species of the budding purple non-sulfur bacteria. J Bacteriol 99(2):597–602
Pfennig N (1975) The phototrophic bacteria and their role in the sulfur cycle. Plant Soil 43(1):11–16
Pfennig N, Truper HG (1973) The Rhodospirillales (phototrophic photosynthetic bacteria). In: Laskin AJ, Lechevalier HA (eds) Handbook of microbiology. CRG Press, Cleveland, pp 17–27
Pfennig N, Truper YG (1992) The family Chromatiaceae. In: Balows A, Truper HG, Dworkin V, Harder W, Schleifer KH (eds) The procaryotes, 2nd edn. Springer, Berlin, pp 3200–3221
Plachy I, Sikyta B (1977) Production of L-aspartic acid from fumaric acid by Alcaligenes metalcaligenes CCEB 312. Folia Microbiol 22(3):440–414
Raymond JC, Sistrom UR (1969) *Ectothiorhodospira halophila*: a new species of the genus Ectothiorhodospira. Arch Microbiol 69(2):121–126
Rudakov OL, Karasevich GA, Pepzikova GA (1978) Acylase activity of mycophilic fungi. Microbiology 3(47):485–487
Sasaki K, Tanaka T, Nishizawa Y, Hayashi M (1991) Enhanced production of 5-aminolevulinic acid by repeated addition of levulinic acid and supplement of precursors in photoheterotrophic culture of *Rhodobacter sphaeroides*. J Ferment Bioeng 71(6):403–406
Sasaki K, Watanabe M, Suda Y, Ishizuka A, Noparatnaraporn N (2005) Applications of photosynthetic bacteria for medical fields. J Biosci Bioeng 100(5):481–488
Sasikala CH, Ramana CV (1995) Biotechnological potentials of anoxygenic phototrophic bacteria. II. Biopolyesters, biopesticide, biofuel, and biofertilizer. In: Advances in applied microbiology. Academic, New York, pp 227–278
Tchan JT (1971) Study of L-asparaginase production by *Rhodopseudomonas capsulate*. J Ferment Technol 49(9):535–539
Tsygankov AA, Gogotov IN (1990) Obtaining biomass of purple non-sulfur bacteria. Appl Biochem Microbiol 26:819–824
Vrati S (1984) Single cell protein production by photosynthetic bacteria grown on clarified effluents of biogas plant. Appl Microbiol Biotechnol 19(1):199–202
Willems A, Monique G, Jozef DL (1992) The prokaryotes: a handbook on the biology of bacteria: ecophysiology, isolation, identification, applications. In: Kondratieva Elena N, Norbert P, Truper Hans G (eds) The phototrophic prokaryotes. Springer, New York, pp 312–330
Woodard SI, Dailey HA (1995) Regulation of heme biosynthesis in *Escherichia coli*. Arch Biochem Biophys 316(1):110–115

Chapter 8
The Genus *Thermus*: A Brief History of Cosmopolitan Extreme Thermophiles: Diversity, Distribution, Biotechnological Potential and Applications

Ani Saghatelyan, Hovik Panosyan, and Nils-Kåre Birkeland

Abstract The genus *Thermus* represents a group of archetypal extreme thermophiles found in both natural and man-made thermal environments. Thermozymes mined from these bacteria have found their use in a broad range of biotechnological and industrial applications, including feed and food production, pharmaceutical and paper-pulp industries, organic synthesis, starch-processing, etc. Moreover, several molecular biological tools and host-vector systems for heterologous expression of extremozymes of thermophiles and hyperthermophiles were developed on the basis of representatives of this genus. Several strains of the genus *Thermus* are also well known as metal reducers with high potency in use for bioremediation. This chapter summarizes the diversity, distribution, main physiometabolic characteristics of members of the genus *Thermus*, as well as their biotechnological capacity and applications.

Keywords *Thermus* · Extreme thermophiles · Thermozymes · Cell factories · Biotechnology

8.1 Introduction

Heat-loving thermophiles are found in many adverse environments discovered on Earth, including hot springs and hydrothermal vents in ocean. Geothermal areas have yielded a broad number of highly diverse thermophilic and hyperthermophilic

---

A. Saghatelyan · H. Panosyan (✉)
Department of Biochemistry, Microbiology and Biotechnology, Yerevan State University, Yerevan, Armenia
e-mail: hpanosyan@ysu.am

N.-K. Birkeland
Department of Biological Sciences, University of Bergen, Bergen, Norway

D. Egamberdieva et al. (eds.), *Microbial Communities and their Interactions in the Extreme Environment*, Microorganisms for Sustainability 32,
https://doi.org/10.1007/978-981-16-3731-5_8

groups of bacteria and archaea. And many of these prokaryotes, including *Thermus*, have a worldwide distribution.

Species belonging to the genus *Thermus* are extreme thermophiles found in neutral and alkaline terrestrial and submarine hot springs and deep-sea thermal vents across the globe. They have been isolated from numerous man-made habitats as well, making them one of the most cosmopolite and abundant of thermophilic organisms (Wilpiszeski et al. 2019; Albuquerque et al. 2018). The type species of the genus, *Thermus aquaticus*, was first isolated in 1969 from a hot spring in Yellowstone National Park in the United States (Brock and Freeze 1969). Since then a number of other *Thermus* spp. have been identified in a variety of habitats across the world (Ming et al. 2020; Albuquerque et al. 2018; Zhou et al. 2018a, b).

Members of the genus *Thermus* are obligate heterotrophs growing on low concentrations of organic material (Albuquerque et al. 2018).

The high growth rates, cell yields of the cultures, and the constitutive expression of a strikingly efficient natural competence apparatus, in combination with other properties, make some strains of the genus perfect laboratory models in structural biology (Cava et al. 2009).

Many strains of *Thermus* spp. are of the essence in research and industrial applications, as they are important sources of biocatalysts and serve almost ideal cell factories for production of biotechnologically relevant proteins (Pantazaki et al. 2002; Cava et al. 2009; Fujino et al. 2020).

The most well-known enzyme mined from the genus *Thermus* is DNA polymerase. Other thermozymes from this genus are also widely used in feed and food production (e.g., amino acid and vitamin synthesis), pharmaceutical, and paper-pulp industries (e.g., xylanases) organic synthesis (e.g., esterases, lipases, proteases), starch-processing (e.g., α-amylases, glucose isomerases (Cava et al. 2009). Some strains were also well-adapted as host cells for homologues and heterologous expression of extremozymes of thermophiles and hyperthermophiles (Cava et al. 2009; Fujino et al. 2020). Several representatives of this genus also show high potency in use for bioremediation (Opperman and van Heerden 2007; Slobodkin et al. 2006; Gounder et al. 2011).

The objective of this chapter is to give a brief overview on diversity, distribution, as well as main physiological, metabolic, and genomic characteristics of the species of the genus *Thermus* and to highlight their biotechnological capacity.

8.2 Taxonomy and Phylogeny

Phylogenetically, *Thermus* is related to the radiation-resistant mesophilic genus *Deinococcus*. The *Deinococcus-Thermus* phylum is one of the most deeply branching bacterial lineages and does not show close affiliation to any other lineage based on analysis of both 16S rDNA and protein sequences (Woese 1987; Williams and Sharp 1995; Griffiths and Gupta 2004). Several other genera have been placed within this lineage, including slightly thermophilic *Meiothermus*, marine thermophiles *Oceanithermus*, *Vulcanithermus*, *Marinithermus*, *Rhabdothermus*, and distantly related mesophilic genus *Truepera* (Albuquerque et al. 2018).

The species of the genus *Thermus* share 85–87% 16S rRNA gene sequence similarity with representatives of the genus *Meiothermus*, and 87–89% with genera *Marinithermus*, *Oceanithermus*, *Rhabdothermus*, and *Vulcanithermus*, demonstrating the clear distinction between these genera (Albuquerque et al. 2018).

The genus *Thermus* to date include 17 species with validly published names that have been recorded in the LPSN website (http://www.bacterio.net/thermus.html). These are: *T. aquaticus* (Brock and Freeze 1969), *T. amyloliquefaciens* (Yu et al. 2015), *T antranikianii* (Chung et al. 2000), *T. arciformis* (Zhang et al. 2010), *T. brockianus* (Williams et al. 1995), *T. calditerrae* (Ming et al. 2014), *T. composti* (Vajna et al. 2012), *T. filiformis* (Hudson et al. 1987), *T. igniterrae* (Chung et al. 2000), *T. islandicus* (Bjornsdottir et al. 2009), *T. oshimai* (Williams et al. 1996), *T. scotoductus* (Kristjánsson et al. 1994), *T. tengchongensis* (Yu et al. 2013), *T. thermophilus* (Manaia et al. 1995), *T. caldifontis* (Khan et al. 2017), *T. tenuipuniceus* (Zhou et al. 2018a), and *T. thermamylovorans* (Ming et al. 2020). Most of them were initially isolated from hot springs with different geographical distributions, some species from geothermally heated soil, oyster mushroom compost and even from hot tap water. Recently, seven new organisms nominated to the genus *Thermus* have been described, but have not been validly published (Table 8.1). These include "*T. caldilimi*" (Li et al. 2019) "*T. anatoliensis*" (Kacagan et al. 2015), "*T. sediminis*" (Zhou et al. 2018b), "*T. rehai*" (Lin et al. 2002), "*T. yunnanensis*" (Gong et al. 2005), "*T. kawarayensis*" (Kurosawa et al. 2005), "*T. parvatiensis*" (Dwivedi et al. 2015). Two strains from Japan, previously designated "*Thermus flavus*" for strain AT-62 and "*Thermus caldophilus*" for strain GK-24, were never validly published and have been shown to belong to the species *T. thermophilus* (Albuquerque et al. 2018).

Phylogenetic analysis demonstrating relationship of species of described *Thermus* spp. based on 16S rDNA sequence comparison (Fig. 8.1) confirms, that the species of *Thermus* with valid names represent a distinct lineage. They share 16S rRNA gene sequence similarities in the range of 91–98%. *T. oshimai* shows the lowest 16S rRNA gene sequence similarity to the other *Thermus* species with similarities in the range of 91–93.5%. In contrast, some species have close relationship based on 16S rRNA gene sequence comparisons. For instance, *T. antranikianii* and *T. scotoductus* (98% sequence similarity) are the closest species among validly published *Thermus* spp. (Albuquerque et al. 2018). *T. caldifontis* and *T. tenuipuniceus* also stand relatively close (97.5% and 97.1% sequence similarity, respectively) to type strain of *T. scotoductus* (Khan et al. 2017; Zhou et al. 2018a).

The species of *Thermus* which are not validly described yet, share very high 16S rRNA gene sequence similarities to other validly described species. Hence, "*T. parvatiensis*," "*T. rehai*," and "*T. yunnanensis*" share >99% similarity to *T. thermophilus*, *T. tengchongensis*, and *T. caliditerrae*, respectively. In the case of "*T. kawarayensis*," the similarity to *T. arciformis* is 98% (Albuquerque et al. 2018). "*T. caldilimi*" is closely related to *T. calditerrae* (97.3% sequence similarity),

Table 8.1 List of Thermus species with some properties (ND—not determined, species which are not validly published yet are represented in brackets)

Location	Species	Source	Growth temperature range (optimum) (°C)	Growth pH range (optimum)	Type strains	G+C content (mol%)	16S rRNA gene	Reference
Niujie hot spring, China	*Thermus amyloliquefaciens*	Sediment	50–70 (60–65)	6.0–8.0 (7.0)	YIM 77409 DSM 25898 KCTC 32024	66.4	KP284528	Yu et al. (2015)
Hot springs at Hruni, Iceland	*Thermus antranikianii*	Water	50–80 (70)	5.0–10.0 (7.5–8.5)	HN3-7 ATCC 700961 DSM 12462	62.5–65.4	Y18411	Chung et al. (2000)
Hot spring in Yellowstone National Park (YNP), USA	*Thermus aquaticus*	Water	40–79 (70)	6.0–9.5 (7.5–7.8)	YT-1 ATCC 2510 DSM 625 JCM 10724	65.5	X58343	Brock and Freeze (1969)
Hot spring in Laibin, China	*Thermus arciformis*	Water	40–77 (70)	6.0–9.5 (7.5–8.0)	TH92 CGMCC 1.6992 JCM 15153	68.3	EU247889	Zhang et al. (2010)
Pine spring in YNP, USA	*Thermus brockianus*	Water	ND (70)	ND	YS38 JCM 11602	62.7	Z15062	Williams et al. (1995)
Hot spring in Tibet, China	*Thermus caldifontis*	Sediment	50–70 (60)	6.0–8.0 (7.0)	YIM 73026 CCTCC AB 2016305 NBRC112415	65.4	KX580314	Khan et al. (2017)
Hot spring, Ngamring Tibet, China	*Thermus caldilimi*	Sediment	45–65 (55)	7.0–8.0 (7.0)	YIM 78456 KCTC 52948 NBRC 113036	65.1	MK681861	Li et al. (2019)
Hydrothermal explosion area in Tengchong, China	*Thermus caliditerrae*	Sediment with water	50–70 (65)	6.0–8.0 (7.0)	YIM 77925 CCTCC2012061 DSM 25901	65.6–65.7	KC852874	Ming et al. (2014)

Compost	*Thermus composti*	Thermophilic phase compost	40–80 (65–75)	5.0–9.0 (7.0)	K-39 DSM 21686 NCAIMB 02340	71.3	EU701067	Vajna et al. (2012)
Hot spring in New Zealand	*Thermus filiformis*	Water	37–80 (70–73)	6.0–8.6 (7.0–7.5)	Wai33 A1 ATCC 43280 DSM 4687 JCM 11600	65.0	X58345	Hudson et al. (1987)
Hot springs in Reykyaflot, Iceland	*Thermus igniterrae*	Water	50–75 (65)	5.5–9.5 (7.5–8.5)	RF-4 ATCC 700962 DSM 12459	68.8–70.3	Y18406	Chung et al. (2000)
Hot springs in Torfajokull, Iceland	*Thermus islandicus*	Biofilm	45–79 (65)	5.5–10.5 (6.0–7.0)	PRI 3838 ATCC BAA-1677 DSM 21543	69.0	EU753247	Bjornsdottir et al. (2009)
Hot springs in the Kawarayu, Japan	*Thermus kawarayensis*	Water	40–73 (68)	5.8–8.9 (7.0)	KW11 DSM 16200 JCM 12314	69.0	AB071811	Kurosawa et al. (2005)
Hot springs São Pedro do Sul, Portugal	*Thermus oshimai*	Water	<80 (70)	ND	SPS17 ATCC 700435 DSM 12092 JCM 11603 NCIMB 13400	63.0	Y18416	Williams et al. (1996)
Hot spring in Manikaran, India	*Thermus parvatiensis*	Water	60–80 (70)	7.0–9.0 (7.2)	RL DSM 21745 MTCC 8932	68.7	EU017402	Dwivedi et al. (2015)
Hot springs at Rehai, China	*Thermus rehai*	Water	40–80 (65–70)	4.5–10.5 (7.5–8.5)	RH99-GF7504 CCTCC-AB200292	63.1	AF331969	Lin et al. (2002)
Selfoss, Iceland	*Thermus scotoductus*	Hot tap water	42–73 (65–70)	ND (7.5)	SE-1 ATCC 51532 DSM 8553 JCM 11601	64.1–64.5	AF032127	Kristjánsson et al. (1994)

(continued)

Table 8.1 (continued)

Location	Species	Source	Growth temperature range (optimum) (°C)	Growth pH range (optimum)	Type strains	G+C content (mol%)	16S rRNA gene	Reference
Geothermal spring in the Long Valley Caldera, California	*Thermus sediminis*	Sediment	45–75 (60–70)	5.0–9.0 (7.0)	L-198 CGMCC1.13590 KCTC XXX	68.21	LC382246	Zhou et al. (2018b)
Rehai National Park, China	*Thermus tengchongensis*	Geothermal soil	55–75 (65)	6.0–8.0 (7.0)	YIM 77924 DSM 25900 KCTC 32025	66.6	JX112365	Yu et al. (2013)
Hot spring in Tengchong, China	*Thermus tenuipuniceus*	Sediment	50–75 (ND)	6.0–9.0 (ND)	YIM 76954 JCM 30350 KCTC 4677	65.5	LC366928	Zhou et al. (2018a)
Jinze hot spring, China	*Thermus thermamylovorans*	Enrichment	37–75 (ND)	6.0–8.0 (ND)	CFH 72773 CCTCC AB2018244 KCTC43129	69.5	MK418796	Ming et al. (2020)
Hot spring in Japan	*Thermus thermophilus*	Water	47–82 (70)	5.1–9.6 (7.5)	HB8 ATCC27634 DSM 579 JCM 10941 NBRC 101084 NCIMB 11244	64.7	X58342	Oshima and Imahori (1974); Williams et al. (1995)
Buharkent geothermal area, Turkey	*Thermus anatoliensis*	Water	45–80 (65)	5.5–10.5 (7.5)	MT1 NCCB 100372 LMG 26290	69.6	HQ419075	Kacagan et al. (2015)
Hot spring in Tengchong, China	*Thermus yunnanensis*	Water	40–78 (65–70)	7.0–10.0 (ND)	RHY12-2 CGMCC 1.3695	ND	AY557603	Gong et al. (2005)

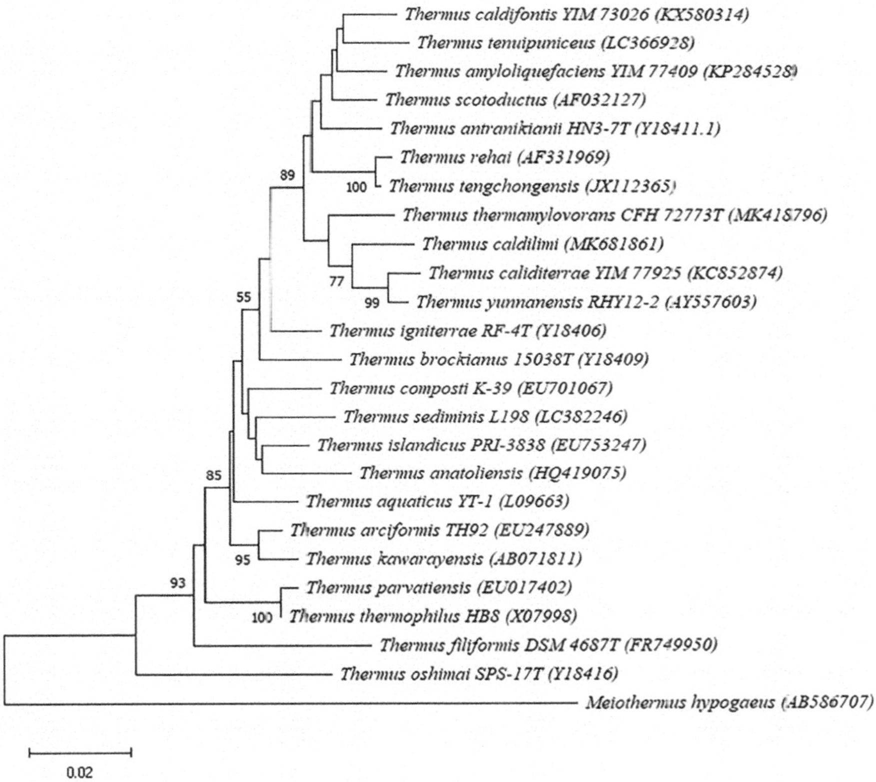

Fig. 8.1 16S rRNA gene sequence-based phylogeny showing the relatedness of the species of the genus *Thermus*. Scale bar indicate two nucleotide substitutions per 100 nucleotides. The optimal tree with the sum of branch length = 0.51927533 is shown. The percentage of replicate trees in which the associated taxa clustered together in the bootstrap test (1000 replicates) are shown next to the branches. Significant bootstrapping values (>50%) are shown on the nodes. This analysis involved 25 nucleotide sequences. All positions containing gaps and missing data were eliminated. There were a total of 1309 positions in the final dataset. Evolutionary analyses were conducted in MEGA X (Kumar et al. 2018)

the similarities of "*T. anatoliensis*" and "*T. sediminis*" with *T. islandicus* are 96.92% and 97.01%, respectively (Li et al. 2019; Kacagan et al. 2015; Zhou et al. 2018b).

Due to extremely variable biochemical and physiological characteristics and fatty acid composition, it is very difficult to define most of the species of the genus *Thermus* based on phenotype (Albuquerque et al. 2018). The phenotypic variability may be caused by natural diversity within each species, as well as could be due to different methods used for assessing phenotypic characteristics. For instance, although the species *T. thermophilus* has variable fatty acid composition, many strains share very high values of DNA-DNA hybridization (Manaia et al. 1995; Nobre et al. 1996). Besides, most species have been described on the basis of a few number of isolates, mostly based on one strain that has not been properly characterized, so that diversity of phenotypic characteristics has not been broadly assessed. For example, the species *T. filiformis* was described based on one strain from New Zealand with a stable filamentous morphology. In contrast, other strains from New Zealand that based on DNA-DNA hybridization values belong to this species, are not filamentous (Albuquerque et al. 2018).

8.3 Morphology, Physiology, Metabolism, and Biochemical Characteristics

Generally, the strains of the genus *Thermus* form rod-shaped cells of variable length and filaments that divide into shorter cells at beginning of the stationary phase of growth. The type strain of *Thermus filiformis* has, in contrast to other strains, a stable filamentous morphology. However, some strains closely related to the type strain of *T. filiformis* do not form extensive filaments. Unusual morphological structures, called "rotund bodies," are occasionally found in many strains. These structures consist of several cells bound longitudinally by a common external layer of the cell envelope enclosing a large space between the cells (Albuquerque et al. 2018; Cava et al. 2009).

The majority of *Thermus* isolates form yellow-pigmented colonies, although the color varies from deep yellow to very pale yellow due to presence of carotenoids. Few reddish-pigmented strains were also described (Albuquerque et al. 2018; Zhou et al. 2018a). Many strains isolated primarily from man-made and dark environments are nonpigmented, although yellow-pigmented strains can also be isolated from these environments. Several nonpigmented *Thermus* strains have been isolated from abyssal hot springs, but even there, some isolates are yellow pigmented (Albuquerque et al. 2018).

The growth temperature range of the species of the genus *Thermus* is between 37 °C and 82–83 °C (Table 8.1). Few strains, closely related to *T. thermophilus* HB8, are able to grow at 80 °C or above. The majority of the *Thermus* strains actually have a maximum growth temperature slightly below 80 °C (Albuquerque et al. 2018). The optimum growth temperature of most strains is near 65–70 °C, but the optimum growth temperature of some strains could be as high as 75 °C (Table 8.1). The type strain of *T. amyloliquefaciens*, was reported to have an optimum growth temperature of 60–65 °C and a maximum growth temperature of 70 °C (Yu et al. 2015). The optimum pH for growth of the strains of the genus *Thermus* is between 7.0 and 8.0, with a pH range for growth between 5.0 and 10.0 (Albuquerque et al. 2018).

Representatives of the genus *Thermus* have a respiratory metabolism and are generally aerobic. Many strains are capable of growth under anaerobic conditions using nitrate as the electron acceptor, some strains also reduce nitrite (Cava et al. 2009).

Complete genome sequences suggest that *Thermus* strains use the Embden–Meyerhof–Parnas pathway for the catabolism of hexoses (Brumm et al. 2015; Henne et al. 2004). The genome sequence of strain *T. aquaticus* Y51MC23 indicates that the organism may be able to perform fermentation with the formation of lactate, ethanol, and acetate (Brumm et al. 2015). However, the ability to fermentation has not been demonstrated experimentally.

Recent whole genome sequence analysis indicate that *Thermus* strains possess a complete tricarboxylic acid cycle and that respiratory chain terminal cytochrome oxidases, namely *caa*3 and *ba*3 are expressed during aerobic growth and low oxygen pressures, respectively (Albuquerque et al. 2018).

Several isolates of *T. thermophilus* reduce nitrate to nitrite anaerobically, while others end up reduction to gaseous nitrogen. Interestingly, the type strain HB8 and strain HB27 do not grow with nitrate as electron acceptor, in contrast to other *T. thermophilus* strains carrying out dissimilatory denitrification (Cava et al. 2009). Other *Thermus* spp. are able to grow anaerobically with nitrogen oxides and polysulfides as electron acceptors (Henne et al. 2004). *T. scotoductus* SA-01 is also capable of dissimilatory iron reduction. This organism also reduces other metal ions, such as gold, chromate, and uranium, during aerobic and anaerobic growth with nitrate (Gounder et al. 2011).

All strains are chemoorganotrophic and are capable of growth on amino acids, peptides and proteins, organic acids, and simple and complex carbohydrates (Albuquerque et al. 2018).

T. scotoductus strain IT-7254 from Iceland, *T. scotoductus* strain SA-01 isolated from fissure water from deep South African gold mine as well as the type strain of *T. islandicus* oxidize thiosulfate to sulfate in the presence of organic carbon sources, indicating that the organisms are facultatively mixotrophic (Balkwill et al. 2004; Gounder et al. 2011; Kieft et al. 1999; Skirnisdottir et al. 2001). Some of these strains can also oxidize sulfur (Bjornsdottir et al. 2009). The presence of a complete *sox* gene cluster in *T. oshimai* strain JL-2 and *T. thermophilus* strain JL-18 suggests that these strains may use thiosulfate as electron donor (Murugapiran et al. 2013b). Other strains of *T. thermophilus* also harbor the *sox* gene cluster, suggesting that thiosulfate oxidation may be widely distributed in *Thermus* (Brumm et al. 2015; Henne et al. 2004).

8.4 Ecology and Distribution

Thermus resides both natural and artificial thermal environments, including terrestrial geothermal areas, hot water taps, self-heating compost manure, and deep mines. *Thermus* strains have been isolated worldwide, mainly from terrestrial hydrothermal areas with water temperature and pH range from 55–70 °C and pH 5.0–10.5, respectively (Albuquerque et al. 2018), also from shallow marine hot springs, deep-sea geothermal areas and even artificial thermal territories (da Costa et al. 2006). Few isolates have been obtained from water with temperatures as high as

95 °C and pH values as low as 3.9 (Albuquerque et al. 2018). The first isolates of the genus *Thermus* were obtained from hydrothermal areas in Yellowstone National Park (Brock and Freeze 1969). Isolates were then recovered from several hydrothermal areas in Japan, followed by isolations from Iceland, New Zealand, New Mexico, continental Portugal, the Island of São Miguel in the Azores, the Australian Artesian Basin (Albuquerque et al. 2018), Armenia (Saghatelyan et al. 2015), Turkey (Kacagan et al. 2015), and Argentina (Navas et al. 2015). Isolates of the genus *Thermus* have also been obtained from deep-sea geothermal areas in the Mid-Atlantic Ridge and in the Guaymas Basin, at depths of 3500 and 2000 m, respectively (Albuquerque et al. 2018), geothermal power plant (Fujino et al. 2017), and 3.5 km deep subsurface gold mine (Kieft et al. 1999).

Interestingly, even species of the genus *Thermus* have been isolated from natural and artificial thermal environments worldwide, they exhibit different patterns of distribution. Geographical isolation may be an important factor in species divergence within the genus *Thermus*. Widely separated locations distinctly differ in species composition, with mixtures of unique endemic lineages and more cosmopolitan species (Hreggvidsson et al. 2006).

Thus, *T. aquaticus* and *T. filiformis* to date have only been encountered in the USA and New Zealand, respectively. *T. brockianus*, *T. oshimai*, and *T. scotoductus* show cosmopolitan distributions, while *T. igniterrae* and *T. antranikianii* have only been isolated from hot springs in Iceland. Strains belonging to *T. thermophilus* show worldwide distribution in marine and coastal hot springs (Hreggvidsson et al. 2006; Bjornsdottir et al. 2009). Halotolerant strains have been isolated from inland hot springs, containing high levels of NaCl, including *T. thermophilus* strains HB8, HB27, AT-62, and GK-24 from Japan and strain B from Iceland. Strains RQ-1 and RQ-2 were also isolated from a hot spring on the island of São Miguel in the Azores with a salinity of about 480 mg/L (Albuquerque et al. 2018). Interestingly, strain TMY, isolated from geothermal power plant in Japan, is unable to grow in presence of high concentrations of NaCl, unlike to other representatives of *T. thermophilus* (Fujino et al. 2017). Isolates of cosmopolite *T. scotoductus* have been obtained from hot springs and geothermal systems in Armenia, Iceland, Portugal, USA (Saghatelyan et al. 2015; Hreggvidsson et al. 2006), a deep South African gold mine (Kieft et al. 1999), and a volcanic hydrothermal region of the flank of Kilauea (Albuquerque et al. 2018). Studies of thermophiles in Icelandic hot springs proofed that *T. scotoductus* was the most genetically diverse and geographically widespread *Thermus* species recovered (Hreggvidsson et al. 2006).

Members of the genus *Thermus* have been found to colonize built environments including geothermal well water (Kristjánsson et al. 1994), gold mine boreholes with up to 3.2 km in depth (Kieft et al. 1999), industrial composting systems, and wet flue-gas desulfurization systems for coal-burning power plants, and, interestingly, *T. scotoductus* was shown to be the dominant thermophile in household water heaters across the USA (Wilpiszeski et al. 2019).

Novel species *Thermus parvatiensis*, has been isolated from the water sample of Manikaran hot water spring (Himachal Pradesh), in India (Dwivedi et al. 2015). Several novel species, *T. rehai*, *T. caliditerrae*, *T. amyloliquefaciens*, *T. caldifontis*,

T. thermamylovorans, *T. caldilimi*, and *T. tenuipuniceus* have been reported from Yunnan and Tibetan hot springs (Lin et al. 2002; Ming et al. 2014, 2020; Yu et al. 2015; Khan et al. 2017; Li et al. 2019; Zhou et al. 2018a).

8.5 Genome Structure and Natural Competence

At the time of writing this chapter, there were 37 *Thermus* genome projects available in Genomes online database (https://gold.jgi-psf.org/index). Complete genome sequences have been reported for *Thermus aquaticus* Y51MC23 (Brumm et al. 2015), *T. thermophilus* HB8 (Bruggemann and Chen 2006), *T. thermophilus* HB27 (Henne et al. 2004), *T. thermophilus* TMY (Fujino et al. 2017), *T. thermophilus* HC11 (Miyazaki 2019), *T. scotoductus* SA-01 (Gounder et al. 2011), *T. oshimai* JL-2, and *T. thermophilus* JL-18 (Murugapiran et al. 2013a, b), *Thermus sp.* strain CCB_US3_UF1 (Teh et al. 2015), *T. brockianus* GE-1 (Schafers et al. 2017), *T. parvatiensis* RL (Tripathi et al. 2017).

Most *Thermus* genomes have been left unfinished in permanent draft status, including *Thermus* sp. NMX2.A1 (Muller et al. 2016), *Thermus* sp. 2.9 (Navas et al. 2015), *T. thermophilus* ATCC 33923 (Jiang et al. 2013), *T. thermamylovorans* CFH 72773 (Ming et al. 2020), *T. caliditerrae* YIM 77777 and *T. tengchongensis* YIM 77401 (Mefferd et al. 2016), *T. amyloliquefaciens* YIM 77409 (Zhou et al. 2016), *T. scotoductus* K1 (Saghatelyan et al. 2015), *T. filiformis* ATCC 43280 (Mandelli et al. 2015).

The genomes of *Thermus* species, in comparison to typical bacterial genomes, are notably different. *Thermus* generally have small genome sizes of less than 2.5 Mb with extrachromosomal elements of common occurrence (Henne et al. 2004; Bruggemann and Chen 2006; Brumm et al. 2015; Gounder et al. 2011; Teh et al. 2015; Tripathi et al. 2017). Analyses revealed that *Thermus* genomes demonstrate high plasticity (Bruggemann and Chen 2006; Gounder et al. 2011). Vast genome rearrangements played a crucial role in sculpturing of genomes of strains of the genus *Thermus* (Kumwenda et al. 2014). Several features of complete genomes of representatives of *Thermus* are shown in Table 8.2.

All of those complete genomes include a chromosome and at least one plasmid. Due to their thermophilic lifestyle, all completed *Thermus* genomes exhibit a high G+C% content varying between 64.9% for *T. scotoductus* SA-01 and 69.5% for *T. thermophilus* HB8. In average, GC content for the *Thermus* genomes is 67.0%.

Plasmids were found in most of the complete genome sequences of *Thermus*. Their size and number vary among strains. For instance, *T. aquaticus* Y51MC23 contains four plasmids with 14.4, 16.6, 69.9, and 78.7 kb in size (Brumm et al. 2015), *Thermus* sp. CCB_US3_UF1 contains only one plasmid of 19.7 kb (Teh et al. 2015), while the genome of *T. oshimai* JL-2 contains one megaplasmid of 0.27 Mb and *T. thermophilus* JL-18 contains a megaplasmid of 0.26 Mb as well as a smaller plasmid of 0.14 Mb, *T. scotoductus* strain SA-1 contains only a small plasmid of 8.3 kb (Murugapiran et al. 2013a; Gounder et al. 2011), *T. brockianus* GE-1 contains

Table 8.2 Some features of complete genomes of representatives of genus *Thermus*. Table was adapted from Teh et al. (2015), Tripathi et al. (2017), Brumm et al. (2015), Miyazaki (2019), Fujino et al. (2017), Muller et al. (2016)

Organism	Chromosome (total genome) size (bp)	Plasmid(s) with size in brackets (bp)	G+C content (%)	Predicted CDS	tRNA	rRNA	NCBI Ac. No
T. scotoductus SA-01	2,346,803 (2,355,186)	pTSC8 (8,383)	64.9	2514	47	6	CP001962
T. thermophilus HB8	1,849,742 (2,197,207)	pTT27 (256,992) pTT8 (9,322) pVV8 (81,151)	69.4	2268	48	6	AP008226
T. thermophilus HB27	1,894,877 (2,127,482)	pTT27 (232,605)	69.4	2224	47	6	AE017221
T. thermophilus JL-18	1,902,595 (2,311,212)	pTTJL1801 (265,886) pTTJL1802 (142,731)	69.0	2424	52	6	CP003252
T. thermophilus TMY	2,121,526 (2,140,665)	pTMY (19,139)	69.0	2528	47	6	AP017920
T. thermophilus HC11	1,910,731 (2,169,490)	pHC11 (258,759)	69.4	-	-	-	AP019801
T. thermophilus SG0.5JP17-16	1,863,201 (2,303,227)	pTHTHE1601 (440,026)	68.6	2405	53	6	CP002777
T. oshimai JL-2	2,072,393 (2,401,329)	pTHEOS01 (271,713) pTHEOS02 (57,223)	68.6	2521	59	6	CP003249
T. sp. CCB_US3_UF1	2,243,772 (2,263,488)	pTCCB09 (19,716)	68.6	2228	48	6	CP003126
T. brockianus GE-1	2,035,182 (2,388,273)	pTB1 (342,792) pTB2 (10,299)	66.9	2789	47	2	CP016312
T. parvatiensis RL	1,872,821 (2,016,098)	pTP143 (143,277)	68.5	2383	54	2	CP014141
T. aquaticus Y51MC23	2,158,963 (2,338,641)	pTA14 (14,448) pTA16 (16,597) pTA69 (69,906) pTA78 (78,727)	68.0	2436	55	3	CP010822

two plasmids, including megaplasmid pTB1 (0.34 Mb) and plasmid pTB2 (10 kb) (Schafers et al. 2017), and *T. thermophilus* HB27 harbors a 0.23-Mb megaplasmid, designated pTT27 (Henne et al. 2004). Interestingly, in *T. thermophilus* JL-18 and *T. oshimai* JL-2 genes involved in denitrification, including *nar* operon, are located on the megaplasmids, as in other strains of *T. thermophilus* (Ramírez et al. 2000), in contrast with *T. scotoductus* SA-01, where these genes are located on the chromosome (Gounder et al. 2011).

Megaplasmids are a common feature among *Thermus* spp. and were identified in *T. parvatiensis* (pTP143; 0.14 Mb), *T. brockianus* GE-1 (pTB1; 0.34 Mb), *T. oshimai* JL-2 (pTHEOS01; 0.27 Mb), *T. thermophilus* HB27 (pTT27; 0.23 Mb), *T. thermophilus* HC11 (pTT27; 0.259Mb), *T. thermophilus* JL-18 (pTTJL1801; 0.266 Mb) (Schafers et al. 2017; Tripathi et al. 2017). Interestingly, megaplasmid is absent in TMY strain, in contrast to other *T. thermophilus* spp. (Fujino et al. 2017). *T. aquaticus* Y51MC23 does not harbor megaplasmid as well. The role of the megaplasmids is believed to act as a storage site for genes related to thermophily (Brumm et al. 2015).

Usually, genomes of different *Thermus* strains belonging to the same species are relatively similar in contrast to plasmids, where numerous genetic rearrangements were observed despite high level of average nucleotide identity (Miyazaki 2019). Generally, high degree of synteny at both genomic and proteomic level is common for closely related *Thermus* strains (Muller et al. 2016). However, in several *Thermus* spp. different metabolic pathways may be detected, which are absent in other, even in closely related strains. For instance, a complete, putative Calvin–Benson–Bassham (CBB) cycle was found in *T. scotoductus* NMX2.A1, that is absent in SA-01. This might be explained by origin of two strains, thus NMX2 originates from HCO_3^-—rich environment, in contrast to SA-01 isolated from deep subsurface (Kristjánsson et al. 1994; Muller et al. 2016).

Among the properties that permitted the development of genetic tools for *Thermus* spp., the most important of all is the constitutive expression of a natural competence system in several strains (Cava et al. 2009).

Competence proteins have an important role in natural transformation. They have been sorted into three groups: DNA-translocator-specific proteins, type IV pili (Tfp)-related proteins, and non-conserved proteins (Averhoff 2009).

The sequence of *T. thermophilus* HB27 genome revealed that at least 16 genes were implicated in natural competence. Three of them (comEA, comEC, dprA) encode proteins similar to components of DNA translocators, four pilin-like proteins (PilA1, PilA2, PilA3, PilA4), a leader peptidase (PilD), a traffic-NTPase protein (PilF), an inner membrane protein (PilC), a PilM-homologue, and a secretin-like protein (PilQ). In addition to these homologues of competence proteins, four additional proteins (ComZ, PilN, PilO, and PilW) were detected with no homologues in the protein data banks. Based on these data, a model for the natural competence system of *T. thermophilus* has been proposed (Cava et al. 2009). Interestingly, PilA is absent in SA-01 and HB8 (Gounder et al. 2011). Strians *T. thermophilus* HB27 and HB8, *T. aquaticus* YT1, *T. flavus* AT62 and *T. caldophilus* GK24 have shown to possess natural competence, and HB27 showed the highest transformation frequency with a maximal uptake rate of about 40 kb/s per cell (Schwarzenlander and Averhoff 2006). Actually, the natural competence system of *T. thermophilus* has been suggested to be responsible for the presence in this genus of genes coding homologues to proteins from Archaea (Omelchenko et al. 2005).

8.6 Biotechnological Potential and Applications of *Thermus*

Representatives of the genus *Thermus* are omnipresent microorganisms among thermophiles. Members of this genus demonstrate a huge potency as a source of thermophilic enzymes of wide industrial and biotechnological interest as well as serve a tool for the overexpression of thermophilic enzymes or for the selection of thermostable mutants from mesophilic proteins by directed evolution (Cava et al. 2009; Fujino et al. 2020).

8.6.1 Thermus *as Source of Extremozymes*

Enzymatic processes involving higher operational temperature demand the thermal stability of the biocatalyst. Enzymes and proteins from the genus *Thermus*, and thermozymes in general are good candidates for biocatalytic processes as they often present higher operational stability. Additionally, purification of these enzymes may be carried out to a great extent by a single step of heat denaturation, when produced in mesophilic hosts (Cava et al. 2009). The other benefits of thermostable proteins and enzymes are the reduced chance of bacterial contamination, especially in food and drug applications and reduced operating costs from enzyme replacement due to thermal denaturation.

A number of enzymes of biotechnological interest, including, but not limited to glucose isomerase, xylose isomerase, proteases, β-glucosidase, *L*-asparaginase, phosphatases, pyrophosphatase and several DNA and RNA processing enzymes have been mined from *Thermus* (Pantazaki et al. 2002). More recently other promising enzymes from *T. thermophilus* have been reported, including NADH-oxidases, mannose-6-P-isomerase and superoxide dismutase (Lopez-Lopez et al. 2015). Several extremozymes sourced from representatives of *Thermus* will be discussed below.

8.6.1.1 Lipolytic Enzymes

Lipolytic enzymes are one of the most important groups of biocatalysts for biotechnological applications (Lopez-Lopez et al. 2015). These include esterases (EC 3.1.1.1, carboxyl ester hydrolases) and lipases (EC 3.1.1.3, triacylglycerol hydrolases), which catalyze the hydrolysis of ester bonds between alcohols and carboxylic acids, and its formation in organic media (Bornscheuer 2002). Lipolytic enzymes mined from thermophiles are preferably applied in several fields of industry, in contrast to their mesophilic analogs, especially due to resistance to various denaturants and higher thermostability (Lopez-Lopez et al. 2015). The presence of lipolytic enzymes in *Thermus* sp. has been known for long time. The strains *T. aquaticus* YT-1 and *T. thermophilus* HB27 were shown to be the two of best producers (Domínguez et al. 2004). The lipolytic activity of protein extracts of these strains showed high thermostability (75–100% activity is remaining after 30 min at 80 °C) demonstrating substrate preference for *p*-nitro-phenyl-esters of medium length fatty acids chains (Lopez-Lopez et al. 2015).

Recombinant esterase Est*Ts*1 was identified from genomic library of *Thermus scotoductus* SA-01. In contrast with usual thermophilic anaerobic lipases with temperature optimum of 70–78 °C, Est*Ts*1 (predicted molecular mass of 28.6 kDa) displayed optimum activity at 80 °C, with a half-life of 48 h at 70 °C. Moreover, EstTs1 showed a preference for p-NP butyrate (C4) and optimum activity at pH 7.0, in contrast to the *T. thermophilus* HB27 and HB8, *T. aquaticus* YT-1, and *Thermus* sp. which prefer p-NP caproate (C8) and have optimum activity at pH 8 (du Plessis et al. 2010).

Other putative lipases/esterases annotated in the *T. thermophilus* genome have been explored. The 7 putative esterase of 329 amino acids and 36.0 kDa in *T. thermophilus* HB2 was annotated, and was classified as a new enzyme family LipT (Lopez-Lopez et al. 2015).

Interestingly, among numerous thermostable lipolytic enzymes which were reported to be active and stable exclusively at alkaline pH values (Choi et al. 2013), only two heat-stable enzymes were described in the literature with slightly acidic pH optima including an esterase from *Thermus thermophilus* (pH optimum at 6.3) and an esterase from *Thermotoga maritima* with a pH optimum at 5.0–5.5 (Tao et al. 2013; Fuciños et al. 2014). There is considerable interest in thermostable lipases such as those obtained from *Thermus aquaticus*, *T. flavus*, and *T. thermophilus*. The latter two were commercially available at Fluka/Sigma-Aldrich.

Phospholipases are enzymes that hydrolyze phospholipids into fatty acids and other lipophilic substances and found use in industrial processes. Thus, the search for new enzymes with differential properties was highly demanded. A gene encoding a novel phospholipase PLP_2.9 identified in the genome of the thermophilic strain *Thermus* sp. 2.9. PLP_2.9 was overexpressed in *E. coli*. PLP_2.9 hydrolyzed *p*-nitrophenylpalmitate at alkaline pH over a wide range of temperatures (55–80 °C), showing high thermostability. PLP_2.9 demonstrated phospholipase A and acyltransferase activities on egg yolk phosphatidylcholine phospholipids and has a significant potential for application in several industries as a catalyst due to its high thermostability (Navas et al. 2018).

8.6.1.2 Proteases

Proteases found many applications in the food and dairy industries (Pantazaki et al. 2002). Besides, proteases have found a novel application as low volume/high value laboratory reagents in nucleic acid isolation (Bergquist et al. 2014).

Many *Thermus* isolates produce extracellular proteases. Recombinant proteases from *Thermus* were described as well. Thus, a gene for an FtsH homologue, an ATP-dependent and temperature-sensitive zinc protease was cloned from *T. thermophilus* HB8 and expressed in *E. coli*. *T. thermophilus* FtsH digests proteins with unfolded structure in an ATP-Zn^{2+}-dependent manner (Pantazaki et al. 2002).

A gene encoding Lon protease, an ATP-dependent serine protease, which is a heat shock protein that degrades denatured or non-functional proteins, was also isolated from *T. thermophilus* HB8. The amino acid sequence of this protease contained several unique motifs conserved in other Lon proteases. Expression of Lon in *E. coli* produced a protein of 89 kDa. The recombinant Lon protease was activated by ATP and α-casein. The results also suggested a difference in substrate specificity between Lon of *T. thermophilus* and that of *E. coli* (Watanabe et al. 1999).

Aqualysin I, a subtilisin-type heat-stable serine protease secreted into the medium by *Thermus aquaticus*. *T. thermophilus* also produced active aqualysin I and was able to grow on a minimal medium containing milk casein as the sole source of carbon and nitrogen (Pantazaki et al. 2002). Aqualysin I from *T. aquaticus* YT-1, with optimum pH of 10.0 at 80 °C, remains active in the presence of 7 M urea, 6 M guanidine hydrochloride and 1% SDS (Matsuzawa et al. 1988).

A protease mined from *Thermus* sp. Rt4A2 was shown to be significantly tolerant to temperature and degradation. Thus, it loses only 25% of its activity at 4 °C, in the presence of 90% acetonitrile. However, when acetonitrile is replaced by butanol at the same concentration, the decrease in activity is twice higher (59%) (Freeman et al. 1993).

The Rt41A protease from *Thermus* sp. is a thermostable alkaline protease that is known commercially as *Pretaq*, being used in the preparation of DNA and mRNA prior to amplification by PCR. Rt41A protease, a member of subtilisin family with molecular mass of 32.5 kDa, is thermostable, showing no loss of activity after 24 h at 70 °C, and a half-life at 90 °C of 20 min. Interestingly, in the absence of Ca^{2+} ions, the half-life of the protease is only 3 min at 70 °C (Toogood and Daniel 2013).

8.6.1.3 Pullulanases, Xylanases, and Other Polymer Degrading Thermozymes

Pullulanases are produced by animals, plants, fungi, and bacteria. Many mesophilic, thermophilic, and hyperthermophilic bacteria and archaea have been reported to produce pullulanases (Messaoud et al. 2002; Chai et al. 2012; Nisha and Satyanarayana 2016). The type I pullulanase is produced by and *Thermus thermophilus* HB8 (Tomiyasu et al. 2001). The gene encoding pullulanase from

Thermus sp. AMD-33 has been cloned in *E. coli* (Sashihara et al. 1988). *Thermus caldophilus* GK-24 produces a thermostable type I pullulanase with 75 °C and 5.5 temperature and pH optima, respectively (Bertoldo and Antranikian 2002). Interestingly, *Thermus aquaticus* YT-1 was shown to produce amylopullulanase using maltose as a carbon source (Plant et al. 1987), although amylopullulanases are distributed mostly among thermophilic anaerobic bacteria (Coleman et al. 1987).

A novel thermoactive xylanase-encoding gene (xyn10) from *T. brockianus* was successfully expressed in *E. coli*. The resulting protein (38.7 kDa), a member of glycoside hydrolase family 10, demonstrates catalytic activity at up to 115 °C and highest its activity was measured at 95 °C and pH 6.0. The protein was extremely thermostable (80% activity remains after incubation at 50–70 °C for 24 h) (Blank et al. 2014). Xyn10 hydrolyzes insoluble and soluble substrates, such as oat spelt xylan, xylan from beech-and birchwood forming xylobiose and xylose. The xylanase exhibited remarkable stability in the presence of various detergents and chaotropic agents, such as guanidine hydrochloride and urea.

The thermozymes involved in the conversion of starch to glucose, maltose and oligosaccharides have been described from thermophiles and hyperthermophiles (Leuschner and Antranikian 1995). Thermostable endoglucanases have been described in *Thermus* (Antranikian and Egorova 2007). Furthermore, in *Thermus* spp. glycosidases also have been described (Xiangyuan et al. 2001). The species *Thermus amyloliquefaciens* was reported for its ability to liquefy starch (Yu et al. 2015).

Amylomaltases catalyze the transfer of a segment of a α-1,4-*d*-glucan to a new 4-position of an acceptor, which may be glucose or another α-1,4-*d*-glucan. Acting upon starch, amylomaltases can produce products of commercial interest, such as cycloamylose, a thermoreversible starch gel, which can be used as a substitute for gelatin. In combination with α-amylase the amylomaltase is used in production of syrups of isomalto-oligosaccharides with reduced sweetness and low viscosity. Amylomaltases from *Thermus* may be used to modify starches. Cyclic glucans can be produced using the thermostable bacterial amylomaltase from *Thermus aquaticus* (Terada et al. 1999; Fujii et al. 2005).

8.6.1.4 Nucleic Acid Manipulation Tools

Proteins and enzymes from *Thermus*, besides their usefulness in a variety of sectors of industry, may be successfully applied to enhance research and diagnostics. Especially, several molecular biological tools have been developed on basis of these organisms (Cava et al. 2009; Song et al. 2020), including DNA/RNA manipulation techniques. These include DNA polymerases, helicases, endonucleases etc.

DNA polymerases represent a family of enzymes which catalyze, in the presence of Mg^{2+} ions, the incorporation of deoxyribonucleoside-5′-monophosphates to the 3′-hydroxyl terminus of a growing DNA strand, using the opposite complementary strand as a template. DNA polymerases are widely used in different in vitro DNA manipulation techniques. These include: DNA cloning, mutagenesis, sequencing,

DNA labeling, etc. Since development of polymerase chain reaction (PCR) technology, thermostable DNA polymerases became more valuable (Yamagami et al. 2014).

DNA polymerases have been reported and patented from various species of *Thermus*, including *T. filiformis*, *T. scotoductus*, *T. antranikianii*, *T. eggertssonni*, *T. kawarayensis*, *T. oshimai*, and *T. igniterrae* (Cava et al. 2009).

The first thermostable DNA polymerase characterized and applied in PCR was *Taq* polymerase isolated from *T. aquaticus* (Chien et al. 1976; Kaledin et al. 1980). *Taq* polymerase has 5′→3′-exonuclease activity, but no detectable 3′→5′exonuclease activity, resulting in low fidelity (Longley et al. 1990). It was demonstrated, that even few mutations in *Taq* polymerase (at positions 742 and 743) can significantly improve its elongation ability, resulting in higher DNA affinity and faster primer extension ability, making mutant *Taq* polymerase more effective for use in PCR (Yamagami et al. 2014).

A thermostable DNA polymerase, isolated from *T. thermophilus* HB8 (*Tth*), was shown to be active under PCR conditions similar to those used for *Taq* polymerase. The molecular weight was estimated to be 67 kDa, and the extension rate 1500 nucleotides per minute (Carballeira et al. 1990). The DNA polymerase from *T. thermophilus* can elongate oligo-DNA with several tandem repeats to very long DNA up to approximately 10,000 bases in vitro at 74 °C with template DNA.

Top DNA polymerase from *T. thermophilus* HB27 was expressed in *E. coli* (Kim et al. 1998). A recombinant DNA polymerase from *T. thermophilus* HB8 containing a polyhistidine tag at the N-terminus was also isolated displaying high polymerase activity, reverse transcriptase activity and better thermostability than the native *Tth* DNA polymerase (Dabrowski and Kur 1998). Recombinant *Tth* DNA polymerase was used for development of detection of new protocol, which simplifies chronic myeloid leukemia diagnosis. The enzyme was catalyzing both coupling enzymic reactions, reverse transcription (RT) and nested polymerase chain reaction (nPCR). Recombinant *Tth* DNA polymerase is one of the more resistant enzymes to the inhibitory effect of K^+ and Na^+ ions in PCR samples. It was shown to be resistant to immunoglobulin G (IgG) in human plasma, which was found to be a major inhibitor of diagnostic PCR (Pantazaki et al. 2002). The ability of *Tth* DNA polymerase to perform both reverse transcription and DNA amplification at elevated temperatures allows this enzyme to be used for RT/PCR, a process valuable for detection, quantification, cloning, and analysis of gene expression at the RNA level. The enzyme can be used for RT/PCR amplification of RNA to a length of at least 1000 bp. It was reported that the RT activity of the polymerase from *T. thermophilus* (*Tth* polymerase) was 100-fold greater than that shown by *Taq* (Gibbs et al. 2009).

A thermostable *Tte* DNA polymerase was isolated from the strain *T. thermophilus* B35. The biochemical properties of this enzyme are similar to those of *Tth* DNA polymerase, but the practical application of *Tte* polymerase seems to be more preferable due to its higher optimal temperature and lack of restriction endonucleases in the initial strain. The polymerase processivity (but not the DNA synthesis efficiency) is Me^{2+}-type independent (Pantazaki et al. 2002).

Several useful DNA polymerases were successfully produced using protein engineering techniques. A cold-sensitive mutant of DNA polymerase from *Thermus aquaticus* was developed with evidently reduced activity at 37 °C, as compared with the wild type (WT) enzyme (Kermekchiev et al. 2003). This mutant polymerase can be used in hot start PCR. Another mutant *Taq* polymerase was developed showing higher resistance to various PCR inhibitors such as whole blood, hemoglobin, plasma, serum IgG, lactoferrin, soil extracts, humic acid (Kermekchiev et al. 2009). A *Taq* polymerases with reduced fidelity was produced (Suzuki et al. 1997, 2000; Tosaka et al. 2001), which might be applicable for error-prone PCR.

DNA ligases play an important role in DNA replication, repair and recombination. The thermostable DNA ligases are valuable as LDR/LCR (Ligase Detection Reaction/Ligase Chain Reaction) enzymes. DNA ligases have been reported from *T. thermophilus*, *T. scotoductus*, *T. filiformis* (Pantazaki et al. 2002; Cava et al. 2009) and are commercially available. A highly thermostable 79 kDa DNA ligase form *T. thermophilus* HB8 was purified. *Tth* DNA ligase from *T. thermophilus* HB8 is NAD^+-dependent, in comparison to mesophilic ATP-dependent DNA ligases, it has higher optimal temperature and, demonstrated higher fidelity than T4 DNA ligase (Pantazaki et al. 2002), enabling the use of *Tth* ligase in ligase-based technologies, such as LCR and multi-locus, site-directed mutagenesis (Cline and Hogrefe 2003; Hames et al. 2005). Two mutant *Tth* ligases K294R and K294P with increased fidelity were obtained using site-directed mutagenesis (Luo et al. 1996). The DNA ligase from *T. thermophilus* (Lauer et al. 1991) was originally used for the detection of a point mutation in the β-globin gene by LCR (Barany 1991). DNA ligases from other *Thermus* species, e.g., *T. scotoductus*, were also utilized for LCR (Housby and Southern 2002). Another NAD^+-dependent DNA ligase AK16D from *Thermus* sp. was cloned and expressed in *E. coli*. It showed similar properties to *T. thermophilus* HB8 ligase with respect to pH, salt, NAD^+, and divalent cation requirements, as well as steady-state kinetics. Both *Thermus* ligases exhibit enhanced mismatch ligation when Mn^{2+} is substituted for Mg^{2-} (Tong et al. 1999).

Other studies demonstrated the presence of thermostable DNA ligases in 23 *Thermus* strains (Pantazaki et al. 2002). Recently, an open reading frame (ORF), from thermophilic bacteriophage TS2126 infecting the *T. scotoductus*, showing homology to T4 RNA ligase was cloned, and recombinant enzyme was purified and characterized (Blondal et al. 2005). The recombinant enzyme ligates single-stranded nucleic acids in an ATP-dependent manner and is moderately thermostable. It exhibits extremely high activity and high ligation efficiency and might be useful for several molecular biological applications, such as RNA ligase-mediated rapid amplification of cDNA ends (RLM-RACE).

DNA helicases are the enzymes that, moving along the helix, uncoil the strands in the direction that they move. The thermostable recombinant *RuvB* from *T. thermophilus* was overexpressed in *E. coli*. The purified protein showed strong double-stranded DNA-dependent ATPase activity at its temperature optimum (>70 °C). In the absence of ATP, *T. thermophilus RuvB* protein bound to linear double-stranded DNA with a preference for the ends. Addition of ATP destabilized the *T. thermophilus RuvB*-DNA complexes. Thermostable *RuvB* proteins displayed helicase activity on supercoiled DNA (Tong and Wetmur 1996). The cloning and characterization of the *UvrD* gene encoding a DNA helicase, which plays an important role in prokaryotic nucleotide excision repair, mismatch repair and DNA replication, from *T. thermophilus* HB8 was also reported (Pantazaki et al. 2002). Thermostable helicases found application in hot start PCR. The addition of a DNA helicase prevents the random annealing of primers and synthesis of non-specific products during the preparation of the reaction mixture and initial heating. The hot start PCR occurs automatically after inactivation of the DNA helicase upon heating of the reaction mixture (Kaboev et al. 1999).

Nucleases include restriction endonucleases and exonucleases. *Thermus* is the one of the most important producers of restriction enzymes. Seventy-six strains of different *Thermus* spp. are known to produce restriction enzymes (Vitkute et al. 2001). *Thermus* represent a rich source of novel restriction endonucleases (REs). Most thermostable type II REs recognizing different DNA sequences have been discovered among species of the genus *Thermus*. REs with 26 different specificities have been discovered within the species of the genus *Thermus* including 10 prototypes: *Taq*I, *TaqII*, *Tfi*I, *Tse*I, *Tsp*4CI, *Tsp*EI, *Tsp*45I, *Tsp*RI, *Tth*111I, and *Tth*111II (Vitkute et al. 2001). A site-specific endonuclease was isolated from *T. thermophilus* strain 111 and named *Tth*111I. The *Tth*111I recognition sequences have symmetry with the twofold axis, as most type II restriction endonucleases do. Another site-specific endonuclease with novel specificity has been purified from the same strain and named *Tth*111II. The enzyme demonstrate activity at up to 80 °C, and Mg^{2+} or Mn^{2+} are required for its endonuclease activity. The DNA sequence that *Tth*111II recognizes as well as the sites that it cleaves were determined (Pantazaki et al. 2002).

A sequence-specific endonuclease (*Taq* I) of novel specificity has been partially purified from *T. aquaticus* (Sato et al. 1977). The enzyme cleaves bacteriophage A DNA at more than 30 sites and bacteriophage OX174 RF DNA at 10 sites. The enzyme is active at temperatures up to 70 °C.

Forty-five type II endonucleases with eighteen different specificities from one hundred and forty isolates of the genus *Thermus* were studied and characterized. Two among them, namely *Tat*I and *Tau*I, represent prototypes which recognize the novel nucleotide sequences 5′-W↓GTACW-3′ and 5′-GCSG↓C-3′, respectively (Vitkute et al. 2001).

Argonaute (Ago) proteins are nucleic acid-guided endonucleases. In contrast to Cas nucleases, Ago nucleases are more flexible, as they do not require the presence of any specific motifs. Under appropriate conditions, Ago from the thermophilic bacterium *Thermus thermophilus* (*Tt*Ago) (Song et al. 2020) cleaves with high efficiency both DNA and RNA complementary to its small interfering DNA guides (siDNA), but spares nucleic acids with a single nucleotide mismatch at and around its catalytic site. Recently, a new assay named NAVIGATER (Nucleic Acid enrichment Via DNA Guided Argonaute from *Thermus thermophilus*) have been developed for increasing the fractions of nucleic acids of clinical interest via DNA-guided Argonaute from *T. thermophilus* (TtAgo) (Song et al. 2020). Use of this technique facilitates 60-fold enrichment of the KRAS G12D cancer biomarker. The sensitivity of Peptide Nucleic Acid and Xenonucleic Acid clamp PCR was shown to be improved about 100 times as well. These allowed to detect low-frequency mutant alleles in blood specimens of pancreatic cancer patients (Song et al. 2020).

Pyrophosphatases are indispensable to cellular energy metabolism. The pyrophosphatase from *T. thermophilus* was purified in a form free of nonspecific RNases and DNases. The enzyme eliminates organic pyrophosphate which is created during incorporation of nucleotide triphosphates and is reported to inhibit DNA polymerase activity and causes DNA and RNA degradation at elevated temperatures. The enzyme exhibits highest activity at 75 °C and is ideally suitable for thermocycling reactions. *T. thermophilus* pyrophosphatase is commercially available and has biotechnological applications in PCR, RT/PCR and DNA sequencing (Pantazaki et al. 2002).

Alkaline phosphatases are nonspecific phosphomonoesterases which hydrolyzes a wide variety of phosphate esters and are classified as alkaline phosphatase according to its optimum pH (Gong et al. 2005). The enzyme is widely used in the diagnostics, immunology and molecular biology, including nonradioactive detection techniques, probing, blotting, and sequencing systems (Pantazaki et al. 2002). Alkaline phosphatases have been described from various *Thermus* species, including *T. aquaticus*, *T. caldophilus*, *T. yunnanensis*, and *T. thermophilus* (Gong et al. 2005), and were applied for labeling of primers, the detection of PCR products and as a reporter in promoter probe vectors (Moreno et al. 2003).

A recombinase A (RecA) protein from the *T. thermophilus* was overexpressed in *E. coli*. RecA is thermostable enzyme that plays important roles in homologous recombination and DNA repair. The purified recombinant RecA could enhance the PCR signals of Hepatitis B Virus (HBV) and improve the detection limit of the HBV diagnosis (viral load of less than 10 IU/ml) (Sundarrajan et al. 2018).

Representatives of the genus *Thermus* were shown to be a valuable source of *single-stranded DNA binding proteins (SSBs)*, which bind and protect single-stranded DNA during replication, recombination and repair. The presence of SSBs during DNA replication has been reported to minimize deletion mutagenesis artifacts and promotes a faster and more specific DNA amplification independently of the polymerase used (Cava et al. 2009).

8.6.1.5 Other Extreomyzemes with Biotechnological and Industrial Relevance

Malate dehydrogenase (MDH), an enzyme involved in the TCA cycle and the malate-aspartate shuttle, plays important roles in glucose metabolism and energy generation. MDH is highly applicable in enzyme immunoassay of a broad range of compounds as a conjugate. The DNA fragment containing the open reading frame of mdh was amplified from the genomic DNA of *T. thermophilus* and cloned and soluble protein was expressed in *E. coli* (Chang et al. 2013).

Catalases are widely distributed in nature. They convert hydrogen peroxide to water and molecular oxygen and might have potential applications in biotechnology and clinical medicine (Pantazaki et al. 2002). The majority of catalases isolated from different organisms contain a heme prosthetic group at the active site. In contrast, Mn-dependent catalases (nonheme or pseudocatalases) contain Mn instead of ferric heme in the active site and have a restricted distribution. Mn-dependent catalases have been purified and biochemically characterized, from thermophiles or hyperthermophiles, including *Thermus* sp. strain YS 8-13, *Thermus thermophilus* (Hidalgo et al. 2004).

Trehalose synthase (TS) converts maltose into trehalose due to intramolecular transglucosilation. A 105 kDa thermostable TS was purified from *Thermus aquaticus* ATCC 33923. The optimum pH and temperature were pH 6.5 and 65 °C, respectively. The enzyme was stable from pH 5.5 to 9.5 and up to 80 °C for 60 min. The activity was inhibited by Cu^{2+}, Hg^{2+}, Zn^{2+}, and Tris (Nishimoto et al. 1996). In other study, thermophilic trehalose synthase from *Thermus antranikianii* (*Ta*TS) was expressed in *E. coli*. The recombinant *Ta*TS showed the highest activity at pH 7.0 and 60 °C. *Ta*TS activity was stable after 6 h of incubation over a broad pH (6.0–10.0) and temperature (4–70 °C) ranges. The enzyme activity was strongly inhibited by Co^{2+}, Cu^{2+}, Zn^{2+}, SDS, and Tris (Lin et al. 2020).

L-Asparaginase (*L*-asparagine-aminohydrolase) converts *L*-asparagine to *L*-aspartic acid and ammonia, and has been used as a chemotherapeutic agent. *L*-Asparaginase was purified to homogeneity from *T. thermophilus*. The antiproliferative activity of the purified *L*- asparaginase of *T. thermophilus* was tested against several human cancer cell lines. The thermostable *L*-asparaginase of *T. thermophilus* does not hydrolyze *L*-glutamine which makes it advantageous for future clinical trials (Pantazaki et al. 2002).

An alanine racemase was purified from *Thermus thermophilus* HB8. This 38 kDa enzyme catalyzed the racemization of *D*- and *L*-alanine. The enzyme was found to be active at 75 °C and pH 8.0 and remained active even after incubation at 80 °C for 30 min (Seow et al. 2000).

Recombinant laccase from *Thermus thermophilus* has been applied to the biobleaching of wheat straw pulp with reduced amounts of hydrogen peroxide required (Zheng et al. 2012).

α-Glucosidases hydrolyze oligosaccharides to glucose. These enzymes have been characterized as maltases. A highly thermostable *α*-glucosidase from *T. thermophilus* HB8 was purified to homogeneity (Pantazaki et al. 2002). The enzyme exhibited high thermostability: after incubation at 90 °C for 10 h the enzyme retained 90% of its activity.

Thermostable xylose isomerases were purified from *T. thermophilus* and *Thermus caldophilus* (Pantazaki et al. 2002). Both enzymes show an optimal reaction temperature of 90 °C.

The photolyase from *Thermus thermophilus* was purified and shown to have flavin adenine dinucleotide as a chromophore. The enzyme showed light-dependent photoreactivation activity in vitro at 35 and 65 °C and was stable when subjected to heat and acidic pH (Kato et al. 1997).

8.6.2 Host-Vector Systems and Cell Factories

The production of active forms of thermozymes is important for both enzymological studies and industrial applications (Fujino et al. 2020). Although thermostable enzymes are commonly produced in mesophilic hosts (Fujino et al. 2020; Hidalgo et al. 2004), a significant number (>40%) of the proteins encoded within thermophilic genomes cannot be expressed or are expressed in an enzymatically inactive form in common mesophilic hosts due to specific requirements of modification machinery, membrane environment, etc. (Cava et al. 2009). Thus, the thermophilic cell factories are highly required to overcome these limitations.

The genetic systems of *Thermus* have been used for improvement of thermozymes and overproduction of thermostable and thermoactive enzymes in *Thermus* spp. which suggests the high potency of use of these organisms as cell factories (Hidalgo et al. 2004; Moreno et al. 2005; Park et al. 2004; Cava et al. 2009). Especially, the extraordinary trait of high competence of natural transformation and the availability of different gene transfer systems together with methodologies for gene expression have facilitated the expression of several heterologous genes in *T. thermophilus* (Park et al. 2004; Cava et al. 2009). Among other species, *Thermus thermophilus* contains an easily manipulable genome, and is therefore one of the best candidate microbes for development of "hot" expression systems (Fujino et al. 2020). In particular, *T. thermophilus* HB27 or its derivatives are most useful for these purposes as it has higher (40 kb/s per cell) transformation efficiency (Schwarzenlander and Averhoff 2006).

There are a number of examples of constitutive and induced expressions of proteins in *T. thermophilus* (Moreno et al. 2003, 2005; Hidalgo et al. 2004; Park et al. 2004).

Constitutive expression was performed frequently using a slpA promoter (Cava et al. 2009) where respiratory complex I (Pnqo) and ribosome protein promoters have been used.

The inducible expression systems available to date in *Thermus* are based on different inducible promoters. These include: (1) the dnaK promoter of *T. thermophilus* which is activated by heat shock (plasmid pTEX2-dnaK, pTEX7), (2) the arginine-inducible promoter Parg (plasmid pTEX8), (3) the carbon-regulated promoter Pscs-mdh (plasmid pTEX9), (4) the nitrate inducible promoter Pnar of the *T. thermophilus* HB8 nitrate reductase operon (Kayser et al. 2001; Park and Kilbane 2004) and (5) silica-inducible promoter (plasmid pSix1) (Fujino et al. 2020).

High level of overproduction in *Thermus* has already been achieved for several enzymes of biotechnological relevance, such as the *T. thermophilus* Mn-dependent catalase and its DNA polymerase (Hidalgo et al. 2004; Moreno et al. 2005).

Kayser and Kilbane (2001) have constructed the *T. thermophilus* Δ*mdh* strain lacking malate dehydrogenase (*mdh*) gene and used it as a host for *Thermus* plasmids expressing an intact *mdh* gene. This host expression vector system is a strong positive selection tool for the introduction of plasmid DNA into *T. thermophilus*, and *mdh* can be used as a reporter gene to quantify promoter strength in *T. thermophilus*.

Moreno et al. (2003) presented plasmid pMKE1 containing replicative origins for *E. coli* and *Thermus* spp., a selection gene encoding a thermostable resistance to kanamycin, and a 720 bp DNA region containing the promoter (Pnar), and the regulatory sequences of the respiratory nitrate reductase operon of *T. thermophilus* HB8. Two genes, encoding a thermophilic β-galactosidase and an alkaline phosphatase were cloned in pMKE1 as cytoplasmic and periplasmic reporters, respectively. The expression of the reporters was specifically induced by the combined action of nitrate and anoxia in facultative anaerobic derivatives of *T. thermophilus* HB27 (*T. thermophilus* HB27::*nar)* to which the gene cluster for nitrate respiration was transferred by conjugation. Overexpressions in the range of 200-fold were obtained for the cytoplasmic reporter, in contrast, that of the periplasmic reporter was limited to 20-fold.

Successful expression at practical levels has been achieved using pMKE2 vector, a variation of pMKE1 that was created by modifying the sequences between *Pnar* and the start codon (Moreno et al. 2005).

An example of homologues expression using *Thermus thermophilus* HB27 strain as a host was expression of α-galactosidase (*Tt*GalA). Interestingly, a soluble and active histidine-tagged enzyme was produced in larger amounts (5 mg/L) in this thermophilic host than in *E. coli* (0.5 mg/L). The purified recombinant enzyme showed an optimal activity at 90 °C and retained more than 40% of activity over a broad range of pH (from 5.0 to 8.0). *Tt*GalA is one of the most thermoactive and thermostable α-galactosidases found to date (Aulitto et al. 2017).

Recently, several genes of the hyperthermophilic archaeon *Pyrococcus horikoshii* OT3 (threonine dehydrogenase, α-mannosidase, and glutamate dehydrogenase) were successfully expressed in *T. thermophilus* HB27 using P31 or Pslp promoter. Moreover, the expression level was comparable or exceeding those in *E. coli.* Notably, α-mannosidase activity was clearly detected with P31 and Pslp promoters, which could not be detected in *E. coli* system (Takayama et al. 2004).

The Mn-dependent catalase genes from *Thermus thermophilus* HB27 and HB8 and a less thermostable mutant carrying two amino acid replacements (M129V and E293G), were overexpressed directly in *T. thermophilus* under the control of the *Pnar* (Hidalgo et al. 2004). Interestingly, upon induction in *T. thermophilus* HB8 and *T. thermophilus* HB27::*nar*, a 20- to 30-fold and 90- to 110-fold increase in catalase specific activity was observed, respectively. Moreover, overexpression in mesophilic *E. coli* resulted in inactive forms of these proteins.

These results demonstrate, that the host-vector system of *T. thermophilus* is useful as cell factory for the recombinant production of biocatalysts from hyperthermophiles active at temperature values over 90 °C, as well as the effectiveness of use of this strain as an alternative cell factory for the overproduction of thermophilic proteins that fail to be expressed in well-known mesophilic hosts.

T. thermophilus was shown to be excellent host bacterium for the detection of novel metagenome borne enzymes that could not readily have been detected by the use of *E. coli* or by in silico analysis (Leis et al. 2015). In their studies, Leis and colleagues constructed mutant strain BL03 with multiple markerless deletions in genes for major extra- and intracellular lipolytic activities. As a result, strain lost ability of growth on defined minimal medium supplemented with tributyrin (as sole carbon source) and might be used as a host for screening of metagenomic DNA pieces that could make possible the growth on tributyrin. Screening of significant number of single fosmid clones from thermophilic metagenomic libraries originating from heated compost and hot spring water samples for esterase activity in both *T. thermophilus* BL03 and *E. coli* EPI300 demonstrated, that a greater number of active esterase clones was detected in the thermophilic bacterium in comparison to the mesophilic *E. coli*. The functional screening implementing *T. thermophilus* BL03 could uncover several lipolytic enzymes from underrepresented species and archaeal origin.

Recently, a new gene expression system based on a silica-inducible promoter for the homologous and heterologous expression of thermostable genes in *Thermus thermophilus* has been developed (Fujino et al. 2020). *Thermus* sp. A4 gene encoding thermostable β-galactosidase was cloned as a reporter gene into the expression vector pSix1, which contains a selection marker that confers thermostable resistance to hygromycin and a 600 bp DNA region containing a putative silica-inducible promoter. Interestingly, truncation of the putative silica-inducible promoter region in *Thermus* expression vector improved the yield of the target protein, possibly by avoiding plasmid instability due to homologous recombination. These resulted in development of an expression vector containing the pSix1 backbone and a 100 bp DNA region corresponding to the silica-inducible promoter. This vector was applied for successful expression of glutamate dehydrogenase of *Pyrobaculum islandicum* (*Pis*GDH) without need of further heat treatment, yielding in 9.5 mg/L of the active protein. This study suggests a high potential of use of silica-inducible expression system as a novel strategy for the intracellular overexpression of thermostable proteins.

8.6.3 **Thermus** *as Metal-Converters and Possible Applications in Bioremediation*

Most of the hydrothermal fluids are found to be enriched with heavy metals (Co, Mo, Cr, and U) at elevated concentrations (Kieft et al. 1999), and expectably, thermophiles possess the ability to reduce these metals, thus providing exciting opportunities for their application in bioremediation.

The genus *Thermus* is well known for bioremediation of heavy metals thus lowering the toxicity at heavy metal contaminated sites. For instance, representatives of the genus *Thermus* shown ability to reduce both selenite and tellurite (Slobodkin et al. 2006; Sokolova et al. 2004; Chiong et al. 1988), an enzymatic uranium reduction has been shown in *Thermus scotoductus* (Kieft et al. 1999). *Thermus scotoductus* was shown to contain a putative peptide ABC transporter, peptide-binding protein capable of U (VI) reduction (Cason et al. 2012).

The ability of aerobic reduction of Cr (VI) was demonstrated for *T. scotoductus* (Opperman and van Heerden 2007). A novel chromate reductase from *Thermus scotoductus* SA-01, originating from South African gold mine, was purified and characterized (Opperman et al. 2008). The chromate reductase is optimally active at a pH of 6.3 and at 65 °C and requires Ca^{2+} or Mg^{2+} for activity. Enzyme activity was also dependent on NADH or NADPH. This chromate reductase was shown to be related to the old yellow enzyme family.

Thermus play a significant role in the cesium assembly. For instance, *Thermus* strain isolated from hot springs in Tibet (China), has been examined for the ability to accumulate cesium from solutions. The accumulation of cesium by this thermophile was rapid with 40–50% accumulation within the first 5 min (Wang et al. 2007).

Recently, the ability of *T. scotoductus* SA-01 to synthesize gold nanoparticles was demonstrated. The strain SA-01, has the ability to reduce Au (III) and produce nanoparticles, making it a suitable candidate for the production of nanoparticles (Erasmus et al. 2014).

The isolate Thermus HR13 from arsenite-rich hot spring in California has been reported to oxidize arsenite without energy gain, but it is able to use arsenate as a terminal electron acceptor in place of oxygen (Gihring and Banfield 2001).

It was demonstrated, that *Thermus thermophilus* HB27 harbors a set of genes involved in arsenic resistance which, in contrast to other microorganisms, were not organized into one single operon. They encode the proteins: arsenate reductase, *Tt*ArsC, arsenic efflux membrane transporter, *Tt*ArsX, and transcriptional repressor, *Tt*SmtB (Antonucci et al. 2018). It was shown that the *Tt*ArsX and *Tt*SmtB proteins are required to provide resistance to arsenic as well as to cadmium. These suggest the potential use of *T. thermophilus* for development of As-Cd biosensors (Antonucci et al. 2018).

Recently, *Tt*ArsC, derived from *T. thermophilus* was bioconjugated with hybrid polyethylene glycol-stabilized gold nanoparticles (AuNPs) to obtain *Tt*ArsC-AuNPs nanobiosystem. Characterization and investigation of this system encourages the use of hybrid biological-AuNPs to improve specific substrate recognition by enzymes in many medical and environmental biosensing applications (Politi et al. 2016).

8.7 Conclusion

The representatives of *Thermus*, showing global distribution in almost all thermally extreme habitats, possess a variety of metabolic capacities and have a high potency for use in bioremediation. Members of this genus have been regarded as models to investigate and enlarge our understanding of the evolution of life at extreme temperatures. *Thermus* is of considerable biotechnological interest as source of thermophilic enzymes. Many molecular biological tools and methods have been developed using proteins and enzymes obtained from these thermophiles. Due to high natural competency and other advantages, several strains of *Thermus* are excellent screening hosts, and newly developed host-vector systems make these organisms perfect cell factories for overexpression of biocatalysts from extreme- and hyperthermophiles and overcome limitations faced during expression in known mesophilic hosts. Thus, present review extends the previous frame of knowledge regarding to diversity and distribution of these extreme cosmopolitans as well as throws a light on their biotechnological potency and recent findings in regard to industrial and research relevance.

Acknowledgments This work was supported by the RA MES State Committee of Science, within the scope of the research projects №. 15T-1F399 and №. 18T-1F261, the Armenian National Science and Education Fund based in New York, USA, to HP (ANSEF-NS-microbio 2493, 3362 and 4676) and partially supported grants from the Norwegian Agency for International Cooperation and Quality Enhancement (CPEA-LT-2016/10095, CPEA-LT-2017/10061).

References

Albuquerque L, Rainey FA, Costa MS (2018) Thermus. Bergey's manual of systematics of archaea and bacteria. Elsevier, London, pp 1–39

Antonucci I, Gallo G, Limauro D, Contursi P, Ribeiro AL, Blesa A, Berenguer J, Bartolucci S, Fiorentino G (2018) Characterization of a promiscuous cadmium and arsenic resistance mechanism in *Thermus thermophilus* HB27 and potential application of a novel bioreporter system. Microb Cell Factories 17(1):78

Antranikian G, Egorova K (2007) Extremophiles, a unique resource of biocatalysts for industrial biotechnology. Physiology and biochemistry of extremophiles. ASM Press, Washington, pp 361–406

Aulitto M, Fusco S, Fiorentino G, Limauro D, Pedone E, Bartolucci S, Contursi P (2017) *Thermus thermophilus* as source of thermozymes for biotechnological applications: homologous expression and biochemical characterization of an α-galactosidase. Microb Cell Factories 16(1):28

Averhoff B (2009) Shuffling genes around in hot environments: the unique DNA transporter of *Thermus thermophilus*. FEMS Microbiol Rev 33:611–626

Balkwill DL, Kieft TL, Tsukuda T, Kostandarithes HM, Onstott TC, Macnaughton S, Bownas J, Fredrickson JK (2004) Identification of iron-reducing *Thermus* strains as *Thermus scotoductus*. Extremophiles 8(1):37–44

Barany F (1991) Genetic disease detection and DNA amplification using cloned thermostable ligase. Proc Natl Acad Sci 88(1):189

Bergquist P, Morgan H, Saul D (2014) Selected enzymes from extreme thermophiles with applications in biotechnology. Curr Biotechnol 3:15

Bertoldo C, Antranikian G (2002) Starch-hydrolyzing enzymes from thermophilic archaea and bacteria. Curr Opin Chem Biol 6(2):151–160

Bjornsdottir SH, Petursdottir SK, Hreggvidsson GO, Skirnisdottir S, Hjorleifsdottir S, Arnfinnsson J, Kristjansson JK (2009) *Thermus islandicus* sp. nov., a mixotrophic sulfur-oxidizing bacterium isolated from the Torfajokull geothermal area. Int J Syst Evol Microbiol 59(Pt 12):2962–2966

Blank S, Schröder C, Schirrmacher G, Reisinger C, Antranikian G (2014) Biochemical Characterization of a recombinant xylanase from *Thermus brockianus*, suitable for biofuel production. JSM Biotechnol Biomed Eng 2:1027

Blondal T, Thorisdottir A, Unnsteinsdottir U, Hjorleifsdottir S, Aevarsson A, Ernstsson S, Fridjonsson OH, Skirnisdottir S, Wheat JO, Hermannsdottir AG, Sigurdsson ST, Hreggvidson

GO, Smith AV, Kristjansson JK (2005) Isolation and characterization of a thermostable RNA ligase 1 from a *Thermus scotoductus* bacteriophage TS2126 with good single-stranded DNA ligation properties. Nucleic Acids Res 33(1):135–142

Bornscheuer UT (2002) Microbial carboxyl esterases: classification, properties and application in biocatalysis. FEMS Microbiol Rev 26(1):73–81

Brock TD, Freeze H (1969) *Thermus aquaticus* gen. n. and sp. n., a nonsporulating extreme thermophile. J Bacteriol 98(1):289–297

Bruggemann H, Chen C (2006) Comparative genomics of *Thermus thermophilus*: Plasticity of the megaplasmid and its contribution to a thermophilic lifestyle. J Biotechnol 124(4):654–661

Brumm PJ, Monsma S, Keough B, Jasinovica S, Ferguson E, Schoenfeld T, Lodes M, Mead DA (2015) Complete Genome Sequence of *Thermus aquaticus* Y51MC23. PLoS One 10(10): e0138674

Carballeira N, Nazabal M, Brito J, Garcia O (1990) Purification of a thermostable DNA polymerase from *Thermus thermophilus* HB8, useful in the polymerase chain reaction. BioTechniques 9:276–281

Cason ED, Piater LA, van Heerden E (2012) Reduction of U(VI) by the deep subsurface bacterium, *Thermus scotoductus* SA-01, and the involvement of the ABC transporter protein. Chemosphere 86(6):572–577

Cava F, Hidalgo A, Berenguer J (2009) *Thermus thermophilus* as biological model. Extremophiles 13(2):213–231

Chai YY, Kahar UM, Salleh MM, Illias RM, Mau Goh K (2012) Isolation and characterization of pullulan-degrading *Anoxybacillus* species isolated from Malaysian hot springs. Environ Technol 33(11):1231–1238

Chang YY, Hung CH, Hwang TS, Hsu CH (2013) Cloning, overexpression, purification and crystallization of malate dehydrogenase from *Thermus thermophilus*. Acta Crystallogr Sect F Struct Biol Cryst Commun 69(Pt 11):1249–1251

Chien A, Edgar DB, Trela JM (1976) Deoxyribonucleic acid polymerase from the extreme thermophile *Thermus aquaticus*. J Bacteriol 127(3):1550–1557

Chiong M, Barra R, González E, Vásquez C (1988) Resistance of *Thermus* spp. to potassium tellurite. Appl Environ Microbiol 54(2):610–612

Choi J-E, Kwon M-A, Na HY, Hahm D-H, Song JK (2013) Isolation and characterization of a metagenome-derived thermoalkaliphilic esterase with high stability over a broad pH range. Extremophiles 17(6):1013–1021

Chung AP, Rainey FA, Valente M, Nobre MF, da Costa MS (2000) *Thermus igniterrae* sp. nov. and *Thermus antranikianii* sp. nov., two new species from Iceland. Int J Syst Evol Microbiol 50:209–217

Cline JM, Hogrefe HH (2003) Multi-site mutagenesis. WO2002US22759 20020718

Coleman RD, Yang SS, McAlister MP (1987) Cloning of the debranching-enzyme gene from *Thermoanaerobium brockii* into *Escherichia coli* and *Bacillus subtilis*. J Bacteriol 169 (9):4302–4307

da Costa MS, Rainey FA, Nobre MF (2006) The Genus *Thermus* and relatives. Prokaryotes 2006:797–812

Dabrowski S, Kur J (1998) Recombinant His-tagged DNA polymerase. I. Cloning, purification and partial characterization of *Thermus thermophilus* recombinant DNA polymerase. Acta Biochim Pol 45(3):653–660

Domínguez A, Sanromán A, Fuciños P, Rúa ML, Pastrana L, Longo MA (2004) Quantification of intra- and extra-cellular thermophilic lipase/esterase production by *Thermus* sp. Biotechnol Lett 26(9):705–708

du Plessis EM, Berger E, Stark T, Louw ME, Visser D (2010) Characterization of a novel thermostable esterase from *Thermus scotoductus* SA-01: evidence of a new family of lipolytic esterases. Curr Microbiol 60(4):248–253

Dwivedi V, Kumari K, Gupta SK, Kumari R, Tripathi C, Lata P, Niharika N, Singh AK, Kumar R, Nigam A, Garg N, Lal R (2015) *Thermus parvatiensis* RL(T) sp. nov., isolated from a hot water spring, located Atop the Himalayan ranges at Manikaran, India. Indian J Microbiol 55 (4):357–365

Erasmus M, Cason ED, van Marwijk J, Botes E, Gericke M, van Heerden E (2014) Gold nanoparticle synthesis using the thermophilic bacterium *Thermus scotoductus* SA-01 and the purification and characterization of its unusual gold reducing protein. Gold Bull 47(4):245–253

Freeman SA, Peek K, Prescott M, Daniel R (1993) Characterization of a chelator-resistant proteinase from *Thermus* strain Rt4A2. Biochem J 295(Pt 2):463–469

Fuciños P, Atanes E, López-López O, Solaroli M, Cerdán ME, González-Siso MI, Pastrana L, Rúa ML (2014) Cloning, expression, purification and characterization of an oligomeric His-tagged thermophilic esterase from *Thermus thermophilus* HB27. Process Biochem 49(6):927–935

Fujii K, Minagawa H, Terada Y, Takaha T, Kuriki T, Shimada J, Kaneko H (2005) Protein engineering of amylomaltase from *Thermus aquaticus* with random and saturation mutageneses. Biol Sect Cell Mol Biol 60:97–102

Fujino Y, Nagayoshi Y, Ohshima T, Ogata S, Doi K (2017) Complete genome sequence of *Thermus thermophilus* TMY, isolated from a geothermal power plant. Genome Announc 5 (5):e01596

Fujino Y, Goda S, Suematsu Y, Doi K (2020) Development of a new gene expression vector for *Thermus thermophilus* using a silica-inducible promoter. Microb Cell Factories 19(1):126

Gibbs MD, Reeves RA, Mandelman D, Mi Q, Lee J, Bergquist PL (2009) Molecular diversity and catalytic activity of *Thermus* DNA polymerases. Extremophiles 13(5):817–826

Gihring TM, Banfield JF (2001) Arsenite oxidation and arsenate respiration by a new *Thermus* isolate. FEMS Microbiol Lett 204(2):335–340

Gong N, Chen C, Xie L, Chen H, Lin X, Zhang R (2005) Characterization of a thermostable alkaline phosphatase from a novel species *Thermus yunnanensis* sp. nov. and investigation of its cobalt activation at high temperature. Biochim Biophys Acta 1750(2):103–111

Gounder K, Brzuszkiewicz E, Liesegang H, Wollherr A, Daniel R, Gottschalk G, Reva O, Kumwenda B, Srivastava M, Bricio C, Berengeur J, van Heerden E, Litthauer D (2011) Sequence of the hyperplastic genome of the naturally competent *Thermus scotoductus* SA-01. BMC Genomics 12:577

Griffiths E, Gupta RS (2004) Distinctive protein signatures provide molecular markers and evidence for the monophyletic nature of the *Deinococcus-Thermus* phylum. J Bacteriol 186 (10):3097–3107

Hames C, Halbedel S, Schilling O, Stülke J (2005) Multiple-mutation reaction: a method for simultaneous introduction of multiple mutations into the glpK gene of *Mycoplasma pneumoniae*. Appl Environ Microbiol 71(7):4097–4100

Henne A, Bruggemann H, Raasch C, Wiezer A, Hartsch T, Liesegang H, Johann A, Lienard T, Gohl O, Martinez-Arias R, Jacobi C, Starkuviene V, Schlenczeck S, Dencker S, Huber R, Klenk HP, Kramer W, Merkl R, Gottschalk G, Fritz HJ (2004) The genome sequence of the extreme thermophile Thermus thermophilus. Nat Biotechnol 22(5):547–553

Hidalgo A, Betancor L, Moreno R, Zafra O, Cava F, Fernandez-Lafuente R, Guisan JM, Berenguer J (2004) *Thermus thermophilus* as a cell factory for the production of a thermophilic Mn-dependent catalase which fails to be synthesized in an active form in *Escherichia coli*. Appl Environ Microbiol 70(7):3839–3844

Housby JN, Southern EM (2002) *Thermus scotoductus* and *Rhodothermus* marinus DNA ligases have higher ligation efficiencies than *Thermus thermophilus* DNA ligase. Anal Biochem 302 (1):88–94

Hreggvidsson GO, Skirnisdottir S, Smit B, Hjorleifsdottir S, Marteinsson VT, Petursdottir S, Kristjansson JK (2006) Polyphasic analysis of *Thermus* isolates from geothermal areas in Iceland. Extremophiles 10(6):563–575

Hudson J, Morgan HW, Daniel RM (1987) *Thermus filiformis* sp. nov., a filamentous caldoactive bacterium. Int J Syst Bacteriol 37:431

Jiang L, Lin M, Li X, Cui H, Xu X, Li S, Huang H (2013) Genome sequence of *Thermus thermophilus* ATCC 33923, a thermostable trehalose-producing strain. Genome Announc 1 (4):e00493

Kaboev OK, Shevelev IV, Luchkina LA, Tret'iakov AN, Shcherbakova OG (1999) Hot start of the polymerase chain reaction using DNA helicase. Bioorg Khim 25(5):398–400

Kacagan M, Inan K, Canakci S, Guler HI, Belduz AO (2015) *Thermus anatoliensis* sp. nov., a thermophilic bacterium from geothermal waters of Buharkent, Turkey. J Basic Microbiol 55 (12):1367–1373

Kaledin AS, Sliusarenko AG, Gorodetskiĭ SI (1980) Isolation and properties of DNA polymerase from extreme thermophylic bacteria *Thermus aquaticus* YT-1. Biokhimiia 45(4):644–651

Kato R, Hasegawa K, Hidaka Y, Kuramitsu S, Hoshino T (1997) Characterization of a thermostable DNA photolyase from an extremely thermophilic bacterium, *Thermus thermophilus* HB27. J Bacteriol 179(20):6499–6503

Kayser KJ, Kilbane JJ (2001) New host-vector system for *Thermus* spp. based on the malate dehydrogenase gene. J Bacteriol 183(5):1792–1795

Kayser KJ, Kwak JH, Park HS, Kilbane JJ 2nd (2001) Inducible and constitutive expression using new plasmid and integrative expression vectors for *Thermus* sp. Lett Appl Microbiol 32 (6):412–418

Kermekchiev MB, Tzekov A, Barnes WM (2003) Cold-sensitive mutants of *Taq* DNA polymerase provide a hot start for PCR. Nucleic Acids Res 31(21):6139–6147

Kermekchiev MB, Kirilova LI, Vail EE, Barnes WM (2009) Mutants of *Taq* DNA polymerase resistant to PCR inhibitors allow DNA amplification from whole blood and crude soil samples. Nucleic Acids Res 37(5):e40

Khan IU, Habib N, Hussain F, Xian WD, Amin A, Zhou EM, Ahmed I, Zhi XY, Li WJ (2017) *Thermus caldifontis* sp. nov., a thermophilic bacterium isolated from a hot spring. Int J Syst Evol Microbiol 67(8):2868–2872

Kieft T, Fredrickson J, Onstott T, Gorby YA, Kostandarithes H, Bailey T, Kennedy D, Li S, Plymale A, Spadoni C, Gray M (1999) Dissimilatory reduction of Fe (III) and other electron acceptors by a *Thermus* isolate. Appl Environ Microbiol 65:1214–1221

Kim JS, Koh S, Kim JJ, Kwon ST, Lee DS (1998) *Top* DNA polymerase from *Thermus thermophilus* HB27: gene cloning, sequence determination, and physicochemical properties. Mol Cells 8(2):157–161

Kristjánsson JK, Hjörleifsdóttir S, Marteinsson VT, Alfredsson GA (1994) *Thermus scotoductus*, sp. nov., a pigment-producing thermophilic bacterium from hot tap water in Iceland and including *Thermus* sp. X-1. Syst Appl Microbiol 17(1):44–50

Kumar S, Stecher G, Li M, Knyaz C, Tamura K (2018) MEGA X: molecular evolutionary genetics analysis across computing platforms. Mol Biol Evol 35(6):1547–1549

Kumwenda B, Litthauer D, Reva O (2014) Analysis of genomic rearrangements, horizontal gene transfer and role of plasmids in the evolution of industrial important *Thermus* species. BMC Genomics 15(1):813

Kurosawa N, Itoh YH, Itoh T (2005) *Thermus kawarayensis* sp. nov., a new member of the genus *Thermus*, isolated from Japanese hot springs. Extremophiles 9(1):81–84

Lauer G, Rudd EA, McKay DL, Ally A, Ally D, Backman KC (1991) Cloning, nucleotide sequence, and engineered expression of *Thermus thermophilus* DNA ligase, a homolog of *Escherichia coli* DNA ligase. J Bacteriol 173(16):5047–5053

Leis B, Angelov A, Mientus M, Li H, Pham VT, Lauinger B, Bongen P, Pietruszka J, Gonçalves LG, Santos H, Liebl W (2015) Identification of novel esterase-active enzymes from hot environments by use of the host bacterium *Thermus thermophilus*. Front Microbiol 6:275

Leuschner C, Antranikian G (1995) Heat-stable enzymes from extremely thermophilic and hyper-thermophilic microorganisms. World J Microbiol Biotechnol 11(1):95–114

Li MM, Xian WD, Zhang XT, Yin YR, Zhou EM, Ding YP, Liu L, Fang BZ, Li WJ (2019) *Thermus caldilimi* sp. nov., a thermophilic bacterium isolated from a geothermal area. Antonie Van Leeuwenhoek 112(12):1767–1774

Lin L, Chen C, Peng Q, Ben K, Zhou Z (2002) *Thermus rehai* sp. nov., isolated from Rehai of Tengchong, Yunnan Province, China. J Basic Microbiol 42(5):337–344

Lin Y-F, Su P-C, Chen P-T (2020) Production and characterization of a recombinant thermophilic trehalose synthase from *Thermus antranikianii*. J Biosci Bioeng 129(4):418–422

Longley MJ, Bennett SE, Mosbaugh DW (1990) Characterization of the 5' to 3' exonuclease associated with *Thermus aquaticus* DNA polymerase. Nucleic Acids Res 18(24):7317–7322

Lopez-Lopez O, Cerdan ME, Gonzalez-Siso MI (2015) *Thermus thermophilus* as a source of thermostable lipolytic enzymes. Microorganisms 3(4):792–808

Luo J, Bergstrom DE, Barany F (1996) Improving the fidelity of *Thermus thermophilus* DNA ligase. Nucleic Acids Res 24(15):3071–3078

Manaia CM, Hoste B, Carmen Gutierrez M, Gillis M, Ventosa A, Kersters K, Da Costa MS (1995) Halotolerant *Thermus* strains from marine and terrestrial hot springs belong to *Thermus thermophilus* (ex Oshima and Imahori, 1974) nom. rev. emend. Syst Appl Microbiol 17 (4):526–532

Mandelli F, Oliveira Ramires B, Couger MB, Paixão DA, Camilo CM, Polikarpov I, Prade R, Riaño-Pachón DM, Squina FM (2015) Draft genome sequence of the thermophile *Thermus filiformis* ATCC 43280, producer of carotenoid-(di)glucoside-branched fatty acid (di)esters and source of hyperthermostable enzymes of biotechnological interest. Genome Announc 3(3): e00475

Matsuzawa H, Tokugawa K, Hamaoki M, Mizoguchi M, Taguchi H, Terada I, Kwon S-T, Ohta T (1988) Purification and characterization of aqualysin I (a thermophilic alkaline serine protease) produced by *Thermus aquaticus* YT-1. Eur J Biochem 171(3):441–447

Mefferd CC, Zhou EM, Yu TT, Ming H, Murugapiran SK, Huntemann M, Clum A, Pillay M, Palaniappan K, Varghese N, Mikhailova N, Stamatis D, Reddy TB, Ngan CY, Daum C, Duffy K, Shapiro N, Markowitz V, Ivanova N, Kyrpides N, Williams AJ, Woyke T, Li WJ, Hedlund BP (2016) High-quality draft genomes from *Thermus caliditerrae* YIM 77777 and *T. tengchongensis* YIM 77401, isolates from Tengchong, China. Genome Announc 4(2):e00312

Messaoud EB, Ammar YB, Mellouli L, Samir B (2002) Thermostable pullulanase type I from new isolated *Bacillus thermoleovorans* US105: Cloning, sequencing and expression of the gene in *E. coli*. Enzym Microb Technol 31:827–832

Ming H, Yin YR, Li S, Nie GX, Yu TT, Zhou EM, Liu L, Dong L, Li WJ (2014) *Thermus caliditerrae* sp. nov., a novel thermophilic species isolated from a geothermal area. Int J Syst Evol Microbiol 64(Pt 2):650–656

Ming H, Zhao ZL, Ji WL, Ding CL, Cheng LJ, Niu MM, Li M, Yi BF, Xia TT, Nie GX (2020) *Thermus thermamylovorans* sp. nov., isolated from a hot spring. Int J Syst Evol Microbiol 70 (3):1729–1737

Miyazaki K (2019) Complete genome sequencing of *Thermus thermophilus* strain HC11, isolated from Mine Geyser in Japan. Microbiol Resour Announc 8:39

Moreno R, Zafra O, Cava F, Berenguer J (2003) Development of a gene expression vector for *Thermus thermophilus* based on the promoter of the respiratory nitrate reductase. Plasmid 49 (1):2–8

Moreno R, Haro A, Castellanos A, Berenguer J (2005) High-level overproduction of His-tagged *Tth* DNA polymerase in *Thermus thermophilus*. Appl Environ Microbiol 71(1):591–593

Muller WJ, Tlalajoe N, Cason ED, Litthauer D, Reva O, Brzuszkiewicz E, van Heerden E (2016) Whole genome comparison of *Thermus* sp. NMX2.A1 reveals principle carbon metabolism differences with closest relation *Thermus scotoductus* SA-01. Genome 6(9):2791–2797

Murugapiran SK, Huntemann M, Wei CL, Han J, Detter JC, Han C, Erkkila TH, Teshima H, Chen A, Kyrpides N, Mavrommatis K, Markowitz V, Szeto E, Ivanova N, Pagani I, Pati A, Goodwin L, Peters L, Pitluck S, Lam J, McDonald AI, Dodsworth JA, Woyke T, Hedlund BP (2013a) *Thermus oshimai* JL-2 and T. *thermophilus* JL-18 genome analysis illuminates pathways for carbon, nitrogen, and sulfur cycling. Stand Genomic Sci 7(3):449–468

Murugapiran SK, Huntemann M, Wei CL, Han J, Detter JC, Han CS, Erkkila TH, Teshima H, Chen A, Kyrpides N, Mavrommatis K, Markowitz V, Szeto E, Ivanova PI, Lam J, McDonald

AI, Dodsworth JA, Pati A, Goodwin L, Peters L, Pitluck S, Woyke T, Hedlund BP (2013b) Whole genome sequencing of *Thermus oshimai* JL-2 and *Thermus thermophilus* JL-18, incomplete denitrifiers from the United States Great Basin. Genome Announc 1:1

Navas LE, Berretta MF, Ortiz EM, Benintende GB, Amadio AF, Zandomeni RO (2015) Draft genome sequence of *Thermus* sp. isolate 2.9, obtained from a hot water spring located in Salta, Argentina. Genome Announc 3:1

Navas LE, Florin-Christensen M, Benintende GB, Zandomeni RO, Berretta MF (2018) Characterization of a novel thermostable enzyme from *Thermus* sp. 2.9 with phospholipase and acyltransferase activities. J Mol Microbiol Biotechnol 28(3):99–106

Nisha M, Satyanarayana T (2016) Characteristics, protein engineering and applications of microbial thermostable pullulanases and pullulan hydrolases. Appl Microbiol Biotechnol 100 (13):5661–5679

Nishimoto T, Nakada T, Chaen H, Fukuda S, Sugimoto T, Kurimoto M, Tsujisaka Y (1996) Purification and characterization of a thermostable trehalose synthase from *Thermus aquaticus*. Biosci Biotechnol Biochem 60(5):835–839

Nobre MF, Carreto L, Wait R, Tenreiro S, Fernandes O, Sharp RJ, Da Costa MS (1996) Fatty composition of the species of the genera *Thermus* and *Meiothermus*. Syst Appl Microbiol 19 (3):303–311

Omelchenko MV, Wolf YI, Gaidamakova EK, Matrosova VY, Vasilenko A, Zhai M, Daly MJ, Koonin EV, Makarova KS (2005) Comparative genomics of *Thermus thermophilus* and *Deinococcus radiodurans*: divergent routes of adaptation to thermophily and radiation resistance. BMC Evol Biol 5:57

Opperman D, van Heerden E (2007) Aerobic Cr (VI) reduction by *Thermus* scotoductus strain SA-01. J Appl Microbiol 103:1907–1913

Opperman DJ, Piater LA, van Heerden E (2008) A novel chromate reductase from *Thermus scotoductus* SA-01 related to old yellow enzyme. J Bacteriol 190(8):3076–3082

Oshima T, Imahori K (1974) Description of *Thermus thermophilus* (Yoshida and Oshima) comb. nov., a nonsporulating thermophilic bacterium from a Japanese thermal spa. Int J Syst Bacteriol 24:102

Pantazaki AA, Pritsa AA, Kyriakidis DA (2002) Biotechnologically relevant enzymes from *Thermus thermophilus*. Appl Microbiol Biotechnol 58(1):1–12

Park HS, Kilbane JJ (2004) Gene expression studies of *Thermus thermophilus* promoters PdnaK, Parg and Pscs-mdh. Lett Appl Microbiol 38(5):415–422

Park HS, Kayser KJ, Kwak JH, Kilbane JJ (2004) Heterologous gene expression in *Thermus thermophilus*: beta-galactosidase, dibenzothiophene monooxygenase, PNB carboxy esterase, 2-aminobiphenyl-2,3-diol dioxygenase, and chloramphenicol acetyl transferase. J Ind Microbiol Biotechnol 31(4):189–197

Plant AR, Clemens RM, Daniel RM, Morgan HW (1987) Purification and preliminary characterization of an extracellular pullulanase from *Thermoanaerobium* Tok6-B1. Appl Microbiol Biotechnol 26(5):427–433

Politi J, Spadavecchia J, Fiorentino G, Antonucci I, De Stefano L (2016) Arsenate reductase from *Thermus thermophilus* conjugated to polyethylene glycol-stabilized gold nanospheres allow trace sensing and speciation of arsenic ions. J R Soc Interface 13:123

Ramírez S, Moreno R, Zafra O, Castán P, Vallés C, Berenguer J (2000) Two nitrate/nitrite transporters are encoded within the mobilizable plasmid for nitrate respiration of *Thermus thermophilus* HB8. J Bacteriol 182(8):2179–2183

Saghatelyan A, Poghosyan L, Panosyan H, Birkeland NK (2015) Draft genome sequence of *Thermus scotoductus* Strain K1, isolated from a geothermal spring in Karvachar, Nagorno Karabakh. Genome Announc 3(6):e01346

Sashihara N, Nakamura N, Nagayama H, Horikoshi K (1988) Cloning and expression of the thermostable pullulanase gene from *Thermus* sp. strain AMD-33 in *Escherichia coli*. FEMS Microbiol Lett 49(3):385–388

Sato S, Hutchinson CA 3rd, Harris JI (1977) A thermostable sequence-specific endonuclease from *Thermus aquaticus*. Proc Natl Acad Sci U S A 74(2):542–546

Schafers C, Blank S, Wiebusch S, Elleuche S, Antranikian G (2017) Complete genome sequence of *Thermus brockianus* GE-1 reveals key enzymes of xylan/xylose metabolism. Stand Genomic Sci 12:22

Schwarzenlander C, Averhoff B (2006) Characterization of DNA transport in the thermophilic bacterium *Thermus thermophilus* HB27. FEBS J 273(18):4210–4218

Seow TK, Inagaki K, Nakamura T, Maeda R, Tamura T, Tanaka H (2000) Purification and some characteristics of a monomeric alanine racemase from an extreme thermophile, *Thermus thermophilus*. J Biosci Bioeng 90(3):344–346

Skirnisdottir S, Hreggvidsson GO, Holst O, Kristjansson JK (2001) Isolation and characterization of a mixotrophic sulfur-oxidizing *Thermus scotoductus*. Extremophiles 5(1):45–51

Slobodkin AI, Tourova TP, Kostrikina NA, Lysenko AM, German KE, Bonch-Osmolovskaya EA, Birkeland NK (2006) *Tepidimicrobium ferriphilum* gen. nov., sp. nov., a novel moderately thermophilic, Fe (III)-reducing bacterium of the order *Clostridiales*. Int J Syst Evol Microbiol 56 (Pt 2):369–372

Sokolova TG, González JM, Kostrikina NA, Chernyh NA, Slepova TV, Bonch-Osmolovskaya EA, Robb FT (2004) *Thermosinus carboxydivorans* gen. nov., sp. nov., a new anaerobic, thermophilic, carbon-monoxide-oxidizing, hydrogenogenic bacterium from a hot pool of Yellowstone National Park. Int J Syst Evol Microbiol 54(Pt 6):2353–2359

Song J, Hegge JW, Mauk MG, Chen J, Till JE, Bhagwat N, Azink LT, Peng J, Sen M, Mays J, Carpenter EL, van der Oost J, Bau HH (2020) Highly specific enrichment of rare nucleic acid fractions using *Thermus thermophilus* argonaute with applications in cancer diagnostics. Nucleic Acids Res 48(4):19

Sundarrajan S, Rao S, Padmanabhan S (2018) Cloning and high-level expression of *Thermus thermophilus* RecA in *E. coli*: purification and novel use in HBV diagnostics. Braz J Microbiol 49(4):848–855

Suzuki M, Avicola AK, Hood L, Loeb LA (1997) Low fidelity mutants in the O-helix of *Thermus aquaticus* DNA polymerase I. J Biol Chem 272(17):11228–11235

Suzuki M, Yoshida S, Adman ET, Blank A, Loeb LA (2000) *Thermus aquaticus* DNA polymerase I mutants with altered fidelity. Interacting mutations in the O-helix. J Biol Chem 275 (42):32728–32735

Takayama G, Kosuge T, Sunamura S, Matsui I, Ishikawa K, Nakamura A, Hoshino T (2004) Use of a *Thermus thermophilus* host-vector system for expression of genes from the hyperthermophilic archaeon *Pyrococcus horikoshii*. J Jpn Soc Extremophiles 3:28

Tao Y, Zhang Y, Wang Y, Cheng Y (2013) Hollow fiber based affinity selection combined with high performance liquid chromatography–mass spectroscopy for rapid screening lipase inhibitors from lotus leaf. Anal Chim Acta 785:75–81

Teh BS, Lau NS, Ng FL, Abdul Rahman AY, Wan X, Saito JA, Hou S, Teh AH, Najimudin N, Alam M (2015) Complete genome sequence of the thermophilic *Thermus* sp. CCB_US3_UF1 from a hot spring in Malaysia. Stand Genomic Sci 10:76

Terada Y, Fujii K, Takaha T, Okada S (1999) *Thermus aquaticus* ATCC 33923 amylomaltase gene cloning and expression and enzyme characterization: production of cycloamylose. Appl Environ Microbiol 65(3):910–915

Tomiyasu K, Yato K, Yasuda M, Tonozuka T, Ibuka A, Sakai H (2001) Cloning and nucleotide sequence of the pullulanase gene of *Thermus thermophilus* HB8 and production of the enzyme in *Escherichia coli*. Biosci Biotechnol Biochem 65(9):2090–2094

Tong J, Wetmur JG (1996) Cloning, sequencing, and expression of ruvB and characterization of RuvB proteins from two distantly related thermophilic eubacteria. J Bacteriol 178 (9):2695–2700

Tong J, Cao W, Barany F (1999) Biochemical properties of a high fidelity DNA ligase from *Thermus* species AK16D. Nucleic Acids Res 27(3):788–794

Toogood HS, Daniel RM (2013) *Thermus* strain Rt41A protease. Handb Proteolyt Enzy 2013:3171–3174

Tosaka A, Ogawa M, Yoshida S, Suzuki M (2001) O-helix mutant T664P of *Thermus aquaticus* DNA polymerase I - Altered catalytic properties for incorporation of incorrect nucleotides but not correct nucleotides. J Biol Chem 276:27562–27567

Tripathi C, Mishra H, Khurana H, Dwivedi V, Kamra K, Negi RK, Lal R (2017) Complete genome analysis of *Thermus parvatiensis* and comparative genomics of *Thermus* spp. provide insights into genetic variability and evolution of natural competence as strategic survival attributes. Front Microbiol 8:1410

Vajna B, Kanizsai S, Keki Z, Marialigeti K, Schumann P, Toth EM (2012) *Thermus* composti sp. nov., isolated from oyster mushroom compost. Int J Syst Evol Microbiol 62 (Pt 7):1486–1490

Vitkute J, Maneliene Z, Janulaitis A (2001) Two new thermostable type II restriction endonucleases from *Thermus aquaticus*: TatI and TauI, which recognize the novel nucleotide sequences 5'-W (downward arrow) GTACW-3' and 5'-GCSG (downward arrow) C-3' respectively. FEMS Microbiol Lett 204(2):253–257

Wang H, Zheng M, Huang X (2007) Cesium accumulation by bacterium *Thermus* sp. TibetanG7: hints for biomineralization of cesium-bearing geyserite in hot springs in Tibet, China. Chin Sci Bull 52(19):2680–2686

Watanabe S, Muramatsu T, Ao H, Hirayama Y, Takahashi K, Tanokura M, Kuchino Y (1999) Molecular cloning of the Lon protease gene from *Thermus thermophilus* HB8 and characterization of its gene product. Eur J Biochem 266(3):811–819

Williams R, Sharp R (1995) The taxonomy and identification of Thermus. In: Sharp R, Williams R (eds) Thermus species. Biotechnology handbooks, vol 9. Springer, Boston. https://doi.org/10.1007/978-1-4615-1831-0_1

Williams RAD, Smith KE, Welch SG, Micallef J, Sharp RJ (1995) DNA relatedness of *Thermus* strains, description of *Thermus brockianus* sp. nov., and proposal to reestablish *Thermus thermophilus* (Oshima and Imahori). Int J Syst Bacteriol 45:495–499

Williams RA, Smith KE, Welch SG, Micallef J (1996) *Thermus oshimai* sp. nov., isolated from hot springs in Portugal, Iceland, and the Azores, and comment on the concept of a limited geographical distribution of *Thermus* species. Int J Syst Bacteriol 46(2):403–408

Wilpiszeski RL, Zhang Z, House CH (2019) Biogeography of thermophiles and predominance of *Thermus scotoductus* in domestic water heaters. Extremophiles 23(1):119–132

Woese CR (1987) Bacterial evolution. Microbiol Rev 51(2):221–271

Xiangyuan H, Shuzheng Z, Shoujun Y (2001) Cloning and expression of thermostable β-glycosidase gene from *Thermus nonproteolyticus* HG102 and characterization of recombinant enzyme. Appl Biochem Biotechnol 94(3):243–255

Yamagami T, Ishino S, Kawarabayasi Y, Ishino Y (2014) Mutant *Taq* DNA polymerases with improved elongation ability as a useful reagent for genetic engineering. Front Microbiol 5:461

Yu TT, Yao JC, Ming H, Yin YR, Zhou EM, Liu MJ, Tang SK, Li WJ (2013) *Thermus tengchongensis* sp. nov., isolated from a geothermally heated soil sample in Tengchong, Yunnan, south-west China. Antonie Van Leeuwenhoek 103(3):513–518

Yu TT, Ming H, Yao JC, Zhou EM, Park DJ, Hozzein WN, Kim CJ, Wadaan MA, Li WJ (2015) *Thermus amyloliquefaciens* sp. nov., isolated from a hot spring sediment sample. Int J Syst Evol Microbiol 65(8):2491–2495

Zhang XQ, Ying Y, Ye Y, Xu XW, Zhu XF, Wu M (2010) *Thermus arciformis* sp. nov., a thermophilic species from a geothermal area. Int J Syst Evol Microbiol 60(Pt 4):834–839

Zheng Z, Li H, Li L, Shao W (2012) Biobleaching of wheat straw pulp with recombinant laccase from the hyperthermophilic *Thermus thermophilus*. Biotechnol Lett 34(3):541–547

Zhou EM, Murugapiran SK, Mefferd CC, Liu L, Xian WD, Yin YR, Ming H, Yu TT, Huntemann M, Clum PM, Palaniappan K, Varghese N, Mikhailova N, Stamatis D, Reddy TB, Ngan CY, Daum C, Shapiro N, Markowitz V, Ivanova N, Spunde A, Kyrpides N, Woyke T, Li WJ, Hedlund BP (2016) High-quality draft genome sequence of the *Thermus amyloliquefaciens* type strain YIM 77409(T) with an incomplete denitrification pathway. Stand Genomic Sci 11:20

Zhou E-M, Xian W-D, Jiao J-Y, Liu L, Li M-M, Ding Y-P, Yin Y-R, Zhao J, Nimaichand S, Xiao M, Li WJ (2018a) Physiological and genomic properties of *Thermus tenuipuniceus* sp. nov., a novel slight reddish color member isolated from a terrestrial geothermal spring. Syst Appl Microbiol 41(6):611–618

Zhou E-M, Xian W-D, Mefferd CC, Thomas SC, Adegboruwa AL, Williams N, Murugapiran SK, Dodsworth JA, Ganji R, Li M-M, Ding YP, Liu L, Woyke T, Li WJ, Hedlund BP (2018b) *Thermus sediminis* sp. nov., a thiosulfate-oxidizing and arsenate-reducing organism isolated from Little Hot Creek in the Long Valley Caldera, California. Extremophiles 22(6):983–991

Chapter 9
Thermoacidophiles for Bioleaching of Copper

Narine Vardanyan and Arevik Vardanyan

Abstract Biohydrometallurgy is a modern, steadily developing alternative metal production technology based on the use of microorganisms and their metabolic products, such as ferric iron, sulfuric acid, etc. for the extraction of metals from ores. Microbiological processing of ores and concentrates has economic, technical and, most importantly, environmental advantages over traditional technologies. Heap leaching is successfully used for recovery of copper from a secondary mineral—chalcocite (Cu_2S). However, the main world reserves of copper are found in the form of chalcopyrite ($CuFeS_2$). Chalcopyrite is the most refractory mineral and undergoes chemical or biological oxidation at a very low rate. One of the most common ways to enhance copper extraction from chalcopyrite is the use of thermophiles. Besides, the intensity of biooxidation of sulfide minerals depends on the pH, redox potential, Fe^{2+}/Fe^{3+} ratio, metals ion concentration and the microorganisms used. It was revealed that the mixed cultures and consortia of moderate thermophilic microorganisms were more efficient and stable in the oxidation of chalcopyrite than pure cultures. From this point of view, developing and optimizing microbial associations for use in commercial copper leaching systems remain an important challenge. In this paper bioleaching of chalcopyrite by pure and mixed cultures of moderate thermophilic bacteria *S. thermosulfidooxidans* and thermotolerant sulfur or iron oxidizing bacteria *L. ferriphilum* CC, as well as the influence of physicochemical factors on this process have been investigated.

Keywords Bioleaching of chalcopyrite · Kinetics of copper recovery · Moderate thermoacidophiles · Associations of iron- and sulfur-oxidizing bacteria

---

N. Vardanyan (✉) · A. Vardanyan
Laboratory of Metal Bioleaching, Institute of Microbiology of the Scientific and Production Center "Armbiotechnology" of National Academy of Sciences of Armenia, Yerevan, Armenia
e-mail: nvard@sci.am

D. Egamberdieva et al. (eds.), *Microbial Communities and their Interactions in the Extreme Environment*, Microorganisms for Sustainability 32,
https://doi.org/10.1007/978-981-16-3731-5_9

9.1 Introduction

Biomining is a modern, steadily developing alternative metal production technology using biohydrometallurgy for the processing of mineral raw materials. Biohydrometallurgy is based on the use of microorganisms and their metabolic products, such as ferric iron, sulfuric acid, etc. for the extraction of metals from ores. Microbiological processing of ores and concentrates has economic, technical and, most importantly, environmental advantages over traditional roasting and autoclave oxidation at high pressures and temperatures (Rawlings 2002; Rawlings et al. 2003).

Biohydrometallurgical processes in the biomining industry are implemented in heap leaching and continuous tank leaching reactors. Leaching has been used to recover copper from ores since ancient times (Ehrlich 1999). However, the development of modern commercial bioleaching processes was associated with the discovery of the first sulfur and iron oxidizing bacteria *Acidithiobacillus ferrooxidans* in the middle of the nineteenth century (Temple and Colmer 1951).

Commercial heap bioleaching was first implemented at the Bingham Kanyon copper mine (Utah, USA) to recover copper from low-grade ores. Subsequently, since 1980, numerous heap bioleaching units of copper have been put into operation in many countries of the world and at the end of the last century, world copper production by biohydrometallurgy reached to 25% (Brombacher et al. 1997).

Tank leaching is widely used in the biomining industry for a number of advantages. Tank leaching is a highly controlled process and provides high bioleaching efficiency in terms of rate and recovery of metals. Constant control of aeration, pH and temperature allows optimizing growth and activity of microorganisms functioning in the tanks. In addition, tank leaching leads to long-term selection and domination of those microorganisms that can grow and function more efficiently under tank leaching conditions.

Tank reactors are used for both bioleaching and biooxidation processes. Bioleaching is the dissolution of insoluble metal sulfides into soluble compounds with further extraction of metals from leaching solutions. Tank bioleaching is used to recover base metals such as copper, zinc, and nickel from the corresponding sulfides.

During biooxidation, the target product (metal) passes into a solid phase. Biooxidation is often used to recover gold and silver. The process is based on the oxidation of pyrite and arsenypyrite using microorganisms in order to release gold in the crystal lattice of minerals, followed by its extraction by traditional cyanidation. The first biooxidation unit for the pretreatment of gold ores was commercialized in 1986 by Gencor at the Fairview mine in South Africa (van Aswegen et al. 1991). Due to the minimal ecological impact on the environment and a number of other advantages, tank biooxidation is successfully used to remove iron and arsenic from gold-bearing ores, gradually replacing the indicated physicochemical and pyrometallurgical technologies (Gahan et al. 2012).

Heap leaching of copper is widely used in the biomining industry, but all existing technologies are based on the extraction of copper from a secondary mineral—

chalcocite (Cu_2S) (Olson et al. 2003; Gahan et al. 2012). Chalcocite is readily leached under the action of protons to form covellite (CuS) (Eq. 9.1) and by Fe (III) (Eq. 9.2) formed as a result of bacterial oxidation of Fe (II).

$$Cu_2S + 2H_2SO_4 + O_2 \rightarrow 2CuS + 2CuSO_4 + 2H_2O \tag{9.1}$$

$$Cu_2S + Fe_2(SO_4)_3 \rightarrow CuSO_4 + 2FeSO_4 + S \tag{9.2}$$

Covellite is also subsequently leached with Fe (III) ions (Eq. 9.3).

$$CuS + Fe_2(SO_4)_3 \rightarrow CuSO_4 + 2FeSO_4 + S \tag{9.3}$$

Chalcopyrite ($CuFeS_2$) is the most abundant sulfide mineral. It is currently estimated that 70% of the world copper reserves now occur in chalcopyrite deposits (Watling 2006). However, chalcopyrite is the most refractory mineral to chemical and biological leaching resulting in low dissolution rates. The main reason for low dissolution rate of chalcopyrite is the formation of passive layer on mineral surface. Reaction products of bioleaching such as elemental sulfur and iron-hydroxide (jarozite) precipitate on the mineral surface and hinder its further dissolution by limiting the flow of bacteria and reaction products to and from the mineral surface (Dopson et al. 2003; Fu et al. 2008; Johnson 2014; Rawlings et al. 1999).

One of the most common solutions to this problem suggested by researchers is to perform bioleaching of chalcopyrite at high temperature using thermophiles. However it worth mentioning that in many cases, instead of extreme thermophiles (70–80 °C) industry has preferred moderate thermophilic microorganisms because they are more resistant to higher pulp densities and higher heavy metal concentrations.

Another alternative for enhancement of chalcopyrite bioleaching is the use of silver as a chemical catalyst. Researchers reported that copper recovery from chalcopyrite was improved through the use of silver as a catalyst in both the presence and the absence of bacteria (Ballester et al. 1990; Gómez et al. 1999; Cancho et al. 2007). It is considered that as a result of a chemical reaction involving an interchange between the silver and the copper and iron from the chalcopyrite lattice silver sulfide is formed on the mineral surface (Eq. 9.4):

$$CuFeS_2 - 4Ag^+ \rightarrow 2Ag_2S + Cu^{2+} + Fe^{2+} \tag{9.4}$$

Silver sulfide dissolves in the presence of ferric ion and the catalyst is regenerated:

$$Ag_2S - 2Fe^{3+} \rightarrow 2Ag^+ + 2Fe^{2+} + S^0 \tag{9.5}$$

The silver effect is enhanced in the presence of iron and sulfur-oxidizing microorganisms. On the one hand, microorganisms contribute to regeneration of Fe^{3+} and, on the other hand, oxidize the elemental sulfur layer produced on the chalcopyrite surface preventing chalcopyrite passivation.

9.2 Bioleaching of Sulfide Minerals

9.2.1 Mechanisms of Oxidation of Sulfide Minerals

Currently, there are three main mechanisms in the bioleaching of sulfide minerals (Tributsch 1999) (Fig. 9.1):

Depending on the type of mineral, two different ways of indirect oxidation of minerals are distinguished (Sand et al. 1995, 2001; Schippers and Sand 1999; Suzuki 2001). Metal sulfides, the valence bonds of which are obtained exclusively from metal orbitals, are oxidized with Fe (III) and cannot be attacked by protons (FeS_2, MoS_2, and WS_2). The dissolution of these minerals, according to the studies by Steudel (1996) proceeds through the formation of thiosulfate (Fig. 9.2).

Other sulfides, in which the orbitals of metals and sulfur participate in the formation of valence bonds, are soluble in acid and are attacked by both protons and Fe (III) ions (ZnS, CdS, NiS, CoS, CuS, and CuS_2). The dissolution of these

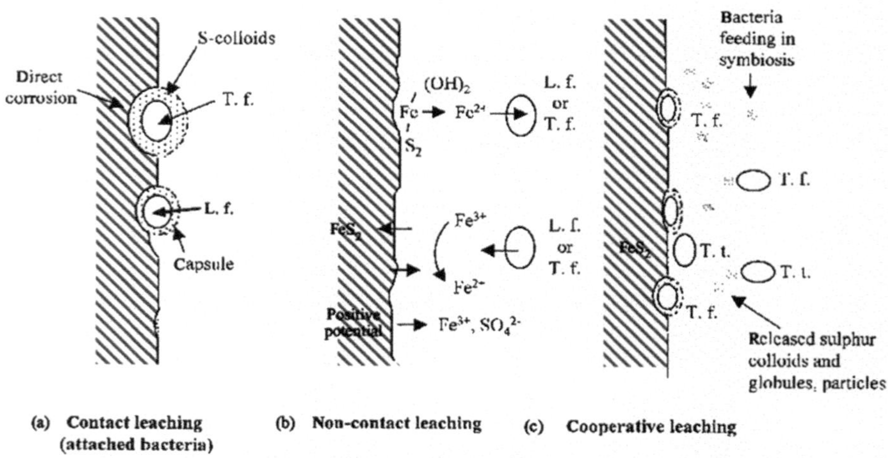

Fig. 9.1 The main mechanisms of bioleaching of sulfide minerals (according to Tributsch 1999): (**a**) Indirect leaching, when the activity of microorganisms is limited by the regeneration of the leaching agent—Fe (III), (**b**) Contact leaching, which assumes the attachment of microorganisms to the surface of the mineral, which creates an environment and facilitates the leaching of the mineral through electrochemical dissolution using Fe (III) ions contained in EPS. (**c**) Cooperative leaching, which assumes cooperation between microorganisms attached to the surface of mineral and free microorganisms in solution. Attached cells, using contact leaching, release and dissolve chemicals that serve as an energetic substrate for free microorganisms

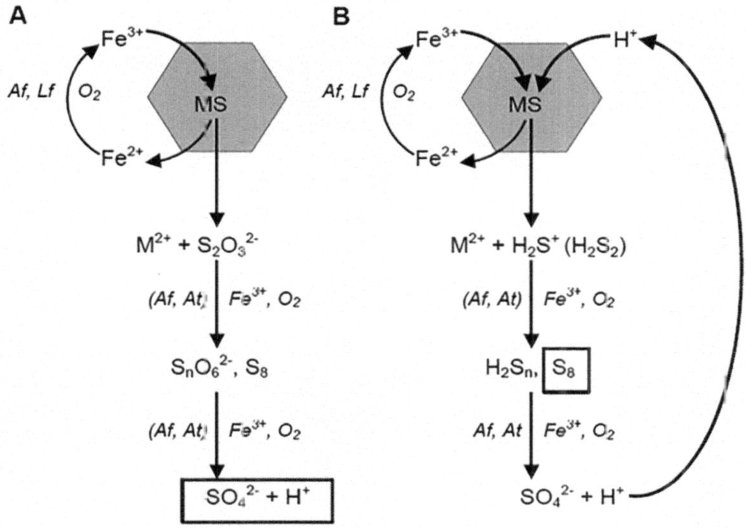

Fig. 9.2 Scheme of thiosulfate (**a**) and polysulfide (**b**) mechanisms of indirect oxidation of sulfide minerals (according to Schippers and Sand 1999)

sulfides proceeds by another mechanism—through the formation of polysulfides (Fig. 9.2).

Polysulfide mechanism:

$$MS + Fe^{3+} + H^+ \rightarrow M^{2+} + 0.5H_2S_n + Fe^{2+} \quad (9.6)$$

$$0.5H_2S_n + Fe^{3+} \rightarrow 0.125S_8 + Fe^{2+} + H^+ \quad (9.7)$$

The resulting ferrous iron (Fe^{2+}) and sulfur (S_8) are further oxidized by chemolithotrophic iron and sulfur-oxidizing bacteria (Eqs. 9.8 and 9.9):

$$Fe^{2+} + 0.5O_2 + H^+ \xrightarrow{At.ferrooxidans} Fe^{3+} + H_2O \quad (9.8)$$

$$0.125S_8 + 1.5O_2 + H_2O \xrightarrow{At.thiooxidans,\ At.caldus} SO_4 + 2H^+ \quad (9.9)$$

Consequently, the role of bacteria in the processes of indirect oxidation of minerals is to supply Fe^{3+} ions (for oxidative attack) and/or protons (for hydrolytic attack).

Contact leaching is based on the attachment of cells to the surface of minerals. The attachment is carried out using extracellular polymer compounds (EPS). Attachment occurs mainly as a result of electrostatic interactions between positively charged cells and a negatively charged mineral (pyrite) at pH 2.0 (Gehrke et al. 1998; Sampson et al. 2000). According to the literature data, the attachment of cells

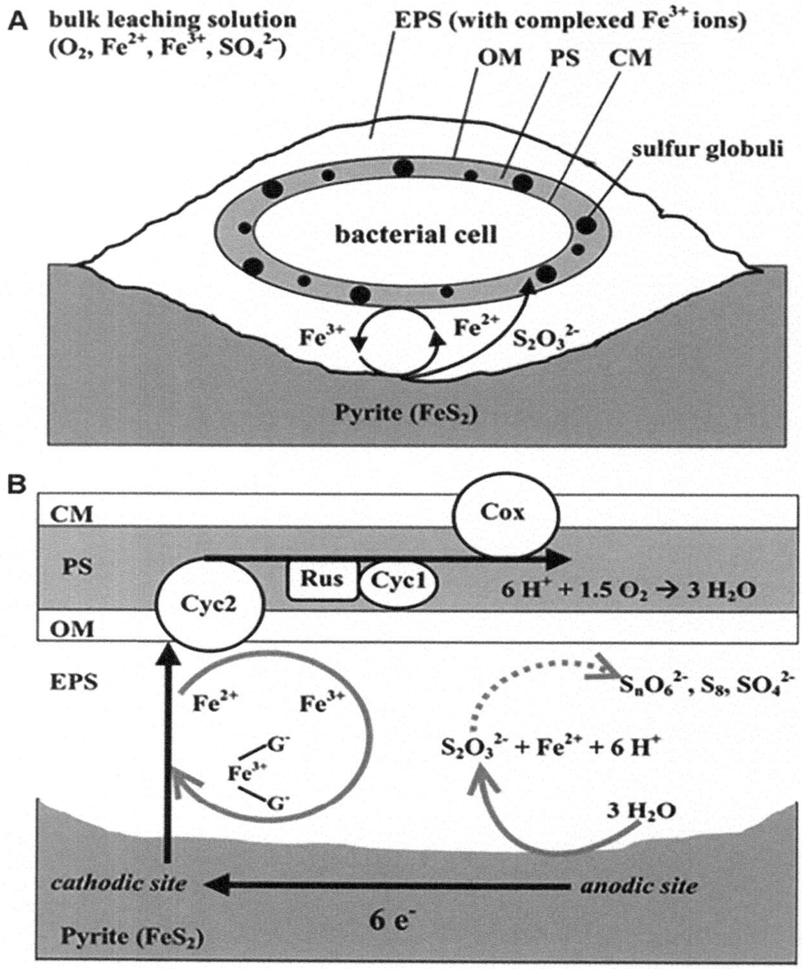

Fig. 9.3 Model of the contact indirect leaching mechanism *At. ferrooxidans* (Rohwerder et al. 2003)

to the mineral does not occur randomly, but preferably to the defective areas of the sulfide mineral (Edwards et al. 1998, 1999; Ohmura et al. 1993; Sanhueza et al. 1999; Gehrke et al. 1998, 2001; Rohwerder et al. 2003). It has been shown that cultures of *At. ferrooxidans* and *Leptospirillum ferrooxidans* have a chemosensory system—chemotaxis, which presents a positive reaction to the Fe (II)/Fe (III) gradient, thiosulfate, etc. (Acuña et al. 1992; Meyer et al. 2002).

It was shown that the rate and intensity of pyrite leaching increase upon direct contact of bacterial cells with the mineral. This is explained by an increase in the concentration of Fe (III) ions in the space between bacterial cells and the surface of pyrite (Fig. 9.3).

This space serves as a reaction medium filled with VPS (Sand et al. 1995; Sand and Gehrke 2006; Rohwerder et al. 2003; Yu et al. 2008, 2011). IPN contain complex ions of Fe (III) in concentrations many times higher than its concentration in the environment. Gehrke et al. (1998) showed the important role of EPS in the pyrite leaching process, since *At. ferrooxidans* cells lacking them are unable to leach pyrite.

9.2.2 Microorganisms Involved in the Leaching of Sulfide Minerals in Technological Processes

Bioleaching by microorganisms is widely used for the production of copper and other non-ferrous and precious metals all over the world (Donati and Sand 2007; Rawlings and Johnson 2007). Stirred continuous leaching reactors are characterized by constant parameters—temperature, pH and aeration. The constancy of conditions in the reactors causes a small number of dominant microorganism species. In general, biodiversity in reactors is limited to 2–4 species (Johnson et al. 2008; Rawlings et al. 1999; Sand et al. 1992). At the same time, it has been found that the composition of microbial consortia largely depends on the nature of the minerals and concentrates subjected to leaching (Table 9.1).

In heap leaching, conditions are not the same in terms of irrigation intensity, temperature, pH, aeration, redox potential, nutrient availability, etc., suggesting a wide variety of mineral-oxidizing microorganisms. Therefore, unlike reactors, leach heaps are characterized by significantly greater biodiversity of microorganisms. In this case, the dominant species can vary randomly, as well as depending on the stage of functioning of the heap (Table 9.2).

Thus, the main leaching bacteria belong to the genus *Acidithiobacillus* (= *Thiobacillus*) (Kelly and Wood 2000, 2005). The first representatives of this genus

Table 9.1 Acidophilic bacteria found in tank bioleaching and biooxidation reactors (Rawlings and Johnson 2007)

Concentrates	T (°C)	Procariotes detected	Source
Zinc/Lead	35–40	*L. ferrooxidans, At. thiooxidans, Acidiphilium cryptum, At. ferrooxidans*	Goebel and Stackebrandt (1994)
Pyrite/arsenopyrite (gold)	40	*L. ferrooxidans, At. thiooxidans, At. ferrooxidans*	Dew et al. (1997)
Polymetallic (copper, zinc, and iron sulfides)	45	*L. ferriphilum, At. caldus, Sulfobacillus* sp., *Ferroplasma acidophilum*	Okibe et al. (2003)
Pyrite, chalcopyrite	45	*At. caldus, S. thermosulfidooxidans, "Sulfobacillus montserratensis"*	Dopson and Lindstrom (2004)
Chalcopyrite	78	*(Sulfolobus shibatae), Metallosphaera Acidianus infernus*	Mikkelsen et al. (2006)

Table 9.2 Acidophilic bacteria found in heap leach reactors

Type of heap, location	Detected microorganisms	Sources
Chalcopyrite (Australia)	*At. ferrooxidans, At. thiooxidans, Ac. cryptum*	Goebel and Stackebrandt (1994)
Copper sulfide/ oxide heap	*Acidithiobacillus* spp., *L. ferrooxidans, Acidiphilium* spp., *Ferrimicrobium acidiphilum*	Bruhn et al. (1999)
Copper sulfide/ oxide heap (USA)	*Sulfobacillus* sp., *Firmicutes, F. acidiphilum, Acidisphaera* sp., *At. ferrooxidans, At. thiooxidans,*	Rawlings and Johnson (2007)
Chalcosite heap (Australia)	*L. ferriphilum, At. caldus, Ferroplasma* spp.	Hawkes et al. (2006)
Copper heap (Chile)	*At. ferrooxidans, L. ferriphilum, F. acidiphilum, Firmicutes*	Demergasso et al. (2005)

are extremely acidophilic iron and/or sulfur-oxidizing mesophilic bacteria *At. ferrooxidans, Acidithiobacillus thiooxidans*, which with the recently isolated and described moderately thermophilic *Acidithiobacillus caldus* belong to gram-negative γ-proteobacteria. Other leaching proteobacteria belong to the genera *Acidiphilium* and *Leptospirillum* (Hippe 2000; Coram and Rawlings 2002; Sand et al. 1992; Vardanyan and Akopyan 2003). Representatives of the genera *Acidimicrobium, Ferroomicrobium,* and *Sulfobacillus* are moderately thermophilic gram-positive bacteria (Clark and Norris 1996; Johnson and Roberto 1997; Norris et al. 1996). Leaching archaebacteria are extreme thermophiles belonging to the genera *Sulfolobus, Acidianus, Metallosphaera* (Fuchs et al. 1995, 1996; Kurosawa et al. 1998; Norris et al. 2000) and recently isolated mesophilic representatives of the genus *Ferroplasma—F. acidiphilum* (Golyshina et al. 2000), *F. acidarmanus* (Edwards et al. 2000).

L. ferrooxidans, Leptospirillum ferriphilum, and *Acidimicrobium ferrooxidans* are characterized by narrow specialization- they are able to oxidize only Fe^{2+}, while *At. ferrooxidans* and *Sulfobacillus* spp. bacteria can grow due to the oxidation of reduced compounds of sulfur, Fe^{2+} and other metal ions. *F. acidophilus* (Johnson and Roberto 1997), *Sulfobacillus* spp. bacteria, as well as thermophilic archaea, are mixotrophs and cannot grow autotrophically (Johnson 1998).

Thus, it can be concluded that the most important bacteria in commercial bioleaching processes are iron and sulfur oxidizing bacteria *Acidithiobacillus ferrooxidans,* sulfur oxidizing *Acidithiobacillus thiooxidans* or *Acidithiobacillus caldus* and iron oxidizing *Leptospirillum* spp. bacteria. (Coram and Rawlings 2002; Okibe et al. 2003). Moreover, the dominant bacteria in the bioleaching processes operated at 45–50 °C are *Leptospirillum ferriphilum* and *At. caldus* (Johnson 2014; Johnson et al. 2005; Rawlings and Johnson 2007; Sand et al. 1992; Schippers and Sand 1999). It has been confirmed that the permanent member of the associations, even under mesophilic conditions, is the species *S. thermosulfidooxidans*.

9.2.3 *Factors Affecting the Intensity of Leaching of Sulfide Minerals in Technological Processes*

9.2.3.1 Influence of the Composition of Microbial Consortia

The intensity of the process of biooxidation of sulfide minerals depends on the temperature, pH, redox potential, and the nature of the microorganisms used. Among these factors, microorganisms are the most important. A number of studies show that mixed cultures and consortia of microorganisms are more efficient and stable in the oxidation of sulfide minerals than pure cultures (Akcil et al. 2007; Falco et al. 2003; Fu et al. 2008; Baker and Banfield 2003; Johnson 2001).

From this point of view, developing and optimizing microbial consortia for use in commercial leaching systems remains an important challenge. Currently, two different "top-down" and "bottom-up" approaches have been proposed to develop the optimal microbial consortia for tank leaching (Rawlings and Johnson 2007).

The "top-down approach uses a highly diverse mixed culture as an inoculum for leaching sulfide ores. This approach is based on the assumption that a stable and efficient consortium of a limited number of bacteria can be formed during the leaching process, while others disappear. Using this approach, many researchers have succeeded in creating efficient and sustainable consortia for bioleaching at pulp densities no greater than 12% (d'Hugues et al. 2002; Zhou et al. 2009; Watling et al. 2013).

A study of the structures of communities formed during the leaching of chalcopyrite showed that their biodiversity was low (Rawlings and Johnson 2007). At the same time, it was noted that the physiological properties and functions of members of the community were complementary. Thus, it has been shown that this community can include not only iron-oxidizing bacteria (*L. ferriphilum, Sulfobacillus acidophilus, F. thermoplasma*) and RISCs oxidizing bacteria (*At. caldus, S. acidophilus*), but also mixotrophs (*S. acidophilus, F. thermoplasma*). Iron-oxidizing bacteria oxidize sulfide minerals using the ferric iron they produce. Sulfur-oxidizing bacteria accelerate the oxidation of minerals by removing the passivating layer of elemental sulfur. Mixotrophs can utilize organic matter contained in exudate or cell lysate, and thus reduce the toxic effect of organic matter on autotrophic bacteria such as *L. ferriphilum*. It is also possible that mixotrophs provide CO_2 for autotrophs. These synergistic interactions between different species increase consortium stability and enhance metal extraction (Johnson 1998; Bacelar-Nicolau and Johnson 1999).

In contrast, a "bottom-up" approach is to create highly efficient, sustainable consortia to leach specific minerals. In such consortia, composite species complement each other in terms of physiological properties, such as the ability to oxidize sulfur and/or iron, to autotrophic or heterotrophic growth, etc. For the creation of such designed consortia, the decisive factors are temperature, pH, concentration of metals, toxic ions, etc. A number of researchers have shown that cultures obtained by this method are most effective in accelerating the oxidation of certain minerals

(Johnson et al. 2008; Akcil et al. 2007; Okibe and Johnson 2004; d'Hugues et al. 2009; Bryan et al. 2011; Mejia et al. 2009).

9.2.4 *Bioleaching of Chalcopyrite*

Studies have shown that mixed cultures of iron and sulfur oxidizing bacteria *At. ferrooxidans* and *At. thiooxidans* are more effective in leaching chalcopyrite ($CuFeS_2$) than the corresponding pure cultures. The presence of sulfur-oxidizing bacteria *At. thiooxidans* increases the rate of dissolution of the mineral and the percentage of copper recovery. However, a mixed culture consisting of the moderately thermophilic bacteria *L. ferrooxidans* and *At. caldus* leaches chalcopyrite more efficiently than the mesophilic bacterium *At. ferrooxidans* in pure and mixed cultures (Fu et al. 2008). On the other hand, it has been noted that after 12–16 days of leaching, the dissolution rate of chalcopyrite *At. ferrooxidans* decreases, which coincides with the formation of jarosite as a passivating layer on the surface of the mineral during bioleaching. The use of a mixed culture consisting of *L. ferriphilum* and *At. caldus* leads to a sharp decrease in the pH of the medium as a result of intense sulfur oxidation, which in turn prevents the formation of jarosite and promotes chalcopyrite leaching (Fu et al. 2008). In addition, *L. ferriphilum* and *At. caldus*, being thermophilic bacteria, leach chalcopyrite more intensively than mesophiles, since the rate of oxidation reactions increases with increasing temperature.

It has been found that cultures containing autotrophic and mixotrophic bacterial species are more effective in stimulating bioleaching of chalcopyrite than mixed cultures containing three or four species, such as *At. caldus*, *L. ferriphilum*, *Sulfobacillus* sp. and *Ferroplasma thermophilum* (Wang et al. 2014).

9.2.4.1 Effect of Growth Conditions

It was previously found that the highest rate of oxidation of elemental sulfur and ferrous iron by moderate thermophilic bacteria *S. thermosulfidooxidans* occurred under mixotrophic conditions in the presence of 0.02% yeast extract (Vardanyan et al. 1990, Vardanyan et al. 2015). The data presented in Table 9.3, showed that the amount of iron and copper transferred into the medium under the conditions of mixotrophic growth of *S. thermosulfidooxidans* str. 69 and str. 86, respectively, were 2.9 and 1.2 times more compared with their autotrophic growth. It should be noted that under autotrophic conditions, the Fe^{2+} ion dominated in the medium, and weak growth of bacteria and oxidation of $CuFeS_2$ were observed. In contrary to autotrophic conditions during the growth of strains on chalcopyrite under mixotrophic conditions in the presence of yeast extract, the Fe^{3+} ion dominated in the medium. Due to the better growth of the strains and oxidation of sulfide sulfur under mixotrophic conditions, a decrease in pH of the medium to pH 1.7 was observed.

Table 9.3 Oxidation of chalcopyrite ($CuFeS_2$) by str. 86 and str. 69 under autotrophic condition and in the presence of 0.02% yeast extract (Duration 10 days, $CuFeS_2$ _5%, initial pH 2.0, *t*—50°)

Strains	Growth conditions	Metals leached, mg/L			Final pH
		Fe^{3+}	Fe^{2+}	Cu^{2+}	
Uninoculated control		84	840	580	3.2
S. thermosulfidooxidans str.69	Autotrophic	112	1092	920	3.2
	Mixotrophic	2856	644	1380	1.7
S. thermosulfidooxidans str.86	Autotrophic	112	924	900	3.1
	Mixotrophic	1092	168	1200	2.25

Table 9.4 Bioleaching of copper and iron by sulfobacilli depending on concentration of chalcopyrite (Duration—3 days, pH 1.8, *t*—50°)

Bacterial strains	$CuFeS_2$, %	Metals bioleached, g/L mg protein	
		Fe_{total}	Cu^{2+}
S. thermosulfidooxidans subsp. asporogenes 41	1	1.08	1.1
	2	2.7	2.7
	3	3.7	4.5
S. thermosulfidooxidans 86	1	6.2	4.8
	2	15.4	10.0
	3	18.1	14.0
S. thermosulfidooxidans 69	1	1.75	2.3
	2	4.37	4.9
	3	6.0	9.1
S. thermosulfidooxidans VKM V-1269	1	1.44	1.2
	2	2.48	2.7
	3	3.6	5.1

9.2.4.2 Effect of Concentration of Substrate

Table 9.4 shows comparative activities of different strains of *S. thermosulfidooxidans* 86 in oxidation of $CuFeS_2$ depending on its concentration. The studies have shown that the higher the concentration of chalcopyrite the more the amount of copper and iron is leached. It should be noted that at all concentrations tested, strains 86 and 69 showed significantly higher activity of $CuFeS_2$ oxidation in comparison with strains VKM V1269 and 41. Thus, str. 86 exceeded strains 1269 and 41 by 4.4–6.3 and 3.2–4.2 times for copper and iron, respectively. In case of str. 69 these values were 1.8–1.9 and 1.2–1.7 times for copper and iron, respectively. Thus, in the oxidation of chalcopyrite among the studied strains, the str. 86 shows the highest activity. This fact can probably be explained by the pronounced high ability of the strain to oxidize sulfur.

The studies have shown that the dependence of the iron and copper leaching rates on chalcopyrite concentrations is described by a typical saturation curve (Fig. 9.4).

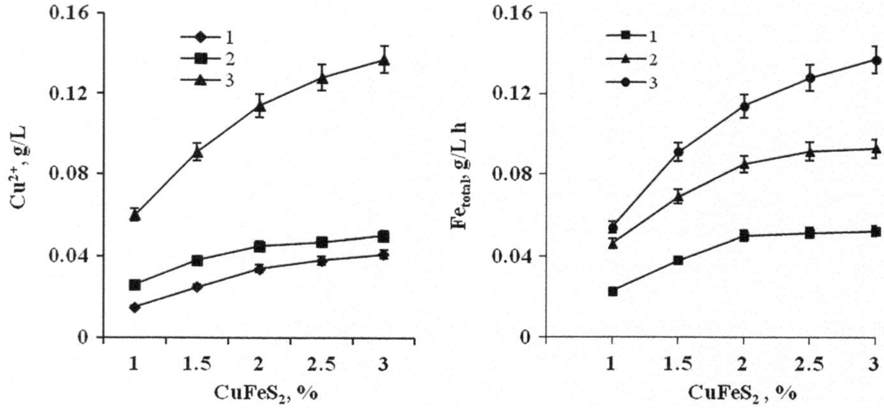

Fig. 9.4 Dependence of copper (**a**) and iron (**b**) bioleaching rates on concentration of $CuFeS_2$: 1—*S. thermosulfidooxidans subsp. asporogenes* str. 41; 2—*S. thermosulfidooxidans* str.69 and 3—str.86

Therefore, by analogy of the kinetics of enzymatic reactions, the bacterial oxidation rates of chalcopyrite can be described by the Michaelis-Menten equation:

$$V = V_{max}[CuFeS_2]/[K_m + [CuFeS_2]] \tag{9.10}$$

The latter in reverse coordinates is known as Lineweaver-Burk the modification:

$$1/V = K_m/V_{max}[CuFeS_2] + 1/V_{max} \tag{9.11}$$

Based on the graph of the dependence of the iron and copper leaching rates on concentration of chalcopyrite in reverse coordinates $1/V$ and $1/S$, the values of K_m and V_{max} for each strain of *S. thermosulfidooxidans* were determined (Fig. 9.5).

According to the data presented in Table 9.5, str. 86 possess with the lowest value of K_m, consequently the highest affinity to the substrate—1.3% $CuFeS_2$. The values of the maximum oxidation rate of chalcopyrite, determined by the leached copper and iron, differ slightly in the studied strains. Nevertheless, relatively high rate of leaching of copper and iron 250 mg/L per hour was observed in str. 86.

9.2.4.3 Effect of pH

The oxidation of chalcopyrite by *S. thermosulfidooxidans* str. 86 was studied in the pH range 1.0–3.0. At the beginning of experiment the increase of pH of the bioleaching solution was observed. According to the mechanism of bioleaching of chalcopyrite the increase in pH occurs due to the consumption of acid during the protonic attack of chalcopyrite (Eq. 9.12):

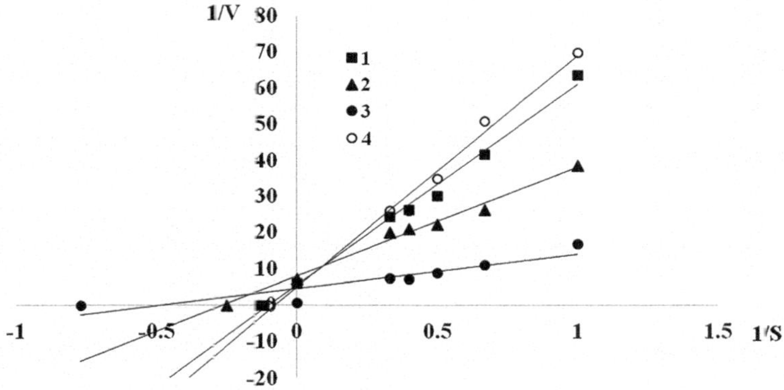

Fig. 9.5 Graphical determination of K_m and V_{max} by the amount of copper leached by *S. thermosulfidooxidans subsp. asporogenes* str. 41- (1); str. 69 (2); str. 86 (3) *S. thermosulfidooxidans* VKM-V 1269 according to Lineweaver-Burk

Table 9.5 K_m and V_{max} values for different strains of sulfobacilli during bioleaching of chalcopyrite (pH 1.8)

Bacterial strains	K_m, %		V_{max}, mg/mL h	
	Fe_{total}	Cu^{2+}	Fe_{total}	Cu^{2+}
S. thermosulfidooxidans subsp. *asporogenes* str. 41	8.0	8.0	0.20	0.17
S. thermosulfidooxidans str. 69	3.6	4.0	0.25	0.14
S. thermosulfidooxidans str. 86	1.8	1.3	0.25	0.2
S. thermosulfidooxidans BKM B-1269	11.8	10.0	0.20	0.17

$$CuFeS_2 + 4H^+ \rightarrow Fe^{2+} + Cu^{2+} + 2H_2S \quad (9.12)$$

After 4–5 days pH started to decrease because of the oxidation of elemental sulfur by *S. thermosulfidooxidans* str. 86 (Eq. 9.13):

$$2S^0 + 3O_2 + 2H_2O \rightarrow 2SO_4{}^{2-} + 4H^+ \quad (9.13)$$

Simultaneously, with pH decrease the extraction of copper and iron from chalcopyrite increased. The data presented in Fig. 9.6 show that the largest amount of copper and iron was leached at pH 1.6. At pH values above 1.8 and below 1.3, the oxidation activity of chalcopyrite decreased, and the process stopped at pH 3.0, the lower pH limit was below 1.0.

9.2.4.4 Effect of Pulp Density

The effect of high concentrations of chalcopyrite on the composition of bacterial consortia during the adaptation of cultures was studied using clone libraries of the

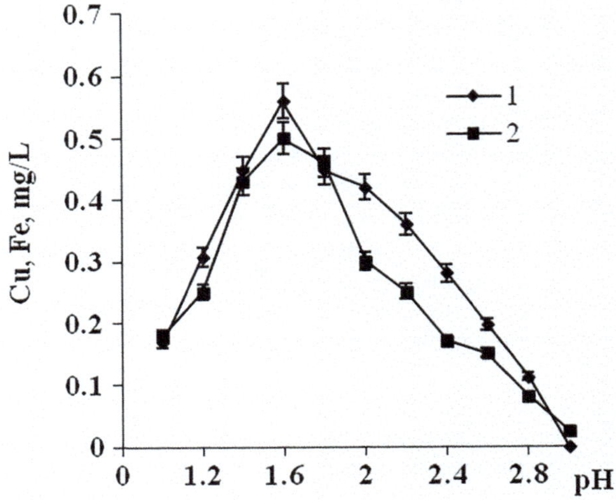

Fig. 9.6 Effect of pH on oxidation of chalcopyrite by 86: 1—Fe leaching, 2—Cu leaching

16S rRNA gene and DGGE analyses. Analyses have shown that with increasing pulp density, the amount of L. ferriphilum in the consortium decreases and is not detected at all when the pulp density exceeds 4% (6 or 8%). *At. caldus* and *S. acidophilus* are found in all pulp densities and adaptation stages. The amount of *F. thermoplasma* changes dramatically during adaptation. So, in the initial stages, the bacterium is hardly detected, and at the end it makes up 30% of the consortium. The percentage of *At. caldus* in the consortium at 20% pulp density gradually decreases from 60% at the beginning to 16% at the end of the process. *S. acidophilus* becomes dominant in the middle stages (66%). Despite *the fact* that the percentage of *F. thermoplasma* in the initial and middle stages was very low, at the end this figure reached 66% (Wang et al. 2014).

In leaching systems functioning under moderately thermophilic conditions, *F. thermoplasma* is the dominant bacterium in the final stages (Zhou et al. 2008; Zhang et al. 2009; Hawkes et al. 2006). It is believed that an increase in pulp density causes high partial pressure, limits the transport of oxygen and carbon dioxide, which leads to inhibition of bacterial growth (Zhou et al. 2009).

An important stage in obtaining highly active cultures of leaching bacteria is their adaptation to high pulp densities. Cultivation under conditions of gradually increasing densities of pulp or metal ions is a well-known method for increasing the bioleaching properties of bacteria and is widely used by many researchers (Astudillo and Acevedo 2008; Zhou et al. 2009; Cameron et al. 2010; Haghshenas et al. 2009; Rawlings 2005).

Studies carried out showed that bioleaching of copper and iron from chalcopyrite enhanced by increasing pulp density (PD) from 2 to 10%. However, at 15% of PD bioleaching of copper and iron by strain *S. thermosulfidooxidans* str. 86 decreased. Maximum extraction of copper and iron by *S. thermosulfidooxidans* str. 86 occurred at 10% of chalcopyrite (Fig. 9.7a, b).

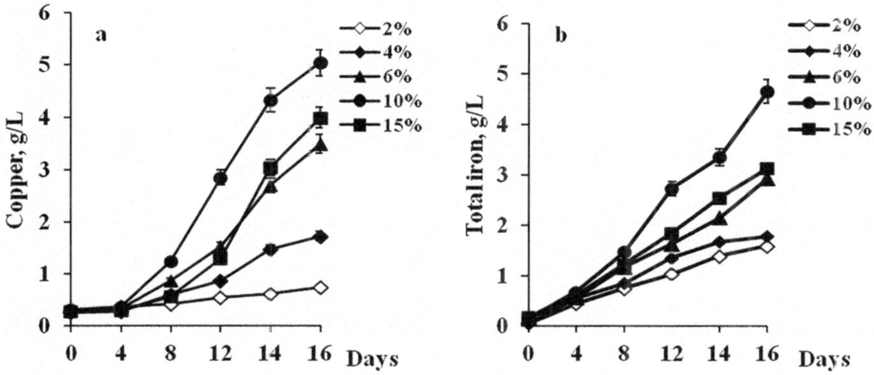

Fig. 9.7 Effect of PD on dynamics of bioleaching of copper (**a**) and iron (**b**) by *S. thermosulfidooxidans* str. 86 (pH 1.7; *t*—35 °C)

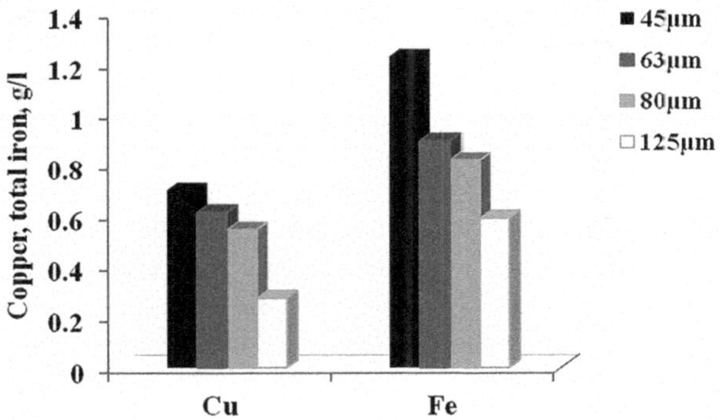

Fig. 9.8 Effect of particle size on bioleaching of copper and iron from $CuFeS_2$ by *S. thermosulfidooxidans* str. 86 (PD—4%, pH 1.8; *t*—35 °C, 180 rpm/min, duration—16 days)

9.2.4.5 Effect of Particle Size

As shown in Fig. 9.7 the rates of copper and iron dissolution increased with decreasing particle size from +125μm to size fraction of +45,—63μm. Maximum yield of copper (0.7 g/L) and total iron (1.23 g/L) reached the size range of +45,—63μm after 13 days of bioleaching of chalcopyrite by *S. thermosulfidooxidans* str. 86 (Fig. 9.8). The explanation for this increase is that smaller particle size has more surface area, enhancing metal dissolution.

9.2.4.6 Effect of Oxidative-Reductive Potential

It has been shown that Fe^{2+} and Fe (III) ions, depending on the concentration, have different effects on the growth and activity of iron-oxidizing bacteria. Das et al. (1999) report that Fe^{3+} ions at low concentrations stimulate oxygen consumption by bacteria, but at high concentrations they inhibit Fe^{2+} oxidation. A decrease in the rate of leaching of chalcopyrite at high concentrations of Fe^{3+} has also been reported by other authors (Howard and Crundwell 1999; Third et al. 2002; Hiroyoshi et al. 2007; Petersen and Dixon 2006; Rodrıguez et al. 2003).

Córdoba et al. (2008b) studied the effect of iron ions on the dissolution of chalcopyrite at low and high potentials and found that, despite the fact that Fe^{3+} ions were responsible for the oxidation of chalcopyrite, Fe^{2+} played an important role in controlling the formation and deposition of jarosiate.

The redox potential (ORP) of the leaching solution depends on the Fe^{3+}/Fe^{2+} ratio and is determined by the Nernst equation (Eq. 9.14):

$$Eh = Eh^{o} + (R.T/n.F) \times \ln\left[Fe(III)/Fe^{2+}\right] \quad (9.14)$$

The ORP in leaching systems increases as a result of the activity of iron-oxidizing microorganisms. The increase in ORP promotes the decomposition of minerals such as chalcocite (Cu_2S), covellite (CuS) and pyrite (FeS_2). However, in the case of chalcopyrite, according to some authors, the maximum dissolution rate of the mineral occurs at low ORP (Hiroyoshi et al. 1999, 2000, 2007; Third et al. 2002; Sandström et al. 2005; Córdoba et al. 2008a; Gericke et al. 2010; Ahmadi et al. 2010, 2011).

Thus, ORP is one of the most important environmental parameters affecting the leaching of chalcopyrite and copper concentrates.

The oxidation of chalcopyrite by the mesophilic bacterium *At. ferrooxidans* becomes more difficult with time due to the passivation of the mineral surface. Studies have shown that slowdown of copper leaching is often accompanied by a decrease in the pH of the medium. Based on this, it was assumed that the reason for the decrease in copper leaching was most likely the formation of jarosite. Iron ions, depending on the reaction of the leaching medium, can precipitate in the form of jarosite (Eq. 9.11).

$$3Fe^{3+} + 2SO_4{}^{2-} + 6H_2O + K^{+} \rightarrow KFe_3(SO_4)_2(OH)_6 + 6H^{+} \quad (9.15)$$

Jarosite forms a passivating layer on the surface of the mineral, which prevents diffusion and hence reduces the rate of chalcopyrite leaching (Stott et al. 2000; Okamoto et al. 2005; Yu et al. 2011).

According to other researchers, the reason for the passivation of chalcopyrite is the high ORP of the leaching medium, which is determined by the ratio of Fe^{3+}/Fe^{2+} ions (Third et al. 2002; Gericke et al. 2010). Therefore, one of the approaches to overcome this effect is to maintain a low level of redox potential in the leach solution. The effect of low ORP on chalcopyrite bioleaching is reported in the early studies of Ahonen and Tuovinen (1993). In order to reduce the redox potential of the solution, the authors propose to suppress aeration and thereby inhibit iron oxidation.

It has been shown that the dissolution of copper from chalcopyrite linearly increases at the ORP from 320 to 370 mV. Under these conditions, oxidation of

F2⁺ by bacteria is inhibited (Bevilaqua et al. 2014). Córdoba et al. (2008b) studied the effect of ORP on the chemical leaching of chalcopyrite at 68 °C. It was shown that with the ORP of 300–400 mV, almost 90% of copper was leached in 6 days, and with the ORP of 500–600 mV, only 30% in 13 days. Based on the data obtained, the authors concluded that the ORP values from 400 to 450 mV were threshold or limiting, above which the oxidative dissolution of chalcopyrite was gradually slowed down due to the passivating effect (Córdoba et al. 2008a). The release of ferrous iron from chalcopyrite and the precipitation of ferric iron resulted in a low ORP. The same authors report that the ORP rises sharply to 650 mV in bacterial cultures and improves the dissolution of copper, while these conditions do not affect chemical leaching.

Other researchers report that in the presence of 40 mM Fe^{2+}, twice as much copper is leached as when using *At. ferrooxidans*, due to the low ORP in the absence of *At. ferrooxidans* (Hiraishi et al. 2000). The authors proposed a model of chalcopyrite leaching, leading to the formation of chalcosite (Cu_2S), which was significantly rapidly and easily oxidized by oxygen and Fe^{3+}. SO_2 can be used as a chemical reducing agent to suppress ORP.

$$CuFeS_2 + 4H^+ + 2e^- \rightarrow Cu^o + 2H_2S + Fe^{2+} \quad (9.16)$$

$$2Cu + H_2S \rightarrow Cu_2S + H^+ + 2e^- \quad (9.17)$$

The leaching rate of chalcopyrite increases with an increase in the ORP of the solution and reaches a maximum value at the optimal ORP. At higher ORPs, the leaching rate decreases.

Similar results were obtained (Third et al. 2000), who studied the effect of stimulating or inhibiting the oxidative activity of iron-oxidizing bacteria in chalcopyrite leaching. The conducted studies allowed them to conclude that the formation of Fe^{3+} as a result of bacterial oxidation of Fe^{2+} inhibited the dissolution of chalcopyrite. The authors proposed to control ORP using oxygen limitation. With controlled ORP, the bacterial oxidation of Fe^{2+} was limited, but at the same time, a sufficient amount of Fe^{3+} was supplied to oxidize chalcopyrite, and as a result, copper extraction increased by 2 times. At the same time, passivation was delayed, but not completely eliminated (Third et al. 2002).

9.2.4.7 Influence of Ferric Iron (Fe^{3+})

Fe^{3+} is a permanent component of chalcopyrite leaching medium. Therefore, the study of the effect of Fe^{3+} ion on its oxidation is of particular importance. Figure 9.9 shows the leaching dynamics of copper and iron during oxidation of chalcopyrite without Fe^{3+} and when Fe^{3+} ion added to the medium.

As can be seen from the data presented in Fig. 9.9, Fe^{3+} ions, at the studied concentrations, stimulate the oxidation of chalcopyrite by str. 36.

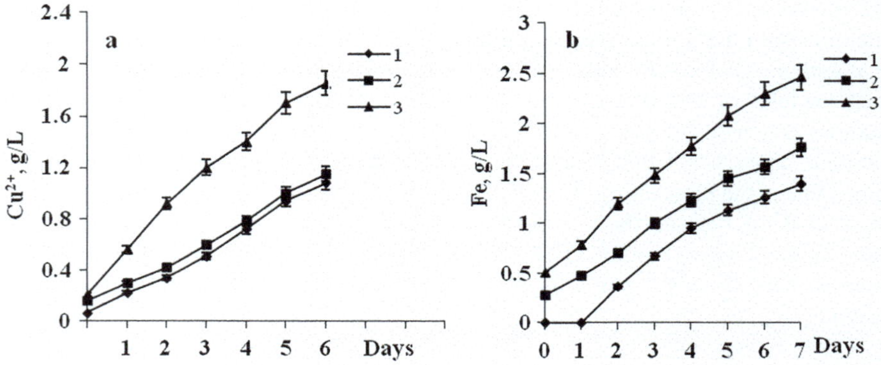

Fig. 9.9 Bioleaching of copper (**a**) and iron (**b**) by *S. thermosulfidooxidans* str.86 during oxidation of chalcopyrite: 1—without addition of Fe^{3+} and in the presence of 2—1.0 g/L Fe^{3+}, 3—1.9 g/L Fe^{3+}

Thus, Fe^{3+} ion, at concentrations of 1.0 and 1.9 g/L stimulates the oxidation of chalcopyrite in str. 86 by 1.3–1.4 and 1.9 times, respectively. The stimulation effect of Fe^{3+} can be explained by its direct involvement in the oxidation of chalcopyrite, known as the indirect route of its leaching. The Fe^{3+} ion interacts with $CuFeS_2$ to form Fe^{2+} and elemental sulfur, which are then subjected to bacterial oxidation and react again with chalcopyrite. The stimulating effect of Fe^{3+} has also been confirmed in *At. ferrooxidans* (Keller and Murr 1982; Lizama and Suzuki 1989).

Chalcopyrite is an acid-soluble metal sulfide and consequently is dissolved by both the ferric ions and proton attack (Rawlings et al. 1999; Watling 2006).

$$CuFeS_2 + 4Fe^{3+} \rightarrow Cu^{2+} + 2S + 5Fe^{2+} \tag{9.18}$$

Ferric iron oxidizes chalcopyrite to copper and ferrous ions in solution and elemental sulfur (Eq. 9.18). The activity of *L. ferriphilum* CC in mixed cultures resulted in regeneration of oxidative agent—Fe(III). However, the presence of sulfur-oxidizing *At. albertensis* SO-2 in the mixed culture, by which the sulfur is oxidized to sulfuric acid prevents the formation of jarosite on the chalcopyrite surface.

9.2.4.8 Tolerance of Microorganisms to Copper Ion

Acidophilic leaching microorganisms occupy the most metal-rich natural and anthropogenic ecosystems, such as acidic drainage water, tailings and other waste products from concentrators. The study of the resistance and adaptation of these microorganisms to high concentrations of metals is of great scientific and practical interest. This interest is primarily caused by the problem of obtaining strains of leaching microorganisms resistant to high concentrations of copper for use in biogeotechnological processes.

Studies have shown that when the concentration of copper inside a bacterial cell exceeds the permissible level, the mechanisms of bacterial resistance are activated (Rensing and Grass 2003; Magnani and Solioz 2007). Thus, in gram-negative bacteria, this is an active excretion of copper from the cytoplasm into the periplasmic space by means of p-type ATP-ases localized in the inner membrane (Rensing and Grass 2003). Certain microorganisms can pump or expel copper from the cytoplasm directly into the extracellular space using the resistant nodule division (RND) system. This type of detoxification is well known and described in *E. coli* (Outten et al. 2001). The ability of some bacteria to bind copper in the periplasmic space using copper chaperones has also been reported (Dopson et al. 2003; Puig and Thiele 2002).

The study of the resistance of acidophilic bacteria to copper is limited only to the gram-negative bacterium *At. ferrooxidans. At. ferrooxidans* is resistant to high concentrations of copper (up to 800 mM $CuSO_4$) and other metals (Dew et al. 1999; Orell et al. 2010). *L. ferrooxidans* is able to grow in the presence of 5 mM copper in the medium (Johnson et al. 1992). The resistance of bacteria to copper is very important from the point of view of their application in biotechnological processes, where the concentration of copper ions can vary in the range from 15 to 100 mM $CuSO_4$ (Watling 2006).

It has been shown that *At. ferrooxidans* ATCC 23270 can function at high copper concentrations due to about 10 genes on the chromosome that are directly related to its copper resistance. They include three genes encoding type p ATPases related to copper transport (copA1Af, copA2Af, and copBAf), three genes related to RND, responsible for the removal of copper from the cell (cusAAf, cusBAf, cusCAf) and two genes, coding periplasmic chaperones for copper (cusFAf and copCAf) (Navarro et al. 2009). Recently, it has been found that the high resistance of some *At. ferrooxidans* strains is due to the presence of additional copper-resistant genes in their genome in the form of genetic islands (GI) (Orellana and Jerez 2011).

Expression of most of these genes was established in *At. ferrooxidans* grown in the presence of high copper concentrations using real-time RT-PCR.

Some of these genes associated with copper resistance were also found in *Leptospirillum* spp. bacteria using metagenomic analyses of nucleotide sequences (Simmons et al. 2008). Two genes were also isolated from *L. ferriphilum* cells, which were responsible for bacterial resistance to arsenic ions. One of them is identical to the gene previously identified in the cells of arsenic-resistant strains *At. caldus* (Tuffin et al. 2006). It has been hypothesized about horizontal gene transfer between leaching and other bacteria, which is an important factor for the transmission of resistance to metals, as well as adaptive and other advantageous properties of these acidophiles (Dopson et al. 2003).

An example of a high tolerance to Fe (III) is the competition between iron-oxidizing bacteria in leaching solutions, which according to Rawlings et al. (1999) flows in favor of *L. ferrooxidans* and *S. themosulfidooxidans* (Vardanyan et al. 1990; Boon et al. 1999). It was shown that Fe (III) ions competitively inhibit the oxidation

Table 9.6 The influence of copper and zinc on oxidation of iron by *S. thermosulfidooxidans* 86 and *L. ferriphilum* CC (for 48 h of cultivation)

S. thermosulfidooxidans 86			*L. ferriphilum* CC	
Cu, mM	Iron oxidized, g/L, 48 h	Inhibition, %	Iron oxidized, g/L, 42 h	Inhibition, %
0	2.95	0	2.44	0
10	3.53	12.5	1.29	47.1
25	3.5	12.0	0.95	60.9
50	2.44	17.4	0.95	60.9
100	1.62	44.9	0.87	64.3
150	–	–	0.82	66.3
200	0.73	75.3	0.67	72.4
250	0.59	80.1	0.62	74.7

of Fe (II) in *At. ferrooxidans, L. ferrooxidans,* and *S. themosulfidooxidans* (Vardanyan et al. 1990; Boon et al. 1999).

It should be noted that the growth of bacteria in the form of a biofilm significantly increases the resistance of bacteria to metals (Costerton et al. 1995; Harneit et al. 2006; Sand and Gehrke 2006; Sutherland 2001).

The influence of copper on oxidation of iron by *S. thermosulfidooxidans* 86 and *L. ferriphilum* CC was studied in the range of concentration from 10 to 250 mM. It was shown that iron oxidation by *L. ferriphilum* CC is inhibited about 50% in the presence of copper in concentration of 10 mM, while inhibition of iron oxidation by *S. thermosulfidooxidans* 86 reached 45% only at 100 mM copper (Table 9.6). Thus, tolerance of *S. thermosulfidooxidans* 86 to copper was much more higher than that of *L. ferriphilum* CC.

Resistance of microorganisms to metals is of great scientific and practical interest which is caused by the problem of obtaining strains of metal leaching bacteria resistant to high concentrations of metals for application in biogeotechnological processes. In this paper the influence of copper and zinc on the oxidation of iron by *L. ferriphilum* CC and *S. thermosulfidooxidans* 86 depending on substrate concentration was studied. It was shown that the tolerance of *S. thermosulfidooxidans* 86 to copper and zinc was much higher than that of *L. ferriphilum* CC. It was revealed that the increase in the concentration of the substrate led to the decrease in the inhibitory effect of copper and zinc on the oxidation of Fe^{2+} by *L. ferriphilum* CC and *S. thermosulfidooxidans* 86. Thus, the increase of substrate concentration will allow in some ways to overcome the inhibitory effect of copper and zinc. It was established that with the growth of bacteria the decrease of inhibitory effect of metal ions was observed. It is assumed that during their growth bacterial cells form biofilm consisting of extracellular polymeric substances (EPS) and create accordingly a less toxic and more favorable environment for the growth of cells in the presence of copper and other ions (Fig. 9.10).

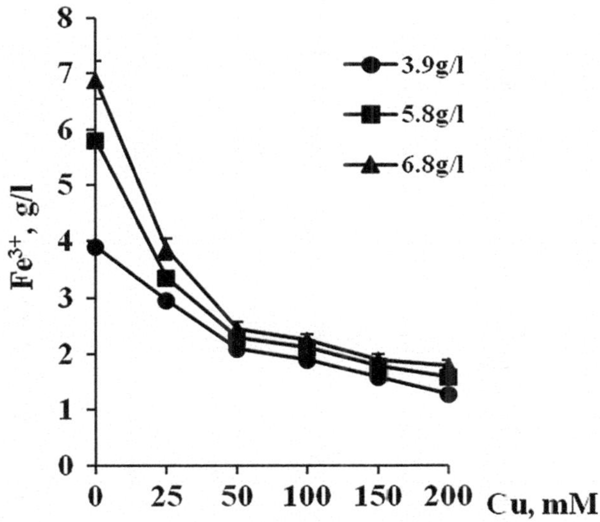

Fig. 9.10 Effect of Cu on Fe oxidation by *S. thermosulfidooxidans* 86 depending on concentration of Fe^{2+} (pH 1.95; T 35, 200 rpm, 72 h)

9.2.4.9 Oxidation of $CuFeS_2$ by Mixed Cultures

It is well-known that the most important bacteria in commercial bioleaching processes are iron and sulfur oxidizing bacteria *Acidithiobacillus ferrooxidans,* sulfur oxidizing *Acidithiobacillus thiooxidans* or *Acidithiobacillus caldus* and iron oxidizing *Leptospirillum* spp. bacteria. (Coram and Rawlings 2002; Okibe et al. 2003). Moreover, the dominant bacteria in the bioleaching processes operated at 45–50 °C are *Leptospirillum ferriphilum, At. caldus,* and *S. thermosulfidooxidans* (Johnson 2014; Johnson et al. 2005; Rawlings and Johnson 2007; Sand et al. 1992; Schippers and Sand 1999).

It is well-known that in natural biocenoses, there are complex relationships between bacteria, including different forms of reciprocal feeding (Dopson and Lindstrom 1999; Bacelar-Nicolau and Johnson 1999; Johnson 1998; Tuovinen et al. 1994). Taking into consideration the above-mentioned associations of sulfobacilli and other iron and/or sulfur oxidizing bacteria have been developed to significantly increase the activity of sulfobacilli in the oxidation of pyrite and chalcopyrite.

The dynamics of oxidation of chalcopyrite by *S. thermosulfidooxidans* str.86 and thermotolerant sulfur or iron oxidizing bacteria and their association is shown in Fig. 9.11. According to the data presented, the highest oxidation activity of $CuFeS_2$ showed *S. thermosulfidooxidans* str.86 growing under mixotrophic conditions in the presence of 0.02% yeast extract. The studies have shown that neither iron oxidizing *L. ferrooxidans* str.72, nor sulfur oxidizing bacteria *At. tandzuti* str.5 in monoculture are capable of performing the oxidation of chalcopyrite. However, when *L. ferrooxidans* str. 72 was grown in association with sulfur oxidizing bacteria str. 5 an increase of oxidation of chalcopyrite was observed (Fig. 9.11).

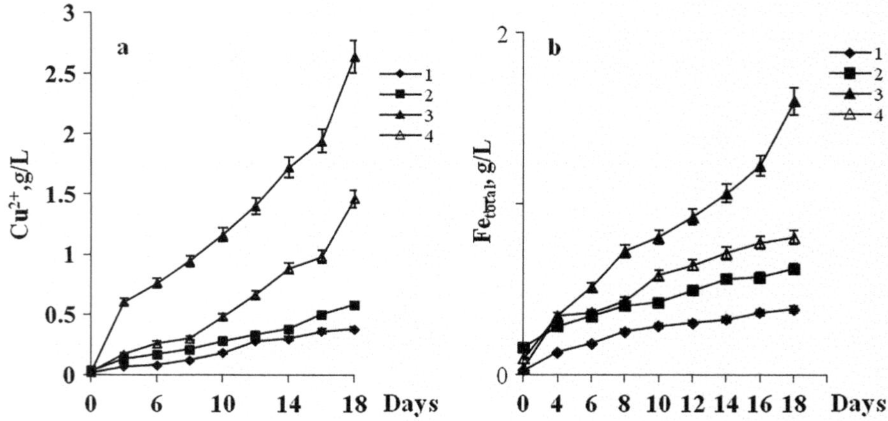

Fig. 9.11 Oxidation of chalcopyrite by: 1—*At. tandzuti* str. 5, 2—*L. ferrooxidans* str.72 3—*S. thermosulidooxidans* str. 86*, 4—*L. ferooxidans* str.72 + *At. tandzuti* str. 5

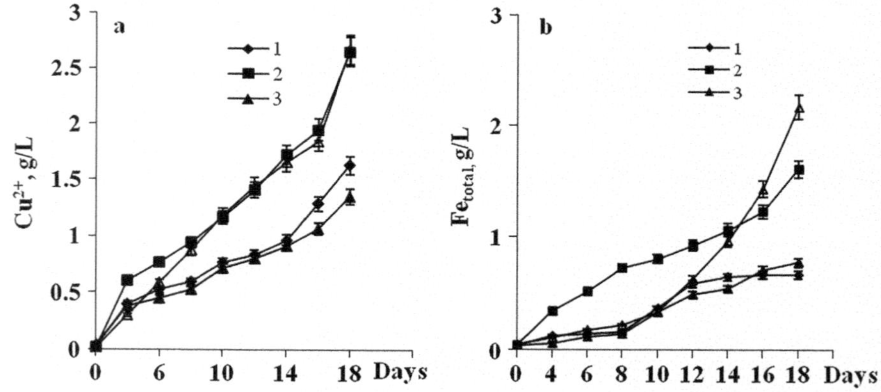

Fig. 9.12 Bioleaching of copper (**a**) and iron (**b**) by sulfur and iron oxidizing bacteria during the oxidation of chalcopyrite by: 1—*S. thermosulidooxidans* str. 86, 2—str. 86*, 3—str. 86 + *L. ferrooxidans* str. 72, 4—str. 86 + str. 5

The same poor oxidation of $CuFeS_2$ was observed during autotrophic growth of *S. thermosulfidooxidans* str. 86 (Fig. 9.12). The studies carried out showed that the activity of autotrophically growing str. 86 significantly increased when grew together with *At. tandzuti* str. 5. Thus, for 8 days of cultivation by the amount of leached copper (0.96 g/L), this association reached the efficiency observed by mixotrophically growing str. 86 (0.94 g/L). During 18 days of growth, the mentioned association by the amount of leached iron (2.16 g/L) significantly exceeds str. 86 grown under mixotrophic conditions in the presence of yeast extract (1.6 g/L). It should be noted that no significant oxidation of $CuFeS_2$ was observed by

S. thermosulfidooxidans growing together with *L. ferrooxidans* str. 72 in the absence of yeast extract with str. 72 (Fig. 9.12).

It can be concluded that the co-cultivation of sulfobacilli with thermotolerant sulfur oxidizing bacteria allows to achieve the efficiency of oxidation of $CuFeS_2$ without adding an organic carbon source to the medium, which is necessary for the growth of sulfobacilli extract (Vardanyan 1998, 2003;Vardanyan and Vardanyan 2016).

Thus, the use of thermotolerant sulfur-oxidizing bacteria with sulfobacilli allows to perform the oxidation of chalcopyrite without the addition of organic substances by intensity, observed during the growth of moderate thermophiles under mixotrophic conditions in the presence of yeast extract. Syntrophic relationships are thought to arise between the strains studied during chalcopyrite oxidation. Probably, strain 86 of moderately thermophilic bacteria provides sulfur oxidizing bacteria with elemental sulfur and its reduced compounds.

9.3 Conclusion

It has been shown that the dependence of the copper and iron leaching rates on chalcopyrite concentrations is described by a typical saturation curve. Therefore, by analogy of the kinetics of enzymatic reactions, the bacterial oxidation rates of chalcopyrite can be described by the Michaelis-Menten equation: *S. thermosulfidooxidans* str. 86 has shown the lowest K_m value, consequently the highest affinity to the substrate-$CuFeS_2$ and highest maximal rate of copper leaching. Decrease of particle size leads to the increase of mineral surface area enhancing copper dissolution. Results obtained have shown that ferric iron (Fe^{3+}) stimulates the oxidation of chalcopyrite in moderate thermophilic sulfobacilli that can be explained by direct involvement of Fe^{3+} in the oxidation of chalcopyrite, known as the indirect route of its leaching. Chalcopyrite is an acid-soluble metal sulfide and consequently is dissolved by both the proton attack and ferric iron (Rawlings et al. 1999; Watling 2006). The Fe^{3+} ion interacts with $CuFeS_2$ to form elemental sulfur and ferrous iron (Fe^{2+}) (Eq. 9.18), which are then subjected to bacterial oxidation and react again with chalcopyrite.

The activity of iron oxidizing bacteria *L. ferriphilum* CC in association resulted in regeneration of oxidizing agent—Fe^{3+}. However, the presence of sulfur-oxidizing *At. tandzuti* in association, by which the sulfur is oxidized to sulfuric acid prevents the formation of passivating sulfur layer and jarosite on the chalcopyrite surface. It can be concluded that the co-cultivation of sulfobacilli with thermotolerant sulfur oxidizing bacteria allows considerably increasing the efficiency of oxidation of $CuFeS_2$ without adding an organic carbon source to the medium, which is necessary for the growth of sulfobacilli extract (Vardanyan 1998, 2003; Vardanyan and Vardanyan 2016). Thus, the use of thermotolerant sulfur-oxidizing bacteria in association with sulfobacilli allows to perform the oxidation of chalcopyrite under autotrophic conditions by intensity, observed during their growth under mixotrophic conditions in the presence of 0.02% yeast extract. Syntrophic relationships are thought to arise between the strains studied during chalcopyrite oxidation. Probably, *S. thermosulfidooxidans* str. 86 provides sulfur oxidizing bacteria with elemental sulfur and its reduced compounds.

References

Acuña J, Rojas J, Amaro A, Toledo H, Jerez C (1992) Chemotaxis of *Leptospirillum ferrooxidans* and other acidophilic chemolithotrophs: comparison with the *Escherichia coli* chemosensory system. FEMS Microbiol Lett 75:37–42

Ahmadi A, Schaffie M, Manafi Z, Ranjbar M (2010) Electrochemical bioleaching of high grade chalcopyrite flotation concentrate in a stirred tank reactor. Hydrometallurgy 104:99–105

Ahmadi A, Schaffie M, Petersen J, Schippers A, Ranjbar M (2011) Conventional and electrochemical bioleaching of chalcopyrite concentrates by moderately thermophilic bacteria at high pulp density. Hydrometallurgy 106:84–92

Ahonen L, Tuovinen O (1993) Redox potential-controlled bacterial leaching of chalcopyrite ores. In: Torma A, Wey J, Lakshmanan V (eds) Biohydrometallurgical technologies, Bioleaching Processes. The Minerals, Metals and Materials Society, Warrendale, PA, pp 571–578

Akcil A, Ciftci H, Deveci H (2007) Role and contribution of pure and mixed cultures of mesophiles in bioleaching of a pyritic chalcopyrite concentrate. Miner Eng 20:310–318

Astudillo C, Acevedo F (2008) Adaptation of *Sulfolobus metallicus* to high pulp densities in the biooxidation of a flotation gold concentrate. Hydrometallurgy 92:11–15

Bacelar-Nicolau P, Johnson D (1999) Leaching of pyrite by acidophilic heterotrophic iron-oxidizing bacteria in pure and mixed cultures. Appl Environ Microbiol 65:585–590

Baker B, Banfield J (2003) Microbial communities in acid mine drainage. FEMS Microbiol Ecol 44:139–152

Ballester A, González F, Blázquez M, Gómez C, Mier J (1990) The influence of several ions in the bioleaching of metal sulphides. Hydrometallurgy 23:221–235

Bevilaqua D, Lahti-Tommila H, Garcia O Jr, Puhakka J, Tuovinen O (2014) Bacterial and chemical leaching of chalcopyrite concentrates as affected by the redox potential and ferric/ferrous iron ratio at 22°C. Int J Miner Process 132:1–7

Boon M, Ras C, Heijnen J (1999) The ferrous iron oxidation kinetics of *Thiobacillus ferrooxidans* in batch cultures. Appl Microbiol Biotechnol 51:813–819

Brombacher C, Bachofen R, Brandl H (1997) Biohydrometallurgical processing of solids: a patent review. Appl Microbiol Biotechnol 48:577–587

Bruhn D, Thompson D, Naoh K (1999) Microbial ecology assessment of a mixed copper oxide/sulfide dump leach operation. In: Amils R, Ballester A (eds) Biohydrometallurgy and the environment. Toward the mining of the 21st century, process metallurgy. Elsevier, Amsterdam, pp 799–808

Bryan C, Joulian C, Spolaore P, El Achbouni H, Challan-Belval S, Morin D, d'Hugues P (2011) The efficiency of indigenous and designed consortia in bioleaching stirred tank reactors. Miner Eng 24:1149–1156

Cameron R, Yeung C, Greer C, Gould W, Mortazavi S, Bédard P, Morin L, Lortie L, Dinardo O, Kennedy K (2010) The bacterial community structure during bioleaching of a low-grade nickel sulphide ore in stirred—tank reactors at different combinations of temperature and pH. Hydrometallurgy 104:207–215

Cancho L, Blázquez M, Ballester A, González F, Muñoz J (2007) Bioleaching of a chalcopyrite concentrate with moderate thermophilic microorganisms in a continuous reactor system. Hydrometallurgy 87:100–111

Clark D, Norris P (1996) *Acidomicrobium ferrooxidans* gen. Nov., sp. nov.: mixed-culture ferrous iron oxidation with *Sulfobacillus* species. Microbiology (UK) 142:785–790

Coram N, Rawlings D (2002) Molecular relationship between two groups of the genus *Leptospirillum* and the finding that *Leptospirillum ferriphilum* sp. nov. dominates south African commercial biooxidation tanks that operate at 40°C. Appl Environ Microbiol 68:838–845

Córdoba E, Muñoz J, Blázquez M, González F, Ballester A (2008a) Leaching of chalcopyrite with ferric ion. Part I: general aspects. Hydrometallurgy 93:81–87

Córdoba E, Muñoz J, Blázquez M, González F, Ballester A (2008b) Leaching of chalcopyrite with ferric ion. Part II: effect of redox potential. Hydrometallurgy 93:88–96

Costerton J, Lewandowski Z, Caldwell D, Korber D, Lappin-Scott H (1995) Microbial biofilms. Annu Rev Microbiol 49:711–745

Das T, Ayyappan S, Chaudhury GR (1999) Factors affecting bioleaching kinetics of sulfide ores using acidophilic microorganisms. Biometals 12:1–10

d'Hugues P, Foucher S, Galle-Cavalloni P, Morin D (2002) Continuous bioleaching of chalcopyrite using a novel extremely thermophilic mixed culture. Int J Miner Process 66:107–119

d'Hugues P, Joulian C, Spolaore P, Morin D, Bryan C, Challan-Belval S, El Achbouni H (2009) Adaptation and evolution of microbial consortia in a stirred tank reactor bioleaching system: indigenous population versus a defined consortium. Adv Mat Res 71:79–82

Demergasso C, Galeguillos P, Escudero L, Zepeda V, Castillo D, Casamayor E (2005) Molecular characterization of microbial population in a low-grade copper ore bioleaching test heap. Hydrometallurgy 80:241–253

Dew D, Lawson E, Broadhurst J (1997) The BIOX process for biooxidation of gold-bearing ores or concentrates. In: Rawlings D (ed) Biomining: theory, microbes and industrial processes. Springer, Berlin, pp 45–80

Dew D, Muhlbauer R, van Buuren C (1999) Bioleaching of copper sulphide concentrates with mesophiles and thermophiles. Alta copper 99 Brisbane, Australia

Donati E, Sand W (2007) Microbial processing of metal sulphides. Springer, Berlin, pp 1–314

Dopson M, Lindstrom EB (1999) Potential role of Thiobacillus caldus in Arsenopyrite Leaching. Appl Environ Microbiol 65:36–40

Dopson M, Lindstrom E (2004) Analysis of community composition during moderately thermophilic bioleaching of pyrite, arsenical pyrite and chalcopyrite. Microb Ecol 48:19–28

Dopson M, Baker-Austin C, Koppineedi P, Bond P (2003) Growth in sulfidic mineral environments: metal resistance mechanisms in acidophilic micro-organisms. Microbiology 149:1959–1970

Edwards K, Schrenk M, Hamers R, Banfield J (1998) Microbial oxidation of pyrite: experiments using microorganisms from an extreme acidic environment. Am Mineral 83:1444–1453

Edwards K, Goebel B, Rodgers T, Schrenk M, Gihring T, Cardona M, Mcguire M, Hamers R, Pace N, Banfield J (1999) Geomicrobiology of pyrite (FeS_2) dissolution: case study at Iron Mountain, California. Geomicrobiol J 16:155–179

Edwards K, Bond P, Gihring T, Banfield J (2000) An archaeal iron-oxidizing extreme acidophile important in acid mine drainage. Science 279:1796–1799

Ehrlich H (1999) Past, present and future of biohydrometallurgy. In: Amils R, Ballester A (eds) Biohydrometallurgy and the environment toward the mining of the 21st century, IBS99. Elsevier, Amsterdam, pp 3–12

Falco L, Pogliani C, Curutchet G, Donati E (2003) A comparison of bioleaching of covellite using pure cultures of *Acidithiobacillus ferrooxidans* and *Acidithiobacillus thiooxidans* or a mixed culture of *Leptospirillum ferrooxidans* and *Acidithiobacillus thiooxidans*. Hydrometallurgy 71:31–36

Fu B, Zhou H, Zhang R, Qiu G (2008) Bioleaching of chalcopyrite by pure and mixed cultures of *Acidithiobacillus* spp. and *Leptospirillum ferriphilum*. Int Biodet Biodegrad 62:109–115

Fuchs T, Huber H, Teiner K, Burggraf S, Stetter K (1995) *Metallosphaera prunae*, sp. nov., a novel metal-mobilizing, thermoacidophilic archaeum, isolated from a uranium mine in Germany. Syst Appl Microbiol 18:560–566

Fuchs T, Huber H, Burggraf S, Stetter K (1996) 16S rDNA-based phylogeny of the archaeal order Sulfolobales and reclassification of Desulfurolobus ambivalens as Acidianus ambivalens comb. nov. Syst. Appl Microbiol 19:56–60

Gahan C, Srichandan H, Kim D, Akci A (2012) Biohydrometallurgy and biomineral processing technology: a review on its past, present and future. Res J Recent Sci 10:85–99

Gehrke T, Telegdi J, Thierry D, Sand W (1998) Importance of extracellular polymeric substances from *Thiobacillus ferrooxidans* for bioleaching. Appl Environ Microbiol 64:2743–2747

Gehrke T, Hallmann R, Kinzler K, Sand W (2001) The EPS of *Acidithiobacillus ferrooxidans* - a model for structure-function relationships of attached bacteria and their physiology. Water Sci Technol 43:159–167

Gericke M, Govender Y, Pinches A (2010) Tank bioleaching of low-grade chalcopyrite concentrates using redox control. Hydrometallurgy 104:414–419

Goebel B, Stackebrandt E (1994) Cultural and phylogenetic analysis of mixed microbial populations found in natural and commercial bioleaching environments. Appl Environ Microbiol 60:1614–1621

Golyshina O, Pivovarova T, Karavaiko G, Kondrateva T, Moore E, Abraham W, Lunsdorf H, Timmis K, Yakimov M, Golyshin P (2000) *Ferroplasma acidiphilum* gen. Nov., sp. nov., an acidophilic, autotrophic, ferrous-iron-oxidizing, cell-wall-lacking, mesophilic member of the Ferroplasmaceae fam. Nov., comprising a distinct lineage of the archaea. Int J Syst Evol Microbiol 50:997–1006

Gómez E, Ballester A, Blázquez M, González F (1999) Silvercatalysed bioleaching of a chalcopyrite concentrate with mixed cultures of moderately thermophilic microorganisms. Hydrometallurgy 51:37–46

Haghshenas D, Alamdari E, Torkmahalleh M, Bonakdarpour B, Nasernejad B (2009) Adaptation of *Acidithiobacillus ferrooxidans* to high grade sphalerite concentrate. Miner Eng 22:1299–1306

Harneit K, Goksel A, Kock D, Klock J, Gehrke T, Sand W (2006) Adhesion to metal sulfide surfaces by cells of *Acidiothiobacillus ferrooxidans, Acidiothiobacillus thiooxidans* and *Leptospirillum ferrooxidans*. Hydrometallurgy 83:245–254

Hawkes R, Franzmann P, Plumb J (2006) Moderate thermophiles including "*Ferroplasma cuprexacervatum*" sp. Nov., dominate an industrial scale chalcocite heap bioleaching operation. Hydrometallurgy 83:229–236

Hippe H (2000) *Leptospirillum* gen. Nov. (ex Markosyan 1972), nom. Rev., including *Leptospirillum ferrooxidans* sp. nov. (ex Markosyan 1972), nom. Rev. and *Leptospirillum thermoferrooxidans* sp. nov. (Golovacheva et al. 1992). Int J Syst Evol Microbiol 50:501–503

Hiraishi A, Matsuzawa Y, Kanbe T, Wakao N (2000) *Acidisphaera rubrifaciens* gen. Nov., sp. nov., an aerobic bacteriophyllcontaining bacterium isolated from acidic environments. Int J Syst Evol Microbiol 50:1539–1546

Hiroyoshi N, Hirota M, Hirajima T, Tsunekawa M (1999) Inhibitory effect of ironoxidizing bacteria on ferrous-promoted chalcopyrite leaching. Biotechnol Bioeng 64:478–483

Hiroyoshi N, Miki H, Hirajima T, Tsunekawa M (2000) A model for ferrous-promoted chalcopyrite leaching. Hydrometallurgy 57:31–38

Hiroyoshi N, Kuroiwa S, Miki H, Tsunekawa M, Hirajima T (2007) Effects of coexisting metal ions on the redox potential dependence of chalcopyrite leaching in sulfuric acid solutions. Hydrometallurgy 87:1–10

Howard D, Crundwell F (1999) A kinetic study of the leaching of chalcopyrite with *Sulfolobus metallicus*. In: Biohydrometallurgy and the environment toward the mining of the 21st century. Elsevier, Amsterdam, pp 209–217

Johnson D (1998) Biodiversity and ecology of acidophilic microorganisms. FEMS Microbiol Ecol 27:307–317

Johnson D (2001) Importance of microbial ecology in the development of new mineral technologies. Hydrometallurgy 59:147–157

Johnson D (2014) Biomining—biotechnologies for extracting and recovering metals from ores and waste materials. Curr Opin Biotechnol 30:24–31

Johnson D, Roberto F (1997) Biodiversity of acidophilic bacteria in mineral leaching and related environments. In: Proceedings of IBS Biomine' 97 Conf., Australian Mineral Foundation, Glenside, Australia, Part 3, pp 1–10

Johnson D, Ghauri M, Said M (1992) Isolation and characterization of an acidophilic, heterotrophic bacterium capable of oxidizing ferrous iron. Appl Environ Microbiol 58:1423–1428

Johnson D, Okibe N, Hallberg K (2005) Differentiation and identification of iron-oxidizing acidophilic bacteria using cultivation techniques and amplified ribosomal DNA restriction enzyme analysis. J Microbiol Methods 60:299–313

Johnson D, Okibe N, Wakeman K, Yajie L (2008) Effect of temperature on the bioleaching of chalcopyrite concentrates containing different concentrations of silver. Hydrometallurgy 94:42–47

Keller L, Murr L (1982) Acid-bacterial and ferric sulfate leaching of pyrite single crystals. Biotechnol Bioeng 24:83–96

Kelly D, Wood A (2000) Reclassification of some species of Thiobacillus to the newly designated genera *Acidithiobacillus* gen. Nov., *Halothiobacillus* gen. Nov., and *Thermithiobacillus* gen. Nov. Int J Syst Evol Microbiol 50:511–516

Kelly D, Wood A (2005) Genus I. *Acidothiobacillus*. In: Brener D, Krieg N, Staley R (eds) Bergey's manual of systematic bacteriology, 2nd edn. Springer, New York, pp 60–62

Kurosawa N, Itoh Y, Iwai T, Sugai A, Uda I, Kimura N, Horiuchi T, Itoh T (1998) *Sulfurisphaera ohwakuensis* gen. Nov., sp. nov., a novel extremely thermophilic acidophile of the order*Sulfolobales*. Int J Syst Bacteriol 48:451–456

Lizama H, Suzuki I (1989) Synergistic competitive inhibition of ferrous iron oxidation by thiobacillus ferrooxidans by increasing concentrations of ferric iron and cells. Appl Environ Microbiol 55:2588–2591

Magnani D, Solioz M (2007) How bacteria handle copper. In: Nies D, Silver S (eds) Bacterial transition metal homeostasis. Springer, Heidelberg, pp 259–285

Mejia E, Ospina J, Marquez M, Morales A (2009) Oxidation of chalcopyrite ($CuFeS_2$) by *Acidithiobacillus ferrooxidans* and a mixed culture of *Acidithiobacillus ferrooxidans* and *Acidithiobacillus thiooxidans* like bacterium in shake flasks. Adv Mat Res 71:385–388

Meyer G, Schneider-Merck T, Böhme S, Sand W (2002) A simple method for investigations on the chemotaxis of *a. ferrooxidans* and *D. vulgaris*. Acta Biotechnol 22:391–399

Mikkelsen D, Kappler U, McEwan A, Sly L (2006) Archaeal diversity in two thermophilic chalcopyrite bioleaching reactors. Environ Microbiol 8:2050–2055

Navarro C, Orellana L, Mauriaca C, Jerez C (2009) Transcriptional and functional studies of *Acidithiobacillus ferrooxidans* genes related to survival in the presence of copper. Appl Environ Microbiol 75:6102–6109

Norris P, Clark D, Owen J, Waterhouse S (1996) Characteristics of *Sulfobacillus acidophilus* sp. nov. and other moderately thermophilic mineral-sulphide-oxidizing bacteria. Microbiology 142:775–783

Norris P, Burton N, Foulis N (2000) Acidophiles in bioreactor mineral processing. Extremophiles 4:71–76

Ohmura N, Kitamura K, Saiki H (1993) Selective adhesion of *Thiobacillus ferrooxidans* to pyrite. Appl Environ Microbiol 59:4044–4050

Okamoto H, Nakayama R, Kuroiwa S, Hiroyoshi N, Tsunekawa M (2005) Normalized redox potential used to assess chalcopyrite column leaching. J MMIJ 121:246–254

Okibe N, Johnson D (2004) Biooxidation of pyrite by defined mixed cultures of moderately thermophilic acidophiles in pH-controlled bioreactors: significance of microbial interactions. Biotechnol Bioeng 87:574–583

Okibe N, Gericke M, Hallberg K, Johnson D (2003) Enumeration and characterization of acidophilic microorganisms isolated from a pilot plant stirred-tank bioleaching operation. Appl Environ Microbiol 69:1936–1943

Olson G, Brierley J, Brierley C (2003) Bioleaching review part B: progress in bioleaching: applications of microbial processes by the minerals industries. Appl Microbiol Biotechnol 63:249–257

Orell A, Navarro C, Arancibia R, Mobarec J, Jerez C (2010) Life in blue: copper resistance mechanisms of bacteria and archaea used in industrial biomining of minerals. Biotechnol Adv 28:839–848

Orellana L, Jerez C (2011) A genomic island provides *Acidithiobacillus ferrooxidans* ATCC 53993 additional copper resistance: a possible competitive advantage. Appl Microbiol Biotechnol 92:761–767

Outten F, Huffman D, Hale J, Ó Halloran T (2001) The independent cue and cus system confer copper tolerance during aerobic and anaerobic growth in Escherichia coli. J Biol Chem 276:30670–30677

Petersen J, Dixon D (2006) Competitive bioleaching of pyrite and chalcopyrite. Hydrometallurgy 83:40–49

Puig S, Thiele D (2002) Molecular mechanisms of copper uptake and distribution. Curr Opin Chem Biol 6:171–180

Rawlings D (2002) Heavy metal mining using microbes. Annu Rev Microbiol 56:65–91

Rawlings D (2005) Characteristics and adaptability of iron- and sulfuroxidizing microorganisms used for the recovery of metals from minerals and their concentrates. Microb Cell Fact 4:13

Rawlings DE, Johnson DB (2007) The microbiology of biomining: development and optimization of mineral-oxidizing microbial consortia. Microbiology 153:315–324

Rawlings D, Coram N, Gardner M, Deane S (1999) *Thiobacillus caldus* and *Leptospirillum ferrooxians* are widely distributed in continuous-flow biooxidation tanks used to treat a variety of metal-containing ores and concentrates. In: Biohydrometrallurgy and the environment toward the mining of the 21st century, part A. Elsevier, Amsterdam, pp 773–778

Rawlings D, Dew D, du Plessis C (2003) Biomineralization of metal-containing ores and concentrates. Trends Biotechnol 21:38–44

Rensing C, Grass G (2003) Escherichia coli mechanism of copper homeostasis in a changing environment. FEMS Microbiol Rev 27:197–213

Rodrıguez Y, Ballester A, Bla'zquez M, Gonza'lez F, Muñoz J (2003) New information on the chalcopyrite bioleaching mechanism at low and high temperature. Hydrometallurgy 71:47–56

Rohwerder T, Gehrke T, Kinzler K, Sand W (2003) Bioleaching review part A: progress in bioleaching: fundamentals and mechanisms of bacterial metal sulfide oxidation. Appl Microbiol Biotechnol 63:239–248

Sampson M, Phillips C, Blake R II (2000) Influence of the attachment of acidophilic bacteria during the oxidation of mineral sulfides. Miner Eng 13:373–389

Sand W, Gehrke T (2006) Extracellular polymeric substances mediate bioleaching/biocorrosion via interfacial processes involving iron (III) ions and acidophilic bacteria. Res Microbiol 157:49–56

Sand W, Rohde K, Sobotke B, Zenneck C (1992) Evaluation of *Leptospirillum ferrooxidans* for leaching. Appl Environ Microbiol 58:85–92

Sand W, Gehrke T, Hellmann R, Schippers A (1995) Sulfur chemistry, biofilm and the indirect attack mechanism a critical evaluation of bacterial leaching. Appl Microbiol Biotechnol 43:961–966

Sand W, Gehrke T, Jozsa P, Schippers A (2001) Biochemistry of bacterial leaching-direct vs. indirect bioleaching. Hydrometallurgy 59:159–175

Sandström A, Shchukarev A, Paul J (2005) XPS characterisation of chalcopyrite chemically and bioleached at high and low redox potential. Miner Eng 18:505–515

Sanhueza A, Ferrer I, Vargas T, Amils R, Jerez C (1999) Attachment of *Thiobacillus ferrooxidans* on synthetic pyrite of varying structural and electronic properties. Hydrometallurgy 51:115–129

Schippers A, Sand W (1999) Bacreial leaching of metal sulfides proceeds by two indirect mechanisms with thiosulfate or via polysulfides and sulfur. Appl Environ Microbiol 65:319–321

Simmons S, DiBartolo G, Denef V, Goltsman D, Thelen M, Banfield J (2008) Population genomic analysis of strain variation in *Leptospirillum* group II bacteria involved in acid mine drainage formation. PLoS Biol 6:1427–1442

Steudel R (1996) Mechanism for the formation of elemental sulfur from aqueous sulfide in chemical and microbiological desulfurization processes. Ind Eng Chem Res 35:1417–1423

Stott M, Walting H, Fransmann P, Sutton D (2000) The role of iron-hydroxy precipitates in the passivation of chalcopyrite during bioleaching. Miner Eng 13:1117–1127

Sutherland I (2001) The biofilm matrix-an immobilized but dynamic microbial environment. Trends Microbiol 9:222–227

Suzuki I (2001) Microbial leaching of metals from sulfide minerals. Biotechnol Adv 19:119–132

Temple K, Colmer A (1951) The autotrophic oxidation of iron by a new bacterium, *Thiobacillus ferrooxidans*. J Bacteriol 62:605–611

Third K, Cord-Ruwisch R, Watling H (2000) The role of iron-oxidizing bacteria in stimulation or inhibition of chalcopyrite bioleaching. Hydrometallurgy 57:225–233

Third K, Cord-Ruwisch R, Watling H (2002) Control of the redox potential by oxygen limitation improves bacterial leaching of chalcopyrite. Biotechnol Bioeng 78:433–441

Tributsch H (1999) Direct versus indirect bioleaching. In: Proceedings of the International biohydrometallurgy symposium IBS'99. Elsevier, Amsterdam, pp 51–60

Tuffin I, Hector S, Deane S, Rawlings D (2006) Resistance determinants of highly arsenic-resistant strain of *Leptospirilum ferriphilum* isolated from a commercial biooxidation tank. Appl Environ Microbiol 72:2247–2253

Tuovinen O, Bhatti T, Bigham J, Hallberg K, Garcia O, Lindstrom E (1994) Oxidative dissolution of arsenopyrite by mesophilic and moderately thermophilic acidophiles. Appl Environ Microbiol 60:3268–3274

van Aswegen P, Godfrey M, Miller D, Haines A (1991) Developments and innovations in bacterial oxidation of refractory ores. Miner Metall Process:188–191

Vardanyan N (1998) The effect of external factors on pyrite oxidation by *Sulfobacillus thermosulfidooxidans* subsp. *Asporogenes*. Biotechnology 6:48–55

Vardanyan N (2003) Oxidation of pyrite and chalcopyrite by mixed cultures of Sulfobacilli and other Sulphur and Iron oxidizing bacteria. Biotechnology 6:79–83

Vardanyan N, Akopyan V (2003) *Leptospirillum*-like bacteria and evaluation of their role in pyrite oxidation. Microbiology 72:438–442

Vardanyan A, Vardanyan N (2016) Bioleaching of pyrite and chalcopyrite by new isolated thermotolerate sulfur-oxidizing bacteria Acidithiobacillus tandzuti sp.nov. Int J Sci Eng Res 7:203–207

Vardanyan N, Karavaiko G, Pivovarova T (1990) The effect of organic substances on the growth and oxidation of inorganic substrates by *Sulfobacillus thermosulfidooxidans*sub sp. *asporogenes*. Microbiology 59:411–417

Vardanyan A, Stepanyan S, Vardanyan N, Markosyan L, Sand W, Vera V, Zhang R (2015) Study and assessment of microbial communities in natural and commercial bioleaching systems. Miner Eng 81:167–172

Wang Y, Zeng W, Qiu G, Chen X, Zhou H (2014) A moderately thermophilic mixed microbial culture for bioleaching of chalcopyrite at high pulp density. Appl Environ Microbiol 80:741–750

Watling H (2006) The bioleaching of sulphide minerals with emphasis on copper sulphids—a review. Hydrometallurgy 84:81–108

Watling H, Collinson D, Shiers D, Bryan C, Watkin E (2013) Effects of pH, temperature and solids loading on microbial community structure during batch culture on a polymetallic ore. Miner Eng 48:68–76

Yu R, Tan J, Yang P, Sun J, Ouyang X, Dai Y (2008) EPS-contact-leaching mechanism of chalcopyrite concentrates by *a. ferrooxidans*. Trans Nonferrous Met Soc Chin 18:1427–1432

Yu R, Zhong D, Miao L, Wu F, Qiu G, Gu G (2011) Relationship and effect of redox potential, jarosites and extracellular polymeric substances in bioleaching chalcopyrite by acidithiobacillus ferrooxidans. Trans Nonferrous Met Soc Chin 21:1634–1640

Zhang R, Wei M, Ji H, Chen X, Qiu G, Zhou H (2009) Application of real-time PCR to monitor population dynamics of defined mixed cultures of moderate thermophiles involved in bioleaching of chalcopyrite. Appl Microbiol Biotechnol 81:1161–1168

Zhou H, Zhang R, Hu P, Zeng W, Xie Y, Wu C, Qiu G (2008) Isolation and characterization of *Ferroplasma thermophilum* sp. nov., a novel extremely acidophilic, moderately thermophilic archaeon and its role in bioleaching of chalcopyrite. J Appl Microbiol 105:591–601

Zhou H, Zeng W, Yang Z, Xie Y, Qiu G (2009) Bioleaching of chalcopyrite concentrate by a moderately thermophilic culture in a stirred tank reactor. Bioresour Technol 100:515–520

Chapter 10
Extreme Thermophilic Microorganisms as an Unique Source of Inspiration for Next Generation Biotechnological Products

Mehmet Karadayi, Burak Alaylar, Sümeyra Gürkök, Gökçe Karadayi, Dilfuza Egamberdieva, and Medine Güllüce

Abstract Thermophilic microorganisms attract the attention of the scientific world with their adaptation to living at high temperatures, far beyond the limits of other living organisms. Initially, research approaches focused on how these organisms could overcome the harmful effects of extreme temperature conditions. However, with the realization that the heat stable biomolecules obtained from these microorganisms have high industrial importance, trends have turned to biotechnology research. As a result of these efforts, a great variety of thermostable biomolecules with high industrial importance such as enzymes, antimicrobial agents, and biosurfactants were discovered and many novel technologies from biorefining to waste treatment were developed. Today, these unique extremophiles are regarded as both heat-resistant biomolecule suppliers for many existing biotechnological processes and a source of inspiration for novel applications to be developed.

In this context, the present chapter has been written to give an outlook to the characteristics, ecology, adaptation mechanisms and biotechnological importance of thermophilic microorganisms.

M. Karadayi (✉) · S. Gürkök · M. Güllüce
Department of Biology, Faculty of Science, Atatürk University, Erzurum, Turkey
e-mail: mkaradati@atauni.edu.tr

B. Alaylar
Department of Molecular Biology and Genetics, Faculty of Science and Art, Agri Ibrahim Cecen University, Ağrı, Turkey
e-mail: balaylar@agri.edu.tr

G. Karadayi
Department of Molecular Biology and Genetics, Faculty of Science, Atatürk University, Erzurum, Turkey
e-mail: gokce.karadayi@atauni.edu.tr

D. Egamberdieva
Faculty of Biology, National University of Uzbekistan, Tashkent, Uzbekistan

D. Egamberdieva et al. (eds.), *Microbial Communities and their Interactions in the Extreme Environment*, Microorganisms for Sustainability 32,
https://doi.org/10.1007/978-981-16-3731-5_10

Keywords Adaptation mechanisms · Biotechnology · Extremophiles · Thermophilic microorganisms

10.1 Introduction

Due to the inspiration they provide to biotechnological applications, the number of studies on microbial diversity of various ecosystems is increasing continuously. Especially, the ecosystem biodiversity research in extreme environments has also increased depending on the developing technology and research methods, and important information has been obtained about the distribution of extremophiles around the world.

Extremophiles, most of which are composed of prokaryotic and eukaryotic microorganisms, are basically defined as organisms that can survive in ecosystems with certain characteristics and are classified according to the nature of the extreme condition (Fig. 10.1).

Among the main groups of extremophiles, thermophiles are known as the group with the most researched and most adapted products to today's technologies. Apart from high temperature, many thermophiles are often well adapted to survive under other extreme environmental conditions, such as wide ranges of pH, salinity, redox potential, and the presence of toxic compounds and solvents. Due to their superior properties, thermophilic microorganisms are of great interest in terms of biotechnological and industrial applications. In these applications, either whole cells or the macromolecules and metabolites derived from thermophiles are used (Elleuche et al. 2014; Mehta et al. 2016; Rathinam et al. 2019). Using thermophiles in industrial and biotechnological processes offers several valuable advantages. First of all, contamination with mesophiles or pathogens is prevented at elevated temperatures. Due to high metabolic activities, product yields of thermophiles are also high. Since thermophiles are stable at extreme conditions, their metabolites and macromolecules such as enzymes and other proteins will also be stable. The cooling steps, required when mesophilic organisms are used, can be bypassed in the processes using the thermophilic organisms. The diffusion rates, ionization, and solubility of chemicals will be better at high temperatures and the rate of reactions will be greater (Urbieta et al. 2015).

A large number of value-added molecules such as thermostable enzymes, biosurfactants, pigments, antimicrobial/antioxidant compounds have been produced by thermophilic microorganisms. In addition, they have been extensively utilized in biomining, bioremediation, dye removal, and production of bioenergy. This chapter aims to compile information on the most common uses of thermophiles.

Fig. 10.1 Main groups of extremophiles that live under certain environmental conditions such as extremely high or low temperatures (thermophiles and psychrophiles), acidic or alkaline pH values (acidophiles and alkaliphiles), high pressures (piezophiles), high ionic strength (halophiles), absence of O_2 (anaerobes) and high radiation (radiation-resistant extremophiles)

10.2 Thermophilic Environments and Their Microbial Biodiversity

Thermophilic microorganisms distribute widely from 45 °C, which is very close to the mesophilic life boundary, to 125 °C, which is the upper limit of life. They are classified according to their optimal growth temperatures as moderate, extreme, and hyperthermophiles (Fig. 10.2).

The dominant groups of microbial biodiversity in extreme habitats with high temperatures exceeding 70 °C are especially consist of bacteria and archaea. Moreover, the existence of many eukaryotic fungal species at lower temperatures is remarkable.

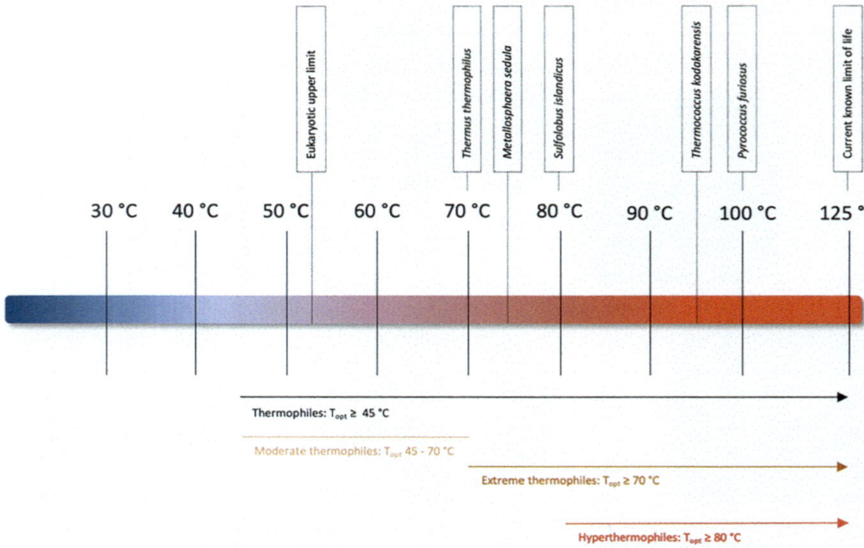

Fig. 10.2 Main groups of thermophiles and their optimum growth temperatures

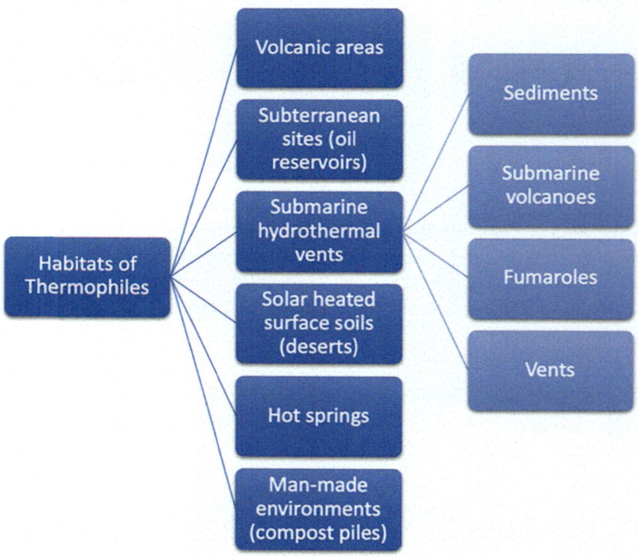

Fig. 10.3 Habitats of thermophilic organisms

Due to the large scale of their habitat temperature, thermophilic organisms are found in many extreme environments on the earth. The natural habitats of these organisms range from volcanic areas to submarine hydrothermal systems (Fig. 10.3).

10.3 Thermophilic Microorganisms as a Unique Source of Inspiration for Next Generation Biotechnological Products

10.3.1 Enzymes

Enzymes used in almost all areas of the industry are generally derived from microorganisms due to their advantages over enzymes derived from plants and animals. These advantages include rapid growth in limited space, ease of genetic manipulation, high catalytic activity and stability, longer shelf life, and high production yield on cheap substrates without forming undesirable by-products (Liu and Kokare 2017; Raveendran et al. 2018). Thermophilic organisms, as rich sources of microbial enzymes, are particularly important for industries that operate at elevated temperatures and harsh conditions (Turner et al. 2007; Elleuche et al. 2014; Rigoldi et al. 2018; Atalah et al. 2019). Optimum growth of thermophiles at elevated temperatures are granted mostly by their thermostable enzymes also called thermozymes. Thermozymes have superior properties over their mesophilic counterparts and have a wide range of applications in industries that require high temperatures, such as the food, feed, brewing, detergent, pulp and paper processing industries.

Thermophilic proteins are very similar to psychrophilic and mesophilic proteins, but there are some differences in their sequences, structures, and thermodynamics, affecting their tolerance to harsh conditions (Kumar and Nussinov 2001). Thermophilic proteins have higher α helix and Β sheet content compared to mesophilic proteins. They show very slow folding rate, which is important in protecting the natural structure against different denaturing conditions. In addition, thermophiles produce special proteins called chaperonins, which restore the proteins after denaturation by re-folding them into their natural active forms (Everly and Alberto 2000).

The conformational stability of proteins is the result of the balance between two opposing factors; flexibility, and stiffness. In fact, thermozymes generally show low catalytic activity at room temperature, and their activity assays should be carried out at temperatures close to their optimum growth temperatures. Thermozymes are more rigid than mesophilic at room temperature, which protects them from denaturation and allows them to maintain catalytically active structure. At high temperatures, they become flexible enough to be fully active and are still sufficiently rigid to protect themselves from denaturation.

Thermostable enzymes also have the mentioned advantages regarding thermophiles in industrial and biotechnological applications. Thermozymes are highly resistant to denaturation in harsh conditions as well as high temperature and have long-term storage. The use of thermostable enzymes in industrial processes reduces the possibility of contamination risk by mesophilic microorganisms. At elevated temperatures, the viscosity of the reaction medium decreases and the bioavailability and solubility of reactants increase, which eventually increases the reaction rate and product formation (Krahe et al. 1996).

Table 10.1 Thermostable enzymes and their applications

Thermostable enzyme	Applications	Reference
α-Amylase	Starch hydrolysis, food (juice, beer, glucose, maltose, and high fructose syrup production) baking, brewing, paper and pulp, textile, laundry and detergents, pharmaceuticals	de Carvalho et al. (2008), Sudan et al. (2018) and Wu et al. (2018)
Chitinase	Food, cosmetics, pharmaceuticals, agrochemicals	Sun et al. (2019)
Cellulase	Cellulose hydrolysis, plant waste treatment, textile, detergent, paper and pulp, color clarification	Peng et al. (2015)
DNA polymerase	DNA amplification in genetic engineering/ PCR	Mullis et al. (1986) and Saiki et al. (1988)
Lipase	Dairy, oleo chemical, laundry and detergent, textile, paper and pulp, pharmaceuticals, cosmetics, biodiesel and leather industry	Choudhury and Bhunia (2015)
Nitrilase	Production of polyacrylonitrile, α- and β-amino acids, pharmaceutical and pesticide, bioremediation	Chen et al. (2015), Han et al. (2015) and Martínková et al. (2017)
Protease	Baking, food (cheese, milk, flavor, beer and brewing), laundry and detergents, leather industry, leather, pharmaceutical, textile, paper and pulp and fuel	Abdul et al. (2019) and Gurkok (2019)
Pullulanase	Baking, starch process, production of glucose syrups and fuel	Curvers et al. (2014)
Xylanase	Baking, pulp and paper, animal feed, starch and fuel viscosity reduction	Paes and O'Donohue (2006) and Basit et al. (2018)

The global enzyme market size is expected to grow at a compound annual growth rate of 7.1% and reach USD 10.047 billion by 2025. The main reason for the growth of this market is the multiple applications of enzymes in diverse industries. Thermozymes have a large share in this market, especially thermostable protease, amylase and lipase are the leading enzymes in the industrial enzyme market with their widespread uses in industries such as food, beverage, biofuels and detergents. Other thermozymes including cellulases, chitinases, DNA polymerases, glucose isomerases, alcohol dehydrogenases, pectinase, phytase, xylanase, and esterases, are used in various industries outlined in Table 10.1 (Kristjansson 1989; Rezanka et al. 2011; Bergquist et al. 2014; Elleuche et al. 2014; Rehman et al. 2017; Rigoldi et al. 2018).

10.3.1.1 Proteases

Proteases also known as peptidases are enzymes that hydrolyze peptide bonds in proteins. Proteases are among the most valuable commercial enzymes and represent

60% of the total enzyme market. They are extensively used in the food, detergent, pharmaceutical, leather, and textile industries (Jain et al. 2012).

Proteases are classified as acid, neutral, and alkaline proteases based on their optimal pH. The alkaline proteases from thermophiles attract more attention biotechnologically and commercially. Their applications are dominated in laundry detergents, brewing, food, leather processing, and pharmaceutical industries. Since 1913 when proteases were first introduced as a detergent additive in laundry by the German chemist Otto Röhm (1913), they are the most widely used detergent enzymes. As alkaline protease enzymes are more suitable for basic washing conditions, they are often used in detergent formulation for degradation of proteinaceous stains such as blood, body secretions, and foods (Jain et al. 2012; Gohel and Singh 2018). Detergent enzymes should be stable and active in the presence of typical detergent ingredients such as surfactants, bleaching agents, builders, and other enzymes (Gurkok 2019).

Thermophilic *Bacillus* species are the major source of alkaline and thermostable proteases and have been isolated and studied for years to be used in detergent formulations (Banik and Prakash 2004; Mukherjee et al. 2007; Hmidet et al. 2009; Rao et al. 2009; Gohel and Singh 2018). During washing, thermostable proteases are needed to remove stubborn stains on cotton fabrics, while cold-adapted protease enzymes that are active at low temperatures are preferred for sensitive fabrics (Gurkok 2019).

Due to the elastolytic and keratinolytic activities, use of proteases is also profound in the leather processing industry that conventionally uses chemicals such as sodium sulfide and lime (Brandelli et al. 2010; Verma et al. 2011). The applications of protease in this industry are related with the bating, soaking, and dehairing phases of preparing skin and hides. In addition, elimination of unwanted pigments by enzymatic methods helps to produce clean leathers. A thermo-halostable protease from *Geobacillus* sp. isolated from undersea fumaroles was tested for its leather dehairing capability. The enzyme resulted in a slight dehairing and produced soft leather at high temperatures and pH, and reduced environmental pollution by eliminating the excessive use of chemicals in the process (Iqbalsyah et al. 2019).

Proteases also have a wide range of applications in the pharmaceutical field. Microbial thermostable and alkaline proteases are extensively used in the production of high nutritional protein hydrolysates and bioactive peptide compounds used in drug formulations for their stress responsive, antifungal, antimicrobial, antioxidant, anti-hypertensive, anticancer, anti-inflammatory, and immune-enhancer activities (Abdul et al. 2019). Protein hydrolysates are also important for infant food formulation and fortification of soft drinks and juices.

10.3.1.2 Amylases

Starch is a complex polysaccharide abundant in plants. It consists of varying amounts of amylose and amylopectin. A mixture of enzymes is required for complete hydrolysis of starch. These enzymes include endoamylases (α-amylases), exoamylases (β-amylase, glucoamylase), and debranching enzymes (isoamylase, pullulanase). Amylases are industrially important enzymes and accounts for approximately 25–30% of the global enzyme market. They have crucial role in starch hydrolysis processes applied in food, paper, brewing, detergent, and textile industries. Many amylolytic enzymes have been isolated and characterized and especially thermostable amylases have attracted attention due to their superior properties (Ardhi et al. 2020).

In starch processing, gelatinization, liquefaction, and saccharification steps are performed at high temperature. During gelatinization, starch granules are dissolved by heating, and if the enzymes used in the process are thermostable, the subsequent hydrolysis step can start without interruption for cooling. In addition to thermostability, acid stability is also a desired feature in liquefaction and saccharification processes. Thermostable and acid-stable amylases are more suitable because they bypass cooling steps, pH adjustment and limit the formation of by-products (Sudan et al. 2018; Wu et al. 2018).

In detergents, amylases have potential for the removal of starch-based stains. They clean by breaking down the α-glycosidic bonds in starch and catalyze the hydrolysis of starch into sugars (de Carvalho et al. 2008). The most important feature of the enzymes to be used in detergent is to be resistant to high temperatures and high pH.

Bacillus spp. play an important role as a source of thermostable amylase enzymes (Asghar et al. 2007). Hmidet et al. (2009) studied thermostable α-amylase produced by Bacillus licheniformis for its potential application as detergent additive in combination with alkali protease. Msarah et al. (2020) reported the production of alpha amylase by thermophilic *B. licheniformis* and *B. subtilis* isolated from hot springs. They applied the enzyme in food waste biodegradation and obtained good reduction in solid content at 65 °C. In another study, *B. licheniformis* was used as the source of extremely thermostable and acid-stable amylase showing maximum activity at 100 °C and pH 5.0 to be used in starch processing. The enzyme was also stable in the presence of inhibitors and surfactants (Wu et al. 2018).

10.3.1.3 Lipases

Lipase enzymes catalyze the hydrolysis of the ester bond of long chain acylglycerols. These enzymes show interesting and unique features that make them excellent biocatalysts, including stereospecificity, regioselectivity, a wide substrate range, and the possibility of catalyzing heterogeneous reactions in an aqueous-apolar interface.

One of the industrial application areas of lipases is their use in detergent formulations. Many thermostable lipases have been isolated and characterized from microorganisms (Hemlata et al. 2016). In cleaning and laundry applications, lipases should be water soluble and stable at alkaline pH, they need to have high tolerance to surfactants and other enzymes such as proteases and amylases, and have low substrate specificity.

10.3.2 Antimicrobial Agents

Antimicrobials are compounds that kill microorganisms or stop their growth. Most of the microorganisms compete with other organisms for resources and/or space in their habitats and develop their own strategies for survival. Many bacterial and fungal species synthesize various antimicrobial substances to eliminate other and mostly closely related species. Antimicrobial agents can be classified based on the target microorganisms. Antibiotics against bacteria, antivirals against viruses, and antifungals against fungi are used. These antimicrobial agents have several clinical and industrial values. The resistance of various pathogenic bacteria to antimicrobial agents appears to be an important problem in clinical practices and necessitates the discovery of new antibiotics. Thermophiles are a promising source in researching new and thermostable antimicrobials for control of microorganisms causing infections. In addition to their uses against pathogenic bacteria in health industry, thermostable antimicrobial substances are especially used against other thermophiles and mesophiles in industries operating at elevated temperatures, such as canned and dairy food industries.

Bacteriocins are peptides having antibacterial activity against other closely related bacteria. Bacteriolysins, known to lyse cell walls of the target bacteria, and non-lytic bacteriocins, some of which are believed to inhibit DNA and protein synthesis, are suggested as an alternative to antibiotics (Cotter et al. 2013; Vaičikauskaitė et al. 2019). A number of *Geobacillus* spp. have been reported to produce bacteriocin against other thermophilic and mesophilic microorganisms. A bacteriocin called geobacillin 26 from a thermophilic bacterium *G. stearothermophilus* 15 isolated from oil well in Lithuania has been reported by Vaičikauskaitė et al. (2019). This bacteriocin has showed antimicrobial activity against closely related thermophilic bacteria species and can be used in the food industry where high temperatures are needed. In another study, Selim et al. (2010) have reported the isolation of thermophilic *Geobacillus* spp. from hot springs in Sinai, Egypt. Extracts of the isolate have showed antimicrobial activity against *B. subtilis* and *Candida albicans*. Alkhalili et al. (2016) reported the thermophilic isolate, *Geobacillus* sp. ZGt-1, from Zara hot spring in Jordan. This strain showed antimicrobial activity against *G. stearothermophilus* and the mesophilic *B. subtilis* and *Salmonella typhimurium*. An antiviral compound produced by *G. stearothermophilus* CECT 43 has been reported by Rivero et al. (2012). Synthesis of 6-chloropurine-2′-deoxyriboside and 6-chloropurine riboside have been achieved by thermophilic bacterium with a conversion of 90% and 68%, respectively.

Many other thermophilic bacteria, such as *Bacillus* species, produce a large number of antimicrobial agents and bacteriocin against bacteria, fungi, protozoa, and viruses (Esikova et al. 2002; Ravot et al. 2006; Hayashi et al. 2013). The thermophilic *Yersinia sp.* isolated from hot spring Jordan Valley have shown antimicrobial activity against *Staphylococcus aureus, Escherichia coli, Pseudomonas aeruginosa, Klebsiella pneumonia, Enterobacter sp.* and *C. albicans* (Khalil et al. 2006).

Skouri-Gargouri and Gargouri (2008) reported a novel thermostable antifungal peptide produced by *Aspergillus clavatus*. This antifungal agent has shown a strong inhibitory activity against mycelial growth of several human and plant pathogenic fungi such as *Fusarium oxysporum, F. solani, A. niger, Botrytis cinerea,* and *Alternaria solani.*

10.3.3 Biorefining

A biorefinery is defined as a refinery converting biomass or renewable raw materials into energy and other beneficial products such as chemicals, biofuels (Turner et al. 2007). Raw materials such as corn (Li et al. 2020), wheat (Zucaro et al. 2018), sugarcane (Edye et al. 2006), cotton (Chen et al. 2016), cellulose (Golecha and Gan 2016) and lignocellulose (Hou et al. 2019; Garlapati et al. 2020) are used as substrates in bioprocess to produce valuable final products.

The biorefinery provides an ecologically sustainable way for obtaining high value-added compounds using metabolic activities and enzymes of microorganisms (Lara et al. 2020). Most of the biorefinery processes occur at high temperature, which allows higher substrate solubility, mixing and mass transfer and lower contamination risk. Additionally, high temperature offers better enzyme penetration and cell-wall dispersion of the raw materials (Turner et al. 2007). Therefore, thermophiles and thermostable enzymes are the viable choices for biorefinery bioprocesses. Compared to mesophiles, thermophiles offer the advantage of resistance to inhibitors that occur during fermentation as well as to high temperature (Mohapatra et al. 2020).

For lignocellulosic biomass, high temperatures (50 °C–70 °C), which are the optimum temperatures of cellulases, are required for saccharification (Jørgensen et al. 2007). Thermostable enzymes have an obvious advantage as catalysts in these processes, as high temperatures often promote better enzyme penetration and cell-wall disorganization of the raw materials (Paes and O'Donohue 2006). The first step in crop biorefining is fractionation, which is carried out by means of physical, chemical, and biological ways. Fractionation usually takes place often at elevated temperatures, which requires thermostable enzymes.

10.3.4 Biofuel

Biofuel is a fuel produced from biomass through modern processes. Unlike fossil fuels such as oil, coal or nuclear fuels produced by very slow geological processes, they are renewable energy sources. Biofuels are considered as a cost-effective and environmentally friendly alternative to petroleum and other fossil fuels, especially in the context of concern about rising oil prices and the contribution of fossil fuels to global warming (Jiang et al. 2017).

Biofuels such as wood can be used directly as a burning raw material to generate heat. Heat can also be used to run generators in a power plant to generate electricity. Some existing power plants burn grass, wood, or other biomass. Liquid biofuels have attracted particular attention in recent years. The most preferred and produced liquid biofuel is ethanol obtained by fermenting starch or sugar. Unlike the "first generation" ethanol biofuel obtained from food crops such as corn and sugar cane, "second generation" ethanol is produced from low-value biomass with high cellulose content, including agricultural and industrial residues, wood chips and other waste materials. Biodiesel, derived from oily plants such as soybean or palm oil, is the second most preferred liquid biofuel and is accepted for use in diesel engines. Methane gas and biogas are other biofuels that can be derived from the decomposition of biomass in the absence of oxygen. Thermophiles play an important role in the biofuel production by introducing novel metabolic pathways and stable enzymes that can act as biocatalysts under harsh industrial conditions. Higher catalytic activity, temperature, and pH stability, and tolerance to the inhibition of the final product are key features needed for efficient biofuel production, and thermophiles offer these conditions by producing stable enzymes (Zhu et al. 2020). Thermozymes are most commonly used in the production of biofuels from starch and lignocellulosic material. The conversion of starch to single fermentable glucose molecules requires high temperature and low pH conditions, with the catalytic action of amylases, glucoamylases, and/or pullulanases.

Lignocellulose is the most abundant plant biopolymer available on earth and is mainly composed of carbohydrate polymers, cellulose and hemicellulose, and the non-carbohydrate phenolic polymer, lignin. It has long been recognized as an attractive resource for producing renewable fuels such as hydrogen gas and ethanol. Being environmentally friendly and renewable, hydrogen is considered as an attractive alternative to future fuel. Cao et al. (2014) reported the hydrogen production directly from conversion of lignocellulosic materials by thermophilic cellulolytic isolates in the genus *Thermoanaerobacterium* growing well at a temperature of 60 °C. Among the tested isolates, *T. thermosaccharolyticum* M18 exhibited greater cellulolytic activity and hydrogen production in a single step process that simultaneously performed cellulase production, cellulose hydrolysis, and fermentation in one step.

10.3.5 Biomining

X–ray films or photographic films contain 1.5–2% silver in its gelatin layers. Thermostable proteases from alkaliphilic *Bacillus* sp. B21–2 (Masui et al. 2004) and *B. cereus* strain S8. (Lakshmi and Hemalatha 2016) have been reported for the successful recovery of silver from used lith film for printing. In this environmentally friendly silver recycling system, gelatin layers on lith film were hydrolyzed enzymatically by thermostable alkaline proteases successfully.

10.3.6 Biosurfactant Production

Biosurfactants are high value-added surface-active biomolecules produced mainly by microorganisms. Two major classes of biosurfactants are glycolipids (rhamnolipids, sophorolipids, trehalose lipids and mannosylerythritol lipids) and lipopeptides (surfactin and lichenysin). They have been receiving great attention due to their structural diversity, low toxicity and biodegradability compared to synthetic surfactants. Biosurfactants are versatile compounds used as emulsifiers, demulsifiers, wetting agents, foaming agents, and spreading agents by reducing the surface and interfacial tension. They have widespread applications in biotechnology and different industries such as food, cosmetic, detergent, pharmaceuticals, agriculture, chemistry, petroleum, mining, metallurgy, oil recovery, and bioremediation (Banat et al. 2010; Camargo et al. 2018).

Production of biosurfactant from thermophiles has also been a topic of interest in recent years. Since, industrial processes involve harsh conditions such as extreme temperature, pH, and salinity, the demand of consumers for biosurfactants showing high tolerance to inhospitable environmental factors has increased (Vijayakumar and Saravanan 2015; Schultz and Rosado 2020). Many biosurfactants from thermophilic microorganisms have been isolated, characterized, and have proven to be compatible with industrial processes.

Mehetre et al. (2019) reported the use of biosurfactant (surfactin) producing thermophilic and thermotolerant bacteria for the biodegradation of polycyclic aromatic hydrocarbons (PAHs) by pure and mixed cultures. McInerney et al. (1990) reported the biosurfactant (lichenysin) from *B. licheniformis* which was resistant to temperature up to 50 °C, in the pH range of 4.5 and 9.0, and NaCl concentrations up to 50 g L^{-1} (McInerney et al. 1990). Singh and Cameotra (2004) reported a biosurfactant from *Arthrobacter protophormiae,* which was highly stable in the wide temperature range between 30 and 100 °C and pH range between 2 and 12 (Singh and Cameotra 2004) and has potential uses in biomedical sciences.

Biosurfactant (rhamnolipid) produced from *P. aeruginosa* strain P-CG3 enhanced the solubilization and desorption of PAHs. The bioavailability and biodegradation of phenanthrene (PHE) increased at 55 °C, implying that rhamnolipid might have the potential to be further applied microbial enhanced oil recovery (Cheng et al. 2004; Wong et al. 2005).

Coronel-León et al. (2015) reported a biosurfactant producing thermophilic *B. licheniformis* strain AL1.1 isolated from Kroner Lake, located at a geothermal area of Deception Island (Maritime Antarctica). As well as its emulsifying properties, the stability of this biosurfactant at high temperatures up to 120 °C, in the pH range of 6.0–11.0 and NaCl concentrations up to 20%, suggest that it has the potential to be applied in the cosmetic industry.

10.3.7 Waste Treatment

Thermophilic wastewater treatment (TWT) is an advanced process that has attracted attention recently. TWT process is used for purification of medium- and high-strength industrial effluents, such as those from food, textile, paper and pulp industries, discharged at elevated temperatures (45 °C–70 °C). Treating these effluents under traditional mesophilic conditions needs pre-cooling. TWT offer several advantages over traditional processes. TWT operates faster and reduces COD to a greater extent than the traditional activated sludge processes. Other advantages include the inactivation of pathogens and mesophiles, low sludge yield, and reduced detention time and operation costs due to operation stability.

In TWT, various substances in wastewater are degraded mainly by the thermophilic bacteria under aerobic or anaerobic conditions. Anaerobic degradation of food wastes is a promising treatment technology as well as production of methane biogas for energy recovery. Complex organic compounds in the wastewater are fermented by microorganisms in thermophilic anaerobic wastewater treatment. The most important output of anaerobic wastewater treatment is biogas, which can be used as fuel. Thermophilic aerobic processes are particularly advantageous for the treatment of high-strength wastewaters that can fully benefit from the rapid biodegradation rates and low sludge yields (LaPara and Alleman 1999; Cheng et al. 2006).

In wastewater treatment systems, nitrification is the main step for biological nitrogen removal. Nitrification is generally not possible at temperatures higher than 40 °C. The warm industrial wastewater above this temperature requires cooling before biological treatment, which increases the energy and operating costs of the facilities for cooling. Lopez-Vazquez et al. (2014) reported a solution for this problem using thermophilic *Nitrosomomas* spp. and moderately thermophilic *Nitrospira sublineage* II for nitritation, nitratation, and denitrification at temperatures as high as 50 °C in an activated sludge wastewater treatment plant treating wastewater from an oil refinery.

Management of domestic and industrial food waste has also become a major problem worldwide. Therefore, reduction and recycling of food waste by microorganisms is extremely important in terms of waste treatment costs and environmental concerns. Food waste is typically composed of organic matter in the form of carbohydrates (cellulose, hemicelluloses, and starch), proteins, lipids, lignin, and organic acids. It also contains low amounts of inorganic substances. The organic part can be fermented and reduced in quantity by microorganisms (An et al. 2018). Thermophilic bacteria are extensively used for food waste treatment. Thermophilic isolates of *Bacillus* spp. have been reported to show effective decomposing activity during food waste reduction occurring at temperature over 45 °C (Yi et al. 2006). An et al. (2018) reported the use of equipment developed for food waste treatment. At temperatures above 50 °C, the machine's operating temperature, thermophilic *Bacillus* species were able to grow and reduce the solid content of food waste. Food waste can also be used by converting it into energy by microorganisms (Kiran et al. 2014). Shin et al. (2004) reported the use of food waste for hydrogen production.

Mesophilic and thermophilic acidogenic bacteria have been compared in terms of hydrogen production yields. Hydrogen production efficiency of thermophilic acidogenic culture was much more than mesophilic culture.

References

Abdul R, Sadia S, Arfan A, Qurban A, Muhammad S, Arif M, Muhammad A (2019) Microbial proteases applications. Front Bioeng Biotechnol 7:110. https://doi.org/10.3389/fbioe.2019.00110

Alkhalili RN, Bernfur K, Dishisha T et al (2016) Antimicrobial protein candidates from the thermophilic *Geobacillus* sp. strain ZGt-1: production, proteomics, and bioinformatics analysis. Int J Mol Sci 17(8):1363. https://doi.org/10.3390/ijms17081363

An B, Park M-K, Oh J-H (2018) Food waste treatment using Bacillus species isolated from food wastes and production of air-dried Bacillus cell starters. Environ Eng Res 23(3):258–264. https://doi.org/10.4491/eer.2017.116

Ardhi A, Sidauruk AN, Suraya N, Pratiwi NW, Pato U, Saryono S (2020) Molecular identification of amylase-producing thermophilic bacteria isolated from Bukit Gadang hot spring, West Sumatra, Indonesia. Biodiversitas 21:994–1000

Asghar M, Asad MJ, Rahman SU, Legge RL (2007) A thermostable alpha amylase from a moderately thermophilic *Bacillus subtilis* strain for starch processing. J Food Eng 79:950–955

Atalah J, Cáceres-Moreno P, Espina G, Blamey JM (2019) Thermophiles and the applications of their enzymes as new biocatalysts. Bioresour Technol 280:478–488. https://doi.org/10.1016/j.biortech.2019.02.008

Banat IM, Franzetti A, Gandolfi I, Bestetti G, Martinotti MG, Fracchia L, Marchant R (2010) Microbial biosurfactants production, applications and future potential. Appl Microbiol Biotechnol 87:427–444

Banik MR, Prakash M (2004) Laundry detergent compatibility of the alkaline protease from *Bacillus cereus*. Microbiol Res 159:135–140

Basit A, Liu J, Rahim K, Jiang W, Lou H (2018) Thermophilic xylanases: from bench to bottle. Crit Rev Biotechnol 38:989–1002

Bergquist LP, Morgan HW, Saul D (2014) Selected enzymes from extreme thermophiles with applications in biotechnology. Curr Biotechnol 3:45–59

Brandelli A, Daroit DJ, Riffel A (2010) Biochemical features of microbial keratinases and their production and applications. Appl Microbiol Biotechnol 85:1735–1750. https://doi.org/10.1007/s00253-009-2398-5

Camargo FP, Prado PF, Tonello PS, Santos ACA, Duarte ICS (2018) Bioleaching of toxic metals from sewage sludge by co-inoculation of Acidithiobacillus and the biosurfactant-producing yeast Meyerozyma guilliermondii. J Environ Manage 211:28–35

Cao GL, Zhao L, Wang AJ, Wang ZY, Ren NQ (2014) Single-step bioconversion of lignocellulose to hydrogen using novel moderately thermophilic bacteria. Biotechnol Biofuels 7:82. https://doi.org/10.1186/1754-6834-7-82

Chen Z, Chen H, Ni Z, Tian R, Zhang T, Jia J, Yang S (2015) Expression and characterization of a novel nitrilase from hyperthermophilic bacterium *Thermotoga maritima* MSB8. J Microbiol Biotechnol 25:1660–1669

Chen J, Liang J, Wu S (2016) Lignin-rich biomass of cotton by-products for biorefineries via pyrolysis. Bioresour Technol 218:402–409. https://doi.org/10.1016/j.biortech.2016.06.122

Cheng KY, Zhao ZY, Wong JWC (2004) Solubilization and desorption of pahs in soilaqueous system by biosurfactants produced from *Pseudomonas aeruginosa* P-CG3 under thermophilic condition. Environ Technol 25(10):1159–1165. https://doi.org/10.1080/09593332508618382

Cheng C, Phipps D, Alkhaddar RM (2006) Thermophilic aerobic wastewater treatment of waste metalworking fluids. Water Environ J 20(4):227–232. https://doi.org/10.1111/j.1747-6593.2005.00010.x

Choudhury P, Bhunia B (2015) Industrial application of lipase: a review. Biopharm J 1(2):41–47

Coronel-León J, de Grau G, Grau-Campistany A, Farfan M, Rabanal F, Manresa A, Marqués AM (2015) Biosurfactant production by AL 1.1, a *Bacillus licheniformis* strain isolated from Antarctica: production, chemical characterization and properties. Ann Microbiol 65:2065–2078. https://doi.org/10.1007/s13213-015-1045-x

Cotter PD, Ross RP, Hill C (2013) Bacteriocins—a viable alternative to antibiotics? Nat Rev Microbiol 11:95–105. https://doi.org/10.1038/nrmicro2937

Curvers S, Koln DE, Svetlichnyi V (2014) Single step bioconversion of lignocellulosic biomass to biofuels using extreme thermophilic bacteria. US Patent 0363869 A1

De Carvalho RV, Côrrea TL, da Silva JC, de Oliveira Mansur LR, Martins ML (2008) Properties of an amylase from thermophilic Bacillus sp. Braz J Microbiol 39(1):102–107. https://doi.org/10.1590/S1517-83822008000100023

Edye LA, Doherty WOS, Blinco JA, Bullock GE (2006) The sugarcane biorefinery: energy crops and processes for the production of liquid fuels and renewable commodity chemicals. Int Sugar J 108:19–27

Elleuche S, Schröder C, Sahm K, Antranikian G (2014) Extremozymes-biocatalysts with unique properties from extremophilic microorganisms. Curr Opin Biotechnol 29:116–123

Esikova TZ, Temirov YV, Sokolov SL, Alakhov YB (2002) Secondary antimicrobial metabolites produced by thermophilic *Bacillus* spp. strains VK2 and VK21. Appl Biochem Microbiol 38:226–231

Everly C, Alberto J (2000) Stressors, stress and survival: overview. Front Biosci 5:780–786

Garlapati VK, Chandel AK, Kumar SPJ, Sharma S, Sevda S, Ingle AP, Pant D (2020) Circular economy aspects of lignin: towards a lignocellulose biorefinery. Renew Sustain Energy Rev 130:109977. https://doi.org/10.1016/j.rser.2020.109977

Gohel SD, Singh SP (2018) Thermodynamics of a Ca^{2+} dependent, highly thermostable and detergent compatible purified alkaline serine protease from *Nocardiopsis xinjiangensis* strain OM-6. Int J Biol Macromol 113:565–574. https://doi.org/10.1016/j.ijbiomac.2018.02.157

Golecha R, Gan J (2016) Cellulosic biorefinery portfolio and diversification: strategies to mitigate cellulosic biorefinery risks in US Corn Belt. Energ Strat Rev 13–14:147–153. https://doi.org/10.1016/j.esr.2016.09.003

Gurkok S (2019) Microbial enzymes in detergents: a review. Int J Sci Eng Res 10(9):75–81

Han C, Yao P, Yuan J, Duan Y, Feng J, Wang M, Wu Q, Zhu D (2015) Nitrilasecatalyzed hydrolysis of 3-aminopropionitrile at high concentration with a tandem reaction strategy for shifting the reaction to β-alanine formation. J Mol Catal B: Enzym 115:113–118

Hayashi MA, Bizerra FC, Da Silva PI Jr (2013) Antimicrobial compounds from natural sources. Front Microbiol 4:195. https://doi.org/10.3389/fmicb.2013.00195

Hemlata B, Uzma Z, Tukaram K (2016) Substrate kinetics of thiol activated hyperthermostable alkaline lipase of *Bacillus sonorensis* 4R and its application in bio-detergent formulation. Biocatal Agric Biotechnol 8:104–111. https://doi.org/10.1016/j.bcab.2016.08.008

Hmidet N, Ali N, Haddar A, Kanoun S, Alya S-K, Nasri M (2009) Alkaline proteases and thermosatble α-amylase co-produced by *Bacillus licheniformis* NH1: characterization and potential application as a detergent additive. Biochem Eng J 47:71–79

Hou W, Kan J, Bao J (2019) Rheology evolution of high solids content and highly viscous lignocellulose system in biorefinery fermentations for production of biofuels and biochemicals. Fuel 253:1565–1569. https://doi.org/10.1016/j.fuel.2019.05.136

Iqbalsyah TM, Malahayati, Atikah, Febriani (2019) Purification and partial characterization of a thermo-halostable protease produced by *Geobacillus* sp. strain PLS A isolated from undersea fumaroles. J Taibah Univ Sci 13(1):850–857. https://doi.org/10.1080/16583655.2019.1650489

Jain D, Pancha I, Mishra SK, Shrivastav A, Mishra S (2012) Purification and characterization of haloalkaline thermoactive, solvent stable and SDS-induced protease from Bacillus sp.: a potential additive for laundry detergents. Bioresour Technol 115:228–236. https://doi.org/10.1016/j.biortech.2011.10.081

Jiang Y, Xin F, Lu J et al (2017) State of the art review of biofuels production from lignocellulose by thermophilic bacteria. Bioresour Technol 245:1498–1506. https://doi.org/10.1016/j.biortech.2017.05.142

Jørgensen HJ, Kristensen B et al (2007) Enzymatic conversion of lignocellulose into fermentable sugars: challenges and opportunities biofuels. Bioprod Biorefin 1:119–134

Khalil AB, Zarqa MA, Al-Qaryouti M (2006) Production of antimicrobial agents from thermophilic *Yersinia* sp.1 and *Aeromonas hydrophila* isolated from hot spring in Jordan Valley. Biotechnology 5:252–256

Kiran EU, Trzcinski AP, Ng WJ, Liu Y (2014) Bioconversion of food waste to energy: a review. Fuel 134:389–399

Krahe M, Antranikian G, Markel H (1996) Fermentation of extremophilic microorganisms. FEMS Microbiol Rev 18:271–285

Kristjansson JK (1989) Thermophilic organisms as sources of thermostable enzymes. Trends Biotechnol 7:349–353

Kumar S, Nussinov R (2001) How do thermophilic proteins deal with heat? A review. Cell Mol Life Sci 58:1216–1233

Lakshmi BKM, Hemalatha KPJ (2016) Eco friendly recovery of silver from used X-ray films by alkaline protease of *Bacillus cereus* strain S8. Front Environ Microbiol 2(6):45–48. https://doi.org/10.11648/j.fem.20160206.14

LaPara TM, Alleman JE (1999) Thermophilic aerobic biological wastewater treatment. Water Res 33(4):895–908. https://doi.org/10.1016/S0043-1354(98)00282-6

Lara A, Rodríguez-Jasso RM, Loredo-Treviño A et al (2020) Chapter 17. Enzymes in the third generation biorefinery for macroalgae biomass. In: Singh SP, Pandey A, Singhania RR, Larroche C, Li Z (eds) Biomass, biofuels, biochemicals. Elsevier, Amsterdam, pp 363–396. https://doi.org/10.1016/B978-0-12-819820-9.00017-X

Li Y-J, Cao X-F, Sun S-N et al (2020) An integrated biorefinery process to comprehensively utilize corn stalk in a MIBK/water/$Al(NO_3)_3 \cdot 9H_2O$ biphasic system: chemical and morphological changes. Ind Crop Prod 147:112173. https://doi.org/10.1016/j.indcrop.2020.112173

Liu X, Kokare C (2017) Chapter 11. Microbial enzymes of use in industry. In: Brahmachari G (ed) Biotechnology of microbial enzymes. Academic Press, San Diego, pp 267–298. https://doi.org/10.1016/B978-0-12-803725-6.00011-X

Lopez-Vazquez CM, Kubare M, Saroj DP et al (2014) Thermophilic biological nitrogen removal in industrial wastewater treatment. Appl Microbiol Biotechnol 98(2):945–956. https://doi.org/10.1007/s00253-013-4950-6

Martínková L, Rucká L, Nešvera J, Pátek M (2017) Recent advances and challenges in the heterologous production of microbial nitrilases for biocatalytic applications. World J Microbiol Biotechnol 33:8

Masui A, Yasuda M, Fujiwara N, Ishikawa H (2004) Enzymatic hydrolysis of gelatin layers on used lith film using thermostable alkaline protease for recovery of silver and PET film. Biotechnol Prog 20:1267–1269. https://doi.org/10.1021/bp030058s

McInerney MJ, Javaheri M, Nagle DP Jr (1990) Properties of the biosurfactant produced by Bacillus licheniformis strain JF-2. J Ind Microbiol Biotechnol 5:95–101

Mehetre GT, Dastager SG, Dharne MS (2019) Biodegradation of mixed polycyclic aromatic hydrocarbons by pure and mixed cultures of biosurfactant producing thermophilic and thermo-tolerant bacteria. Sci Total Environ 679:52–60. https://doi.org/10.1016/j.scitotenv.2019.04.376

Mehta R, Singhal P, Singh H, Damle D, Sharma AK (2016) Insight into thermophiles and their wide-spectrum applications. 3 Biotech 6(1):81

Mohapatra S, Jena S, Jena PK et al (2020) Partial consolidated bioprocessing of pretreated *Pennisetum* sp. by anaerobic thermophiles for enhanced bioethanol production. Chemosphere 256:127126. https://doi.org/10.1016/j.chemosphere.2020.127126

Msarah MJ, Ibrahim I, Hamid AA, Aqma WS (2020) Optimisation and production of alpha amylase from thermophilic Bacillus spp. and its application in food waste biodegradation. Heliyon 6(6): e04183. https://doi.org/10.1016/j.heliyon.2020.e04183

Mukherjee KA, Adhikari H, Rai KS (2007) Production of alkaline protease by a thermophilic *Bacillus subtilis* under solid-state fermentation (SSF) condition using *Imperata cylindrica* grass and potato peel as low-cost medium: characterization and application of enzyme in detergent formulation. Biochem Eng J 39:353–361

Mullis K, Faloona F, Saiki R, Horn G, Erlich H (1986) Specific enzymatic amplification of DNA in vitro: the polymerase chain reaction. Cold Spring Harb Symp Quant Biol 51:263–273

Paes G, O'Donohue MJ (2006) Engineering increased thermostability in the thermostable GH-11 xylanase from *Thermobacillus xylanilyticus*. J Biotechnol 125:338–350

Peng X, Qiao W, Mi S, Jia X, Su H, Han Y (2015) Characterization of hemicellulase and cellulase from the extremely thermophilic bacterium *Caldicellulosiruptor owensensis* and their potential application for bioconversion of lignocellulosic biomass without pretreatment. Biotechnol Biofuels 8(131):1–14

Rao CS, Sathish T, Ravichandra P, Prakasham R (2009) Characterization of thermo-and detergent stable serine protease from isolated *Bacillus circulans* and evaluation of eco-friendly applications. Process Biochem 44:262–268. https://doi.org/10.1016/j.procbio.2008.10.022

Rathinam NK, Bibra M, Salem DR, Sani RK (2019) Thermophiles for biohydrogen production in microbial electrolytic cells. Bioresour Technol 277:171–178

Raveendran S, Parameswaran B, Ummalyma SB, Abraham A, Mathew AK, Madhavan A, Rebello S, Pandey A (2018) Applications of microbial enzymes in food industry. Food Technol Biotechnol 56:16–30

Ravot G, Masson JM, Lefèvre F (2006) 34 applications of extremophiles: the industrial screening of extremophiles for valuable biomolecules. Methods Microbiol 35:785–813. https://doi.org/10.1016/S0580-9517(08)70037-0

Rehman R, Ahmed M, Siddique A, Hasan F, Hameed A, Jamal A (2017) Catalytic role of thermostable metalloproteases from *Bacillus subtilis* KT004404 as dehairing and destaining agent. Appl Biochem Biotechnol 181:434–450. https://doi.org/10.1007/s12010-016-2222-5

Rezanka T, Siristova L, Sigler K (2011) Rhamnolipid-producing thermophilic bacteria of species *Thermus* and *Meiothermus*. Extremophiles 15:697–709

Rigoldi F, Donini S, Redaelli A, Parisini E, Gautieri A (2018) Review: engineering of thermostable enzymes for industrial applications. APL Bioeng 2(1):011501. https://doi.org/10.1063/1.4997367

Rivero CW, De Benedetti EC, Sambeth JE, Lozano ME, Trelles JA (2012) Biosynthesis of anti-HCV compounds using thermophilic microorganisms. Bioorg Med Chem Lett 22 (19):6059–6062. https://doi.org/10.1016/j.bmcl.2012.08.045

Röhm O (1913) German Patent DE 283 923

Saiki R, Gelfand D, Stoffe S, Scharf S, Higuchi R, Horn G, Mullis K, Erlich H (1988) Primer directed enzymatic amplification of DNA with thermostable DNA polymerase. Science 239:487–491

Schultz J, Rosado AS (2020) Extreme environments: a source of biosurfactants for biotechnological applications. Extremophiles 24:189–206. https://doi.org/10.1007/s00792-019-01151-2

Selim S, Sherif ME, El-Alfy S (2010) Phenotypic and genotypic diversity of thermophilic bacteria isolated from hot springs in Sinai, Egypt and testing their ability to produce antimicrobial agents and extracellular enzymes. J Biotechnol 150(Supplement):270–271. https://doi.org/10.1016/j.jbiotec.2010.09.181

Shin H, Youn J, Kim S (2004) Hydrogen production from food waste in anaerobic mesophilic and thermophilic acidogenesis. Int J Hydrogen Energy 29(13):1355–1363

Singh P, Cameotra SS (2004) Potential applications of microbial surfactants in biomedical sciences. Trends Biotechnol 22:142–146

Skouri-Gargouri H, Gargouri A (2008) First isolation of a novel thermostable antifungal peptide secreted by *Aspergillus clavatus*. Peptides 29(11):1871–1877. https://doi.org/10.1016/j.peptides.2008.07.005

Sudan SK, Kumar N, Kaur I, Sahni G (2018) Production, purification and characterization of raw starch hydrolyzing thermostable acidic α-amylase from hot springs, India. Int J Biol Macromol 117:831–839. https://doi.org/10.1016/j.ijbiomac.2018.05.231

Sun X, Li Y, Tian Z, Qian Y, Zhang H, Wang L (2019) A novel thermostable chitinolytic machinery of *Streptomyces* sp. F-3 consisting of chitinases with different action modes. Biotechnol Biofuels 12:136. https://doi.org/10.1186/s13068-019-1472-1

Turner P, Mamo G, Karlsson EN (2007) Potential and utilization of thermophiles and thermostable enzymes in biorefining. Microb Cell Fact 6:9. https://doi.org/10.1186/1475-2859-6-9

Urbieta MS, Donati ER, Chan K-G, Shahar S, Sin LL, Goh KM (2015) Thermophiles in the genomic era: biodiversity, science, and applications. Biotechnol Adv 33(6):633–647

Vaičikauskaitė M, Ger M, Valius M, Maneikis A, Lastauskienė E, Kalėdienė L, Kaunietis A (2019) Geobacillin 26—high molecular weight bacteriocin from a thermophilic bacterium. Int J Biol Macromol 141:333–344. https://doi.org/10.1016/j.ijbiomac.2019.09.047

Verma A, Pal HS, Singh R, (2011) Agarwal S. Potential of alkaline protease isolated from Thermoactinomyces sp. RM4 as an alternative to conventional chemicals in leather industry dehairing process. Int J Agric Environ Biotechnol 4:173–178

Vijayakumar S, Saravanan V (2015) Biosurfactants-types, sources and applications. Res J Microbiol 10:181–192. https://doi.org/10.3923/jm.2015.181.192

Wong JWC, Fang M, Zhao Z, Xing B (2005) Effects of a surfactants on solubilization and degradation of phenanthene under thermophilic conditions. J Environ Qual 33(6):2015–2025

Wu X, Wang Y, Tong B, Chen X, Chen J (2018) Purification and biochemical characterization of a thermostable and acid-stable alpha-amylase from *Bacillus licheniformis* B4-423. Int J Biol Macromol 109:329–337. https://doi.org/10.1016/j.ijbiomac.2017.12.004

Yi H, Jeong J, Park Y, Deul K, Ghim S (2006) Effect of thermophilic bacteria on degradation of food wastes. Korean J Microbiol Biotechnol 34:363–367

Zhu D, Adebisi WA, Ahmad F, Sethupathy S et al (2020) Recent development of extremophilic bacteria and their application in biorefinery. Front Bioeng Biotechnol 8:483. https://doi.org/10.3389/fbioe.2020.00483

Zucaro A, Forte A, Fierro A (2018) Life cycle assessment of wheat straw lignocellulosic bio-ethanol fuel in a local biorefinery prospective. J Clean Prod 194:138–149. https://doi.org/10.1016/j.jclepro.2018.05.130

Chapter 11
Biodiversity, Ecological, and Commercial Importance of Psychrophilic Microorganisms

Gökçe Karadayi, Burak Alaylar, Mehmet Karadayi, Sümeyra Gürkök, Şeyma Aksu, Dilfuza Egamberdieva, and Medine Güllüce

Abstract Cold habitats are defined as extreme ecosystems with annual temperature averages below 0 °C, and contrary to popular belief, they cover a fairly large area on the earth. A notable part of the cold habitats includes deep oceans, snow cover, permafrost, sea ice, glaciers, cold water lakes, cold underground soils, cold deserts, and caves. These cold regions of the Earth are hosting psychrophilic archaea, bacteria, yeasts, molds, protozoa, photosynthetic cyanobacteria, and algae. Nowadays, cold habitats and their microbial biodiversity have become one of the important research subjects due to the high industrial application potential of the inhabitant microorganisms and their products. They are commonly accepted as rich sources of numerous valuable tools for application in a broad spectrum of innovative technologies. Thus, the number of researches on their biodiversity, ecology and commercial importance is increasing continuously. In this context, common characteristics of the cold habitats, their microbial biodiversity, adaptive features of psychrophilic microorganisms and their commercial importance of thermophilic microorganisms are presented and discussed in this chapter.

G. Karadayi
Department of Molecular Biology and Genetics, Faculty of Science, Atatürk University, Erzurum, Turkey
e-mail: gokce.karadayi@atauni.edu.tr

B. Alaylar
Department of Molecular Biology and Genetics, Faculty of Science and Art, Agri Ibrahim Cecen University, Ağrı, Turkey
e-mail: balaylar@agri.edu.tr

M. Karadayi (✉) · S. Gürkök · M. Güllüce
Department of Biology, Faculty of Science, Atatürk University, Erzurum, Turkey
e-mail: mkaradayi@atauni.edu.tr

Ş. Aksu
Institute of Natural and Applied Sciences, Atatürk University, Erzurum, Turkey

D. Egamberdieva
Faculty of Biology, National University of Uzbekistan, Tashkent, Uzbekistan

D. Egamberdieva et al. (eds.), *Microbial Communities and their Interactions in the Extreme Environment*, Microorganisms for Sustainability 32,
https://doi.org/10.1007/978-981-16-3731-5_11

Keywords Cold adaptations · Cold habitats · Biotechnology · Extremophiles · Psychrophilic microorganisms

11.1 Introduction

Contrary to common belief, Earth is a cold planet and about 85% of the biosphere is permanently exposed to temperatures below 5 °C. This part of the biosphere, extends from the Arctic to the Antarctic, from high mountains to the deep oceans, is named as cold environments or cold habitats. A notable part of the cold habitats consists of deep oceans (64% of the oceans), snow cover (35% of land surface), permafrost (24% of land surface), sea ice (13% of the earth's surface), and glaciers (10% of terrestrial environment). Moreover, cold water lakes, cold underground soils, cold deserts, and caves can be given as examples of other cold environments (Fig. 11.1) (Singh et al. 2006; Margesin and Miteva 2011; Hassan et al. 2016).

Temperature is one of the most important factors determining the limits of life. It has a strong influence on the survival and/or growth of an organism, both indirectly by its effect on water where all the biochemical reactions occur and by its direct effect on organic molecules of the living cells (Feller 2013). In this respect, cold habitats with unusual characteristics contain many challenging conditions both in

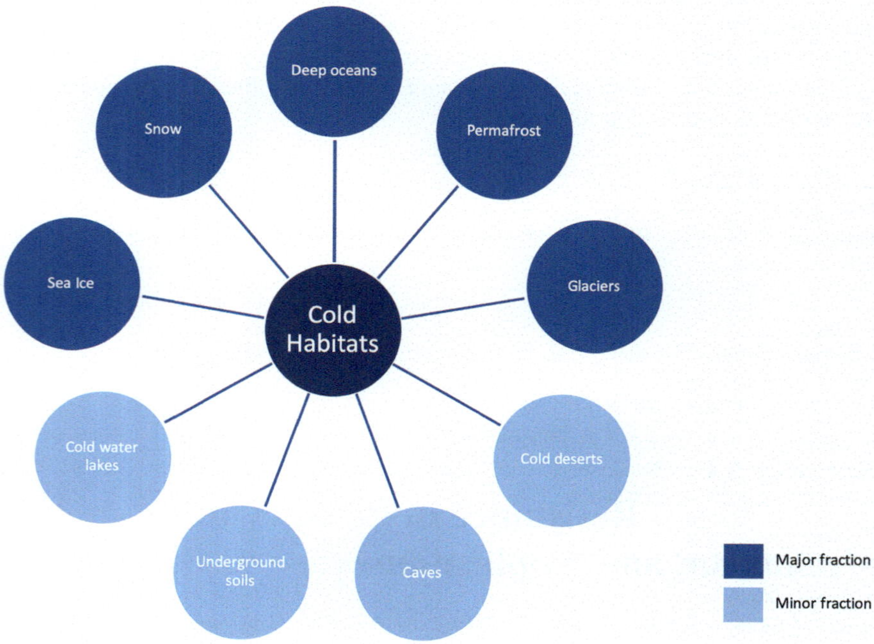

Fig. 11.1 Cold habitats of the Earth

keeping water in liquid state and in maintaining metabolic activities. Surprisingly, a great variety of cold-adapted organisms, mainly including bacteria, archaea, yeasts, filamentous fungi, algae, and protozoans, can survive and grow in these habitats by overcoming the adverse effects of cold environments. These organisms, most of which consist of microorganisms, have developed a series of complex structural, physiological, metabolic, and functional adaptations to extreme conditions of cold ecosystems such as limited availability of nutrients, low water activity, low enzymatic reaction rate, extreme pH, and salinity. Besides their critical role for the maintenance of ecological cycles in low temperature habitats, these adaptations are also a unique source of inspiration for the development of biotechnological processes that can facilitate our daily life.

11.2 Cold Habitats

11.2.1 Atmosphere and Clouds

The stratosphere and mesosphere layers (41–77 km) where temperatures can reach $-100\ ^\circ$ C are generally dominated by Gram-positive bacteria (Griffin 2008; Pearce et al. 2009; Wainwright et al. 2004). Highly specialized adaptation mechanisms against high UV radiation, oxidative stress, low nutrient availability, and drought factors determine maintaining the microbial life in these extremely cold environments.

Microbial cells originating from terrestrial or aquatic ecosystems or organisms living in these ecosystems are transported vertically and horizontally by aerosol particles for short and long distances depending on the meteorological and seasonal conditions. This information led to the creation of a global atmospheric bacteria transport model in the early 2000s. Accordingly, it is estimated that the mass of bacteria found annually in the atmosphere and transported around the world is 40–1800 Gg (1 Gg $= 10^9$ g) (Burrows et al. 2009). The microbial mass remains in the atmosphere for varying lengths of time (from days to weeks), travels great distances, covers very large areas and eventually precipitates back into terrestrial or aquatic ecosystems. This wide distribution of microorganisms in the atmosphere has brought up the view that they can have an active role in atmospheric processes and especially in cloud formation (Morris et al. 2008a).

Clouds are an essential part of the atmosphere, where water vapor cools and condenses into water droplets or ice crystals. Cloud water is considered a more favorable microbial habitat than dry air because cloud droplets remain liquid at temperatures below $0\ ^\circ$ C that affects atmospheric chemistry (Deguillaume et al. 2008). Thus, a great variety of microbial cells can maintain their lives by metabolizing organic compounds under this condition (Sattler et al. 2001). Scientific research has revealed that there are a large number of bacterial and fungal cells (10^3–10^5 living-cell/mL), including known and new species, within the tropospheric cloud layers. Some important examples of microbial species detected in atmosphere and clouds are *Bacillus aerius*, *Bacillus aerophilus*, *Bacillus aryabhattai*, *Bacillus altitudinis*, *Bacillus isronensis*, *Bacillus stratosphericus*, *Deinococcus aetherius*, *Deinococcus aerius*, *Erwinia billingiae*, *Janibacter hoylei*, *Pseudomonas fluorescens*, *Pseudomonas graminis*, *Pseudomonas poae*, *Pseudomonas reactans*, *Pseudomonas syringae*, *Pseudomonas trivialis*, *Pseudomonas veronii*, *Pseudomonas viridiflava*, *Xanthomonas campestris,* and *Pseudoxanthomonas* sp. (Margesin and Miteva 2011; Vaïtilingom et al. 2012; Joly et al. 2013). As biological aerosols, these microbial cells could play important roles by nucleating ice in supercooled clouds (Lohmann and Feichter 2005). Hence, it was proposed that ice nucleating microbial cells in the atmosphere could contribute to the formation of clouds (Bauer et al. 2002; Joly et al. 2013). Recently, it has been suggested that microbial glycoprotein and lipoprotein biosurfactants have a possible role in increasing cloud condensation efficiency (Ekstrom et al. 2010).

11.2.2 Snow

Snow is a large component of the cryosphere, which covers 35% of the Earth's land surface permanently or seasonally, mostly in the Northern Hemisphere. Its particularly important ecological features are seasonal temperature fluctuation, aerobic conditions and very high light and UV irradiation (Jones 1999; Cockell and Cordoba-Jabonero 2004). As a habitat, snow is closely related to the atmosphere due to its continuous dynamic aeolian nature. It is also the primary source for the formation of glaciers. Besides, it temporarily covers other soil surfaces in warmer regions during winter and affect the ecological features of these habitats (Hodson et al. 2008; Pearce 2009; Pearce et al. 2010).

In the first studies conducted to determine the biodiversity of the snow habitat, photosynthetic snow algae, which cause snow coloration with their high bioactivity as primary producers, was the heavily examined group (Hoham and Duval 2001). In recent studies, other groups of microorganisms have begun to be included and a large number of bacteria and fungus species have been identified in snow habitats (Lopatina et al. 2013; Cameron et al. 2015; Lopatina et al. 2016). In the study conducted in 2013, the diversity of bacterial communities in the samples collected from superficial areas around two Russian stations in Antarctica, named Druzhnaja and Leningradskaja, was investigated and Various species belonging to the genera *Variovorax*, *Flavisolibacter*, *Corynebacterium*, *Methylobacterium*, *Sphingomonas*, *Pseudomonas*, *Stenotrophomonas*, *Aquabacterium*, *Aquaspirillum*, *Comamonas*, *Janthinobacterium*, *Caulobacter*, *Ralstonia*, *Burkholderia*, *Curvibacter*, *Polaromonas*, *Bacillus*, *Pedobacter*, *Arthrobacter*, *Microbacterium*, *Flavobacterium*, *Acinetobacter*, *Mitsuria*, *Phaselicistis*, *Grimontella,* and *Arcicella* have been shown on the surface snow (Lopatina et al. 2013). In addition, in a more recent study conducted by Lopatina et al. 2016, samples were taken from the surface snow around four Russian stations in East Antarctica and the bacterial communities in these samples were analyzed by metagenomic methods. According to the results of this study, it was determined that the groups that dominate the bacterial population of the surface snow in the sampled areas are Alphaproteobacteria, Betaproteobacteria, Gammaproteobacteria, Sphingobacteriia, Flavobacteriia, Cytophagia, Actinobacteria, Cyanobacteria, and Bacilli (Lopatina et al. 2016).

Beyond these bacterial communities, over 30 Archaeal taxa also were identified in research studies (Lutz et al. 2015).

Studies have reported that the microbial density and diversity in the snow cover changes with altitude and latitude. Different studies have identified microbial cell numbers ranging from 10^3 to 10^5 per mL in molten snow, which often correlate positively with Ca^{+2} concentrations and acts as a trap for dust.

11.2.3 Cryoconite Holes

Cryoconite (ice dust) holes are water filled ice pits with underlying earthy dark material commonly found on the surface of all snowless glaciers. These extreme habitats, which vary in size, can exceed 1 m in width and half a meter in depth. The habitable zone in these structures is formed by soil depressions in the air that absorb solar radiation and melt down in the ice. While most of the cryoconite holes are open to the atmosphere, the cryoconite formations in Antarctica (such as those commonly seen in McMurdo Dry Valleys) are covered with an ice cap due to harsh seasonal conditions. It has also been discovered that these formations can in some cases even connect to the underground flow systems (Hodson et al. 2008; MacDonell and Fitzsimons 2008). Many studies on polar and alpine cryoconite holes have shown that the biodiversity in these habitats includes a wide variety of viruses, archaea, bacteria, cyanobacteria, eukaryotic algae, yeast, and protozoa species (Margesin et al. 2002; Vincent 2007; Zhakia et al. 2008). In an example study conducted by Christner et al. (2003), samples taken from a cryoconite hole in Antarctica were analyzed for biodiversity and various species belonging to eight bacteria (Acidobacterium, Actinobacteria, Cyanobacteria, Cytophagales, Gemmimonas, Planctomycetes, Proteobacteria, and Verrucomicrobia), three metazoan (nematodes, tardigrades, and rotifers), one fungus (Choiromyces), one ciliate (Spathidium), and one green algae (Pleurastrium) groups have been identified. In another similar study, the biodiversity of the cryoconite holes in the Antarctic McMurdo Dry Valley glaciers was investigated, and at the end of the study, cyanobacteria (*Chlorococcus, Chroococcus, Crinalium, Oscillatoria, Nostoc,* and *Spirulina*), rotifer (*Philodina gregaria* and *Cephalodella catellina*), tardigrades (*Acutuncus antarcticus* and *Hypsibius* spp.) and many species of ciliates were recorded (Porazinska et al. 2004). In a recent study conducted in East Antarctica, the biodiversity of the cryoconites in the Queen Maud Land region was investigated and species belonging to the groups of Proteobacteria, Cyanobacteria and Actinobacteria were found in the samples. In this study, the presence of species belonging to the genus *Nitrososphaera*, a genus of the Archaea, was also recorded (Lutz et al. 2019).

11.2.4 Glaciers

Glaciers are considered to be the most difficult environment for living organisms for reasons such as freezing temperatures ranging from −56 ° C to −10 ° C, high hydrostatic pressure, insufficient availability of nutrients, water and light. However, the polar and non-polar ice sheets covering 15,861,766 km^2 are an enormous reservoir of microbial life that has been uniquely preserved in chronological layers for thousands of years (Priscu and Christner 2004; Priscu et al. 2007).

In biodiversity studies conducted with these unique resources, it was determined that large bacterial phylogenetic groups such as Actinobacteria, Firmicutes, Proteobacteria, and CFB (Sitophaga-Flavobacterium-Bacteroides) dominate glacial ecosystems. In addition, many studies have shown the presence of psychrophilic microbial eukaryotes (fungi, yeasts), some plant and bacterial viruses, and several Archaea species (Miteva 2008; Simon et al. 2009).

11.2.5 Permafrost

Permafrost habitats are surface land parts that remain below 0 °C for two or more consecutive years. These habitats are generally characterized by having more than one stress factor (freezing, salinity, drought, oligotrophy) that directly affect microbial life. Microorganisms in these habitats have been studied with culture-dependent and culture-independent methods, and the existence of various life forms, the majority of which are bacteria, have been determined (Steven et al. 2006; Gilichinsky et al. 2008).

Bacterial communities in permafrost include aerobes and anaerobes. Reduction conditions in permafrost support the preservation of anaerobes such as acetoclastic and hydrogenotrophic methanogens, sulfate reducers, Fe(III) reducers and denitrificators (Rivkina et al. 1998; Trotsenko and Khmelenina 2005; Gilichinsky et al. 2008). It has been determined that the Archaea species belong to the Euryarchaeota (Methanomicrobiaceae, Methanosarcinaceae, Methanosaetaceae) and Crenarchaeota groups. Some of the methanogens have also been reported to show halophilic properties (Ganzert et al. 2007; Rivkina et al. 2007; Yang et al. 2008).

11.2.6 Deep Oceans

Approximately 71% of the earth's surface is covered by the ocean, and more than half of this area is over 3000 m deep. The deep sea is defined as the lowest layer below the thermocline in the ocean at a depth of 1828 m or more.

The deep sea, whose most characteristic feature is the extreme cold, is characterized by having many challenging conditions for life at the same time. Therefore, microorganisms living here have developed a few extraordinary features that allow them to thrive in their environment (Abe 2004; Deming 2009). Most of the isolated microorganisms are psychropiezophilic (low-temperature- and high-pressure-adapted), which cannot be cultured at temperatures higher than 20 °C and at pressures lower than 0.1 MPa (Nogi 2008).

Communities of bacteria, archaea, protists, and yeast make up most of the biomass in the ocean and are responsible for 98% of primary production (Whitman et al. 1998). Microbial density is low in deep seas; however, there is an enormous phylogenetic diversity that cannot be clearly identified by culture-based research (Sogin et al. 2006). Among the psychrophilic and piezophilic bacteria in the deep seas, members of the Gammaproteobacteria class are dominant. *Colwellia, Moritella, Photobacterium, Psychromonas, Marinomonas,* and *Shewanella* can be listed as examples for the dominant genera in deep oceans.

11.3 Adaptation in Psychrophiles

Depending on the environment, numerous biotic (e.g., predation by herbivores and viruses, antibiotics, cell-cell interactions), abiotic (e.g., pH, salinity, oxygen, nutrient flow) and broader ecological factors (e.g., sea ice versus sea water, free life against added particles) can greatly influence the selection and growth characteristics of individual microorganisms. Also, the variety of microorganisms that colonize the Earth's biosphere, most of which is cold, is enormous. As a result, various physiotypes have been developed to successfully colonize cold environments. In addition, very few classes of microorganisms have evolved that can successfully colonize both low and high temperature habitats. Methanogens, members of Archaea, are the only known group to have individual species whose growth temperature range is below zero to 122 °C (Saunders et al. 2003; Cavicchioli et al. 2006; Reid et al. 2006; Takai et al. 2008). Therefore, there are limited opportunities to compare the adaptive characteristics of psychrophiles and hyper/thermophiles belonging to the same genus or family.

11.3.1 Cellular Mechanism of Cold Adaptation

Low temperature can hinder transcription and translation due to the increased stability of the random secondary structures of the transcripts. Prevention or degradation of inhibitory secondary structures of RNA can be achieved with RNA chaperones. Cold shock proteins (Csps) are small proteins that bind to RNA to maintain its single-stranded structure (Jones and Inouye 1994). DEAD box RNA helicases can resolve secondary structures in an ATP-dependent manner and are upregulated during cold growth in some psychrophiles (Lim et al. 2000). Psychrophiles vary greatly in the number of *csp* genes found in their genome. The *csp* genes contain a nucleic acid binding domain known as the cold shock domain (CSD) and have additional roles in addition to serving as RNA chaperones. Proteins containing individual CSDs can regulate the cold shock response or play an important role in growth following low temperatures in mesophiles (Hebraud and Potier 1999). Thus, most of the Csps function as cold-adaptive proteins in psychrophiles, as they are expressed continuously rather than transiently at low temperatures (D'Amico et al. 2006). Overexpression of the *csp*A of *Psychromonas arctica* has been shown to increase the resistance of *Escherichia coli* to low temperatures (Jung et al. 2010). In addition, it appears that one of the three Csp is important for the growth of *Shewanella oneidensis* at low temperatures (Gao et al. 2006).

Not all bacteria and archaea that can grow at low temperatures have known Csp homologues. For example, *Rhodoferax (Albidoferax) ferrireducens* lacks identifiable *csp* genes, although *csp* genes are present in other members of *Burkholderiales (Betaproteobacteria)*, including *Polaromonas* strains. The *csp* genes are found in archaea *Methanogenium frigidum* (stenopsychrophile) and *Halorubrum lacusprofundi* (eurypsychrophile), but are absent in *Methanococcoides burtonii*, an eurypsicrophilic archeon isolated from the same Antarctic lake as *M. frigidum* (Giaquinto et al. 2007). For *M. burtonii*, small proteins consisting of a single RNA-binding TRAM domain are upregulated at low temperatures and have been proposed to serve as RNA chaperones similar to Csps (Williams et al. 2010a, 2011). These putative RNA chaperones have been termed Ctr (cold sensitive TRAM domain) proteins and are specific to an archaea subgroup. The amount of Ctr proteins in *M. burtonii* is particularly high at very low growth temperature (2 °C), and a role in facilitating cell function during cold stress has been suggested.

Small RNA-binding proteins (Rbp) can facilitate cold adaptation, but similar to Csps, they may also have other functions in the cell (Maruyama et al. 1999; Christiansen et al. 2004). These Rbps accumulate with exposure to cold stress and play important roles in regulating the termination of transcription (Mori et al. 2003). Rbps are small proteins that contain a single glycine-rich RNA-binding motif. It is also common in cyanobacteria but rare in other bacteria (Maruyama et al. 1999; Ehira et al. 2003). Mesophilic cyanobacterium *Anabaena variabilis* has eight *rbp* genes, all but one cold regulated (Maruyama et al. 1999). With osmotic stress, *rbp* gene expression is increased in *Anabaena sp*. PCC7120 strain. The responses to cold and osmotic stresses overlap because both reduce the availability of free water (Mori et al. 2003).

Enzymes that play a role in the breakdown of RNA and proteins, such as RNases and proteases found in *Psychrobacter arcticus* (Bergholz et al. 2009) and *M. burtonii* (Williams et al. 2010b), which are permafrost bacteria, are upregulated during growth at low temperature in some psychrophilic bacteria and archaea. This has been interpreted as a strategy to preserve biosynthetic precursors (Bergholz et al. 2009) or improved quality control of irreparably damaged RNA and proteins (Williams et al. 2010b).

At sufficiently low temperatures where ice is formed, cells are exposed to additional stress factors such as ice damage, oxidative damage, and osmotic imbalance (Tanghe et al. 2003; Williams et al. 2010b, 2011). Extracellular polymeric substances (EPS) can offer protection against mechanical degradation of the cell membrane caused by ice. Sea ice bacteria such as *Colwellia psychrerythraea* produce polysaccharide-rich EPS (Thomas and Dieckmann 2002; Junge et al. 2004). The resulting biofilms can protect against invasive ice crystal damage and facilitate uptake of nutrients inside channels formed in sea ice.

Low temperatures reduce membrane fluidity and permeability. In response, the elastic liquid crystal structure of the cell membrane is replaced by the gel phase, which can disrupt the biological functions of the membrane, including transport (Phadtare 2004). This can be offset by increasing the proportion of unsaturated fatty acids in the lipid bilayer, resulting in a more loosely packed array (Russell 2008). Increasing the proportion of unsaturated fatty acids can be achieved by reducing the saturation of pre-existing fatty acids or by de novo synthesis of less saturated fatty acids. In a metagenomic analysis, a microbial community in glacial ice was found to be relatively rich for genes involved in the maintenance of membrane fluidity (Simon et al. 2009). The changes in membrane lipids appear to be a generally conserved characteristic for cellular adaptation to cold.

11.3.2 Protein Adaptation to Cold

Many types of proteins, including various classes of enzymes (e.g., glucanases, hydrolases, oxidoreductases, hydrogenases, isomerases, nucleic acid modifying enzymes), have evolved to function effectively at temperatures ranging from below zero to well above 100 °C (Adams and Kelly 1994; Demirjian et al. 2001; Siddiqui and Cavicchioli 2006). By comparing the structure, activity, and stability of the same type of proteins (preferably orthologues with high sequence identity) from different thermal classes, the researchers gained useful insight into how proteins evolved and what properties seem important to provide certain thermal characteristics.

In recent years, genomics has been applied to microbial communities from all environmental samples (metagenomics), thus providing DNA sequence information for proteins from uncultured microorganisms. The metagenomics of samples from cold environments include the generation of large datasets obtained by shotgun sequencing (Lopez-Bueno et al. 2009, Lauro et al. 2011, Varin et al. 2012,) and functional screening of clones for cold-active enzymes (Elenda et al. 2007). Genomics and metagenomics facilitate subsequent targeted analyses to evaluate specific properties of individual proteins (e.g., site-directed mutagenesis). Broad-spectrum modification (e.g., mutagenesis by directed evolution, chemical modification of certain amino acid side groups) and evaluation of changes in thermal properties of individual enzymes have also been used to identify structural properties that play a role in providing thermal activity and stability (Cavicchioli et al. 2006; Siddiqui et al.

2006). Collectively, such studies have revealed a lot about the adaptation of proteins to temperature.

In general terms, properties associated with adaptation (e.g., ratio of specific amino acids, hydrophobicity of exposed surfaces) tend to have opposite actions between psychrophilic and hyper/thermophilic proteins (Siddiqui and Cavicchioli 2006; Feller 2008). Proteins from psychrophiles have higher activity and thermal solubility compared to mesophilic and thermophilic homologs (Demirjian et al. 2001; Siddiqui and Cavicchioli 2006). For example, α-amylases obtained from the psychrophilic bacterium *Pseudoalteromonas haloplanktis* and the thermophilic bacterium *Bacillus amyloliquefaciens* have optimal activity temperatures of 28 °C and 84 ° C, respectively (D'Amico et al. 2003). The unique properties of cold-active enzymes have attracted both academic and commercial interest, as low temperature environments pose important problems for the function of enzymes and proteins from mesophiles.

11.3.3 Enzyme Mechanism in Cold Adaptation

In low temperature environments, there is insufficient kinetic energy to overcome enzyme activation barriers, thus resulting in very slow chemical reaction rates. For a biochemical reaction occurring in a mesophyll at 37 °C, a decrease in temperature from 37 °C to 0 °C results in a 20–80-fold reduction in enzyme activity. This is the main factor inhibiting growth at low temperatures. However, organisms adapted to low temperatures have developed various ways to overcome this limitation; these include the energy-costly enhanced enzyme production strategy (Crawford and Powers 1992) and the seasonal expression of isoenzymes (Somero 1995). However, the most common adaptation characteristic of cold-active enzymes is a reaction rate (kcat) that is largely independent of temperature. Most psychrophilic enzymes reduce the activation energy barrier between the ground state (substrate) and the activated state, resulting in kcat that is not temperature sensitive. To assist in substrate binding at low energy cost, the active sites of cold-active enzymes tend to be larger and more accessible to substrates. As a result, the binding affinity of substrates for cold-active enzymes is generally lower (higher Km) than their thermophilic counterparts.

High catalytic rates at low temperatures are usually achieved by the flexible structure and the accompanying low stability of cold-active enzymes, which is called the activity-stability exchange (Siddiqui and Cavicchioli 2006). Many cold-active enzymes have a more unstable and flexible catalytic region, i.e., localized flexibility, than the rest of the protein structure (Siddiqui et al. 2005; Feller 2008). Accordingly, in an environment characterized by low kinetic energy and delayed molecular motion, activity of cold-active enzymes rely on molecular dynamics and thus greater disorganization for maintaining function (Feller 2007).

Genomic analysis of psychrophilic archaea revealed proteins characterized by a higher content of uncharged polar amino acids (especially Gln and Thr), a lower content of hydrophobic amino acids (especially Leu), increased exposure of hydrophobic residues and low charge associated with destabilizing the surface of psychrophilic proteins (Saunders et al. 2003). The evolutionary choice of amino acid usage has made such an adaptation possible (Allen et al. 2009). Through the genome analyses of marine Gammaproteobacteria, it has been reported that cold-adapted strains have higher contents of Ile, Lys and Asn as well as lower Ala, Arg, and Pro contents. (Zhao et al. 2010). Among them, Pro and Arg are associated with the ability to provide increased stability by limiting backbone rotations and forming multiple hydrogen bonds and salt bridges (Feller and Gerday 2003).

Psychrophilic proteins are characterized by reduced core hydrophobicity, increased surface hydrophobicity, increased surface hydrophilicity, a lower arginine / lysine ratio, weaker intermediate domains and subunit interactions, increasingly longer cycles, reduced secondary structure content, more glycine residues, less proline in cycles, prolines in α-helices, fewer and weaker metal bonding sites, less disulfide bridges, less electrostatic interaction (H-bonds, salt bridges, cation-pi interactions, aromatic-aromatic interactions), reduced oligomerization and increased unfolded state in conformational entropy and more (Siddiqui and Cavicchioli 2006).

11.3.4 Other Factors Affecting Enzyme Adaptation

Limited number of heat-labile enzymes may be labile enzymes at subzero or near-zero temperatures (D'Amico et al. 2003; Xu et al. 2003) and some oligomer or cofactor-requiring enzymes (e.g., tryptophanase) are reversibly inactivated at lower temperatures as a result of subunit and cofactor decomposition (Kogan et al. 2009). Therefore, if an essential cellular enzyme is inactivated in the cold or cold denatured, it can define the lower temperature limit for growth rather than the freezing point of the aqueous medium in which the organism grows.

11.3.5 Cold-Adapted Enzymes and Climate Change

Global warming has a particularly strong impact on polar and mountain environments where 30% of the global soil carbon pool is located. The reduction of cellulose, hemicellulose and humic substances in soil organic matter (SOM) to dissolved organic compounds by extracellular enzymes (e.g., glucanases, ligninases) represents the rate-limiting step in carbon release (Weedon et al. 2011; German et al. 2012). The kinetic and thermodynamic properties of extracellular enzymes, including their responses to environmental factors (e.g., nutrient source, nitrogen and oxygen availability, phenolic and substrate concentration, soil moisture, permafrost melting and temperature) are now beginning to be included in predictive models explaining the effects of global warming on the carbon cycle (Davidson and Janssens 2006; Weedon et al. 2011; German et al. 2012).

Considering such issues related with global warming, it is important to acknowledge that cold-adapted enzymes work efficiently at low temperatures and therefore help reduce CO_2 emissions by reducing electricity consumption associated with heating (Cavicchioli et al. 2002, 2011). For example, washing machines use a large portion of a home's electricity budget, and 80% of electricity is used to heat water (Nielsen 2005). By using cold-active enzymes, washing temperatures can be reduced from 40 °C to 30 °C, resulting in a 30% reduction in electricity consumption. More importantly, by setting wash temperatures 10 °C lower, CO_2 emissions associated with burning fossil fuels for energy production are reduced by 100 g per wash (Nielsen 2005). The application of cold-adapted enzymes in a number of other industries such as textiles, food, wastewater treatment, and pulp also helps to reduce toxic by-products, electricity consumption and CO_2 emissions (EuropaBio Rep. 2009; Cavicchioli et al. 2011).

11.3.6 Ice Binding Proteins

Ice binding proteins (IBPs), such as antifreeze proteins (AFPs) and ice nucleation proteins (INPs), have been identified in a variety of cold-adapted organisms and their potential applications in biotechnology have been recognized in various fields.

AFPs bind to ice crystals, causing thermal hysteresis (TH) and ice recrystallization inhibition (IRI) (Lorv et al. 2014). TH is a non-collective effect defined by the difference between a solution's freezing and melting points (Raymond and DeVries 1977). In IRI, AFPs prevent the formation of large ice crystals by the boundary migration of smaller ice crystals (Yu et al. 2010). Another type of IBP is ice nucleation proteins (INPs) that induce ice formation at subzero temperatures (Lindow et al. 1982). Water freezing is a stochastic process determined by the temperature and direction of water molecules. Spontaneous or homogeneous ice nucleation in pure water occurs at a temperature of −38.5 °C and bacterial INPs can act as water molecule regulators that promote ice nucleation at higher temperatures (−2 °C to −10 °C) (Lee et al. 1995).

11.3.6.1 Antifreeze Proteins (AFP)

AFPs were first identified in Antarctica teleost fish (DeVries and Wohlschlag 1969). However, AFPs are not limited to fish and they have been found in a wide variety of living organisms from cold environments such as plants (Bravo La. Griffith M. 2005; Middleton et al. 2009), insects (Graether and Sykes 2004; Kristiansen et al. 2011), fungi (Hoshino et al. 2003; Xiao et al. 2010) microalgae (Janech et al. 2006; Raymond et al. 2009), yeast (Lee et al. 2010) and archaea (Saunders et al. 2003).

According to the ice binding patterns, AFPs are divided into two main categories: active and hyperactive AFPs. The difference between active and hyperactive AFPs is still under investigation. The main difference may be their selectivity for different ice faces; where hyperactive AFPs bind to the base and planes of ice and block growth along the c-axis (Pertaya et al. 2008). In a mutational study of an insect protein (*Tenebrio molitor* AFP, TmAFP), modification from threonine to valine results in lower AFP activity for the basal plane of the ice due to the protein's lower binding affinity (Liou et al. 2000).

Active AFPs simply bind to the prism plane, leaving the basal plane free (Scotter et al. 2006). In addition, ice surface coating by hyperactive AFPs increases TH activity in *T. mollitor* up to 6 °C at sub-millimolar concentrations (Graham et al. 1997). By comparison, active AFPs generally found in fish are characterized by a lower TH of 0.5 °C –1.0 °C at millimolar protein concentrations (Hanada et al. 2014).

11.3.6.2 Ice Nucleation Proteins (INP)

Molecules that act as ice nucleating agents are diverse in nature (such as dust and pollen) and produce ice nuclei between −8 °C and − 15 °C (Margaritis and Bassi 1991). However, INPs have been found in insects living in cold environments (Duman 2001), fungi (Tsumuki et al. 1992) and plants (Brush et al. 1994). Most INP studies have been performed on bacteria with active ice nucleus in temperature ranges between 0 °C and − 10 °C (Cochet and Widehem 2000). In this regard, Schmid et al. (1997) provided a classification method used to identify cultured bacterial subpopulations; this classification is based on the heterogeneous threshold temperatures of INP activity. The most effective bacterial INPs are called type I and can initiate ice formation from 0 °C to −4 °C. Type II INPs initiate ice formation from −5 °C to 7 °C, type III INPs initiate ice formation from −8 °C to −10 °C. The basis of these subpopulations of INPs is their monomer, trimer, and oligomer grouping (Burke and Lindow 1990). This assumption has been demonstrated through radiation inactivation of these proteins. This study showed that 150 kDa is the minimum molecular mass required for ice nucleation at temperatures between −12 °C and − 13 °C (Govindarajan and Lindow 1988).

11.4 Biotechnological Applications

IBPs are mainly used in food industry (Yeh et al. 2009; Zhang et al. 2010) and medicine (Koushafar and Rubinsky 1997; Muldrew et al. 2001; Amir et al. 2003; Hirano et al. 2008; Lee et al. 2012). For example, AFPs are used in ice cream production (Warren et al. 1992; Feeney and Yeh 1998) to maintain their quality, and INPs are used for artificial snow production (Cochet and Widehem 2000). However, several IBPs from bacteria have been identified to date, and their applications have only been demonstrated at the laboratory level. In light of these facts, a better understanding of current and future biotechnological applications of bacterial IBPs is required. Therefore, this small review summarizes the available information on bacterial IBPs (both AFPs and INPs) and their applications in biotechnology.

Many industrial and biotechnological processes use cold-active biomolecules from organisms adapted to cold environments (Cavicchioli et al. 2011). Natural and recombinant AFPs have been primarily applied in various fields such as the food industry and medicine (Amir et al. 2003; Kontogiorgos et al. 2007; Hirano et al. 2008). Recombinant AFPs from fish are used in industry to improve food preservation in freezing (Yeh et al. 2009). Recombinant AFPs derived from yeast *Leucosporidium sp.* has been used for cryopreservation of red blood cells (Lee et al. 2012). In cryo-surgery, the production of needle-like ice crystals at high AFP concentration ($\geq$5 mg/mL) helps to remove cells (Koushafar and Rubinsky 1997). Only fish AFPs have been used successfully to destroy subcutaneous rat tumor cell lines. However, although these studies have been conducted at the laboratory level, a great application potential for the medical industry is predicted (Koushafar and Rubinsky 1997; Muldrew et al. 2001).

Bacterial INPs have also been applied in various fields, particularly the ones with the highest temperature (type I) activity (Gurian-Sherman and Lindow 1992). Several food technology studies have reported positive results when bacterial INPs were used to increase ice nucleation temperature, resulting in reduced freezing times and ice crystal size and improved frozen solid food quality (Zhang et al. 2010). As a result of ice formation at higher temperatures, lower energy costs are required to freeze food. It has also been shown that freezing food with less cooling is possible when using biological ice cores produced by bacteria (i.e., from *P. syringae* and *E. herbicola*). Widehem and Cochet (2003) showed that freezing processes can be improved by the addition of lyophilized cells of *P. syringae*, which can act as ice nuclei, resulting in a significant reduction in supercooling point by forming more ice crystals. In particular, INP-producing bacteria in samples collected from rain and snow are closely related to the water cycle (Morris et al. 2008b). In fact, the presence of bacteria in cloud water has revealed that this transport mechanism makes biological ice nucleation a common phenomenon in the atmosphere, affecting meteorological processes such as rain and hail storm (Christner et al. 2008). In addition, one of the most important applications in the industry is the current use of Snomax (Telemet, Inc., NewYork, USA), a freeze-dried bacterial INP for artificial snow production (Cochet and Widehem 2000).

Although there are few engineering or genetic modifications to improve the function of IBPs, improvements in recombinant protein production have been achieved by linking an identified protein function to the encoding gene. This technology has increased our ability to modify certain peptides to achieve the desired properties in vitro. As an example, protein engineering through modification of certain residues in isoforms of AFPs resulted in antifreeze activity gain, as reported by Garnham et al. (2012).

11.4.1 Applications in Agriculture

Frost damage is responsible for more economic loss than any other weather-related phenomenon in the United States and many other parts of the world (Pearce 2001; Chevalier et al. 2012). Sessile organisms such as plants have adopted ice recrystallisation inhibition (IRI) mechanisms primarily to tolerate cold temperatures (Thomashow 1998; Lorv et al. 2014). Studies have shown that frost-tolerant plants such as winter rye (*Secale cereale*) and ryegrass (*Lolium perenne*) secrete AFPs in intercellular spaces and apoplast that affect freezing point and ice crystal growth (Griffith et al. 1992; Middleton et al. 2012). Frost-tolerant plants such as *L. perenne* accumulate fructans as a source of carbohydrates, and this is thought to play an important role in freezing tolerance. (Valluru and Van Den Ende 2008; Valluru et al. 2008). In contrast, other plant species regulate freezing point depression by the presence of cryoprotectants such as trehalose and sucrose that allow plants to overcool, reaching lower temperatures without ice formation in tissues. (Kawahara 2008). Studies on blueberries (*Vaccinium spp.*) have also shown that controlled ice nucleation is a beneficial mechanism for preventing overcooling in the protoplasm, inducing extracellular freezing, and / or accommodating the ice crystal in certain tissues (body shell) (Kishimoto et al. 2014). However, frost-sensitive plants may be affected by the presence of INP-producing bacteria that cause ice formation and damage and dehydration in plant cells (Lindow et al. 1982; Cambours et al. 2005).

Current methods of preventing frost damage in agroecosystems are primarily physical, i.e., heaters, wind machines, artificially generated fog, and sprinkling water directly on the plants. However, these methods are costly. As a different method to reduce the effects of frost damage, control of INP-producing bacterial populations is performed to increase the supercooling ability of plants (Lindow 1983). In this context, bactericides (such as oxytetracycline and streptomycin) have been applied as a preservative to prevent the formation of INP-producing bacteria in plants (Lindow 1984). However, the use of antibiotics in foodstuffs is undesirable, as the remaining antibiotics may affect the intestinal flora, cause allergic reactions, and also facilitate bacterial resistance to antibiotics (Jing et al. 2009). Therefore, it has been suggested that the inoculation of antagonistically working bacteria against INP-producing bacteria also reduces the frequency of biological ice nucleation damage on plants as shown by Lindow (1984, 1987). Moreover, active bacterial INPs have been shown to be effective in pest control by reducing the capacity of potato beetles to resist freezing temperatures (Castrillo et al. 2001; Lee et al. 2001). Environmental scientists have also suggested INP-producing bacteria as a cloud condensation inducer to produce artificial rain (Levin et al. 1986; Joly et al. 2013). Obviously, great efforts will be required to make these applications effective and economical for use in the field.

Another mechanism discovered is the use of bacteria that promote plant growth. The only example reported in the literature is the inoculation of *P. putida* GR12–2 isolated from the rhizosphere of Canadian High Arctic plants secreting an extracellular AFP (Sun et al. 1995). Inoculation of spring and winter canola seeds with

P. putida GR12–2 resulted in an increase in root elongation at 5 °C. Bacteria have also been found to colonize intercellular and apoplast spaces in leaves as epiphytic organisms living in the phyllosphere of plants (Lindow and Brandl 2003), so inoculation of AFP-producing epiphytic bacteria can be applied to reduce frost damage in susceptible plants. This possibility is supported by the finding that AFPs isolated from fish have been experimentally shown to inhibit the ice nucleation activity of an *E. herbicola* producing INP, a common organism living in the plant phyllosphere (Parody-Morreale and Murphy 1988).

Similar results have been found using type III AFP, which prevents the ice nucleation process by adsorbing both on the growing ice nucleus surface and on dust particles (Du et al. 2003). On the other hand, a recent article has shown that a plant's AFP can also control ice growth and act as a defence strategy against bacterial ice nucleation (Tomalty and Walker 2014). Recent advances in metagenome analyses have also suggested that the survival of Antarctic algae is driven by the accumulation of bacterial AFP activity (Raymond 2016). This is due to the identification of homologues in the same domain (DUF3494) in bacterial gene sequences discovered in the metagenome analysis of moss leaves (Davies 2016). The same area (DUF3494) has been reported to be transferred by horizontal gene transfer between *Colwellia sp* SLW05 (ColAFP) and prokaryotic organisms such as *Flavobacteriaceae* bacteria and eukaryotic microorganisms such as fungi and copepods (Hanada et al. 2014). The high expression of AFP domains by eukaryotic microbial communities indicates that this protein plays an important role in Arctic and Antarctic sea ice survival (Uhlig et al. 2015).

In summary, in the context of the significant economic losses caused by frost events in agriculture, it provides a strategy needed to reduce the damage of IBP-producing bacteria or IBPs from ice crystal formation and to stimulate growth in seasonal cold temperatures.

References

Abe F (2004) Piezophysiology of yeast: occurrence and significance. Cell Mol Biol 50:437–445

Adams MWW, Kelly RM (1994) Thermostability and thermoactivity of enzymes from hyperthermophilic archaea. Bioorg Med Chem 2:659–667

Allen M, Lauro FM, Williams TJ, Burg D, Siddiqui KS et al (2009) The genome sequence of the psychrophilic archaeon, Methanococcoides burtonii: the role of genome evolution in cold adaptation. ISME J 3:1012–1035

Amir G, Rubinsky B, Kassif Y et al (2003) Preservation of myocyte struc- ture and mitochondrial integrity in subzero cryopreservation of mammalian hearts for transplantation using antifreeze proteins—an electron microscopy study. Eur J Cardio Thorac 24:292–297

Bauer H, Kasper-Giebl A, Loflund M, Giebl H, Hitenberger R, Zibuschka F, Puxbaum H (2002) The contribution of bacterial and fungal spores to the organic carbon content of cloud water, precipitation and aerosols. Atmos Res 64:109–119

Bergholz PW, Bakermans C, Tiedje JM (2009) Psychrobacter arcticus 273–4 uses resource efficiency and molecular motion adaptations for subzero temperature growth. J Bacteriol 191:2340–2352

Bravo La. Griffith M. (2005) Characterization of antifreeze activity in Antarctic plants. J Exp Bot 56:1189–1196

Brush R, Griffth M, Mlynarz A (1994) Characterization and quantifcation of Intrinsic ice nucleators in winter rye (secale cereale) leaves. Plant Physiol 104:725–735

Burke M, Lindow S (1990) Surface properties and size of the ice nucleation site in ice nucleation active bacteria: theoretical considerations. Cryobiology 27:80–84

Burrows SM, Elbert W, Lawrence MG, Poschl U (2009) Bacteria in the global atmosphere e part 1: review and synthesis of literature data for different ecosystems. Atmos Chem Phys 9:9263–9280

Cambours M, Nejad P, Granhall U et al (2005) Frost-related dieback of willows. Comparison of epiphytically and endophytically isolated bacteria from different Salix clones, with emphasis on ice nucleation activity, pathogenic properties and sea- sonal variation. Biomass Bioenergy 28:15–27

Cameron KA, Hagedorn B, Dieser M, Christner BC, Choquette K, Sletten R (2015) Diversity and potential sources of microbiota associated with snow on western portions of the Greenland ice sheet. Environ Microbiol 17:594–609

Castrillo L, Rutherford S, Lee R et al (2001) Enhancement of ice- nucleating activity in Pseudomonas fluorescens and its effect on efficacy against overwintering Colorado potato beetles. Biol Control 21:27–34

Cavicchioli R, Siddiqui KS, Sowers KR, Andrews D (2002) Low-temperature extremophiles and their applications. Curr Opin Biotechnol 13:253–261

Cavicchioli R, Curmi PMG, Siddiqui KS, Thomas T (2006) Proteins from psychrophiles. In: Rainey FA, Oren A (eds) Methods in microbiology: extremophiles, vol 35. Academic, London, pp 395–436

Cavicchioli R, Charlton T, Ertan H, Mohd Omar S. Siddiqui KS. Williams TJ. (2011) Biotechnological uses of enzymes from psychrophiles. J Microbial Biotechnol 4:449–460

Chevalier R, Hoogenboom G, McClendon R et al (2012) A web-based fuzzy expert system for frost warnings in horticultural crops. Environ Model Softw 35:84–91

Christiansen JK, Larsen MH, Ingmer H, Søgaard-Andersen L, Kallipolitis BH (2004) The RNA-binding protein Hfq of Listeria monocytogenes: role in stress tolerance and virulence. J Bacteriol 186:3355–3362

Christner BC, Kvitko BH, Reeve JN (2003) Molecular identification of Bacteria and Eukarya inhabiting an Antarctic cryoconite hole. Extremophiles 7:177–183

Christner B, Morris C, Foreman C et al (2008) Ubiquity of biological ice nucleators in snowfall. Science 319:1214

Cochet N, Widehem P (2000) Ice crystallization by Pseudomonas syringae. Appl Microbiol Biotechnol 54:153–161

Cockell CS, Cordoba-Jabonero C (2004) Coupling of climate change and biotic UV exposure through changing snoweice covers in terrestrial habitats. Photochem Photobiol 79(1):26–31

Crawford DL, Powers DA (1992) Evolutionary adaptation to different thermal environments via transcriptional regulation. Mol Biol Evol 9:806–813

D'Amico S, Marx JC, Gerday C, Feller G (2003) Activity-stability relationships in extremophilic enzymes. J Biol Chem 278:7891–7896

D'Amico S, Collins S, Marx JC, Feller G, Gerday C (2006) Psychrophilic microorganisms: challenges for life. EMBO Rep 7:385–389

Davidson EA, Janssens IA (2006) Temperature sensitivity of soil carbon decomposition and feedback to climate change. Nature 440:165–173

Davies P (2016) Antarctic moss is home to many epiphytic bacteria that secrete antifreeze proteins. Environ Microbiol Rep 8:1–2

Deguillaume L, Leriche M, Amato P, Ariya PA, Delort AM, Pöschl U, Chaumerliac N, Flossmann AI BH, Morris CE (2008) Microbiology and atmospheric processes: chemical interactions of primary biological aerosols. Biogeosciences 5:1073–1084

Deming JW (2009) Extremophiles: cold environments. Encyclopedia of microbiology. Elsevier, Oxford, pp 147–158

Demirjian DC, Moris-Varas F, Cassidy CS (2001) Enzymes from extremophiles. Curr Opin Chem Biol 5:144–151

DeVries A, Wohlschlag D (1969) Freezing resistance in some Antarctic fshes. Science 163:1073–1075

Du N, Liu XY, Hew CL (2003) Protein structure and folding: ice nucleation inhibition: mechanism of antifreeze by antifreeze protein. J Biol Chem 278:36000–36004

Duman JG (2001) Antifreeze and ice nucleator proteins in terrestrial arthropods. Annu Rev Physiol 63:327–357

Ehira S, Hamano T, Hayashida T, Kojima K, Nakamoto H et al (2003) Conserved temperature-dependent expression of RNA-binding proteins in cyanobacteria with different temperature optima. FEMS Microbiol Lett 225:137–142

Ekstrom S, Noziere B, Hultberg M, Alsberg T, Magner J, Nilsson ED, Artaxo P (2010) A possible role of ground-based microorganisms on cloud formation in the atmosphere. Biogeosciences 7:387–394

Elenda C, Schmeissera C, Hoebenreich H, Steelea HL, Streit WR (2007) Isolation and characterization of a metagenome-derived and cold-active lipase with high stereospecificity for (R)-ibuprofen esters. J Biotechnol 130:370–377

EuropaBio Rep (2009) How industrial biology can tackle climate change. EuropaBio, Jan. 28, Bruss., Belg. http://www.europabio.org/industrial/positions/how-industrial-biotechnology-can-tackle-climatechange

Feeney R, Yeh Y (1998) Antifreeze proteins: current status and possible food uses. Trends Food Sci Technol 9:102–106

Feller G (2007) Life at low temperatures: is disorder the driving force? Extremophiles 11:211–216

Feller G. (2008) Enzyme function at low temperatures in psychrophiles. Siddiqui KS, Thomas T 35–69

Feller G (2013) Psychrophilic enzymes: from folding to function and biotechnology. Scientifica (Cairo) 2013:1–28

Feller G, Gerday C (2003) Psychrophilic enzymes: hot topics in cold adaptation. Nat Rev Microbiol 1:200–208

Ganzert L, Jurgens G, Münster U, Wagner D (2007) Methanogeniccommunities in permafrost-affected soils of the Laptev Sea coast, Siberian arctic, characterized by 16S rRNA gene fingerprints. FEMS Microbiol Ecol 59:476–488

Gao H, Yanh ZK, Wu L, Thompson DK, Zhou J (2006) Global transcriptome analysis of the cold shock response of Shewanella oneidensis MR-1 and mutational analysis of its classical cold shock proteins. J Bacteriol 188:4560–4569

Garnham CP, Nishimiya Y, Tsuda S et al (2012) Engineering a naturally inactive isoform of type III antifreeze protein into one that can stop the growth of ice. FEBS Lett 586:3876–3881

German DP, Marcelo KRB, Stone MM, Allison SD (2012) The Michaelis-Menten kinetics of soil extracellular enzymes in response to temperature: a cross-latitudinal study. Glob Chang Biol 18:1468–1479

Giaquinto L, Curmi PM, Siddiqui KS, Poljak A, DeLong E et al (2007) Structure and function of cold shock proteins in archaea. J Bacteriol 189:5738–5748

Gilichinsky D, Vishnivetskaya T, Petrova M, Spirina E, Mamykin V, Rivkina E (2008) Bacteria in permafrost. In: Margesin R, Schinner F, Marx JC, Gerday C (eds) Psychrophiles: from biodiversity to biotechnology. Springer Verlag, Berlin, pp 83–102

Govindarajan A, Lindow S (1988) Size of bacterial ice-nucleation sites measured in situ by radiation inactivation analysis. Proc Natl Acad Sci U S A 85:1334–1338

Graether S, Sykes B (2004) Cold survival in freeze-intolerant insects: the structure and function of beta-helical antifreeze proteins. FEBS J 271:3285–3296

Graham LA, Liu YC, Walker VK, Davies P (1997) Hyperactive antifreeze protein from beetles. Nature 388:727–728

Griffin DW (2008) Non-spore forming eubacteria isolated at an altitude for 20,000 m in Earth's atmosphere: extended incubation periods needed for culture-based assays. Aerobiologia 24:19–25

Griffith M, Ala P, Yang DSC et al (1992) Antifreeze protein produced endogenously in winter rye leaves. Plant Physiol 100:593–596

Gurian-Sherman D, Lindow S (1992) Ice nucleation and its application. Curr Opin Biotechnol 3:303–306

Hanada Y, Nishimiya Y, Miura A et al (2014) Hyperactive antifreeze protein from an Antarctic Sea ice bacterium Colwellia sp. has a compound ice-binding site without repetitive sequences. FEBS J 281:3576–3590

Hassan N, Rafiq M, Hayat M, Shah AA, Hasan F (2016) Psychrophilic and psychrotrophic fungi: a comprehensive review. Rev Environ Sci Biotechnol 15:147–172

Hebraud M, Potier P (1999) Cold shock response and low temperature adaptation in psychrotrophic bacteria. J Mol Microbiol Biotechnol 1:211–219

Hirano Y, Nishimiya Y, Matsumoto S et al (2008) Hypothermic preser- vation effect on mammalian cells of type III antifreeze proteins from notched-fin eelpout. Cryobiology 57:46–51

Hodson A, Anesio AM, Tranter M, Fountain A, Osborn M, Priscu J, Laybourn-Parry J, Sattler B (2008) Glacial ecosystems. Ecol Monogr 78:41–67

Hoham RW, Duval B (2001) Microbial ecology of snow and freshwater ice with emphasis on snow algae. In: Jones HG, Pomeroy JW, Walker DA, Hoham RW (eds) Snow ecology. Cambridge University Press, Cambridge, pp 168–228

Hoshino T, Kiriaki M, Ohgiya S et al (2003) Antifreeze proteins from snow mold fungi. Can J Bot 81:1175–1181

Janech MG, Krell A, Mock T et al (2006) Ice-binding proteins from sea ice diatoms (Bacillariophyceae). J Phycol 42:410–416

Jing T, Gao X, Wang P et al (2009) Determination of trace tetracycline antibiotics in foodstuffs by liquid chromatography-tandem mass spectrometry coupled with selective molecular-imprinted solid-phase extraction. Anal Bioanal Chem 393:2009–2018

Joly M, Attard E, Sancelme M, Deguillaume L (2013) Ice nucleation activity of bacteria isolated from cloud water. Atmos Environ 70:392–400

Jones HG (1999) The ecology of snow-covered systems: a brief overview of nutrient cycling and life in the cold. Hydrol Process 13:2135–2147

Jones PG, Inouye M (1994) The cold-shock response: a hot topic. Mol Microbiol 11:811–818

Jung YH, Yi JY, Jung HJ, Lee YK, Lee HK et al (2010) Overexpression of cold shock protein A of Psychromonas arctica KOPRI 22215 confers cold resistance. Protein J 29:136–142

Junge K, Eicken H, Deming JW (2004) Bacterial activity at −2 to −20°C in Arctic wintertime sea ice. Appl Environ Microbiol 70:550–557

Kawahara H (2008) Cryoprotectants and ice-binding proteins. In: Margesin R et al (eds) Psychrophiles: from biodiversity to biotechnology. Springer-Verlag, Berlin, pp 229–246

Kishimoto T, Yamazaki H, Saruwatari A et al (2014) High ice nucleation activity located in blueberry stem bark is linked to primary freeze initiation and adaptive freezing behavior of the bark. AoB Plants 6:1–17

Kogan A, Gdalevsky GY, Cohen-Luria R, Goldgur Y, Phillips RS et al (2009) Conformational changes and loose packing promote E. coli Tryptophanase cold lability. BMC Struct Biol 9:65

Kontogiorgos V, Regand A, Yada RY et al (2007) Isolation and characterization of ice structuring proteins from cold-acclimated winter wheat grass extract for recrystallization inhibition in frozen foods. J Food Biochem 31:139–160

Koushafar H, Rubinsky B (1997) Effect of antifreeze proteins on frozen primary prostatic adeno-carcinoma cells. Urology 49:421–425

Kristiansen E, Ramløv H, Højrup P et al (2011) Structural characteristics of a novel antifreeze protein from the longhorn beetle Rhagium inquisitor. Insect Biochem Mol 41:109–117

Lauro FM, Allen M, Wilkins D, Williams TJ, Cavicchioli R (2011) Genetics, genomics and evolution of psychrophiles. In: Horikoshi K, Antranikian G, Robb F, Stetter K, Bull AT (eds) Extremophiles handbook. Springer, Heidelberg, pp 1179–1199

Lee R, Warren G, Gusta L, (eds). (1995) Biological ice nucleation and its applications. American Phytopathological Society, St. Paul, MN, p 1995

Lee RE, Castrillo LA, Lee MR, Wyman JA, Costanzo JP (2001) Using ice-nucleating bacteria to reduce winter survival of Colorado potato beetles: development of a novel strategy for

biological control. In: Denlinger DL, Giebultowicz JM, Saunders DS (eds) Insect timing: circadian rhythmicity to seasonality. Elsevier Science B.V., Amsterdam, pp 213–227
Lee J, Park K, Park S et al (2010) An extracellular ice-binding glycoprotein from an Arctic psychrophilic yeast. Cryobiology 60:222–228
Lee SG, Koh HY, Lee JH et al (2012) Cryopreservative effects of the recombinant ice-binding protein from the arctic yeast Leucosporidium sp. on red blood cells. Appl Biochem Biotechnol 167:824–834
Levin Z, Yankofsky S, Pardes D et al (1986) Possible application of bacterial condensation freezing to artificial rainfall enhancement. J Clim Appl Meteorol 26:1188–1197
Lim J, Thomas T, Cavicchioli R (2000) Low temperature regulated DEAD-box RNA helicase from the Antarctic archaeon, Methanococcoides burtonii. J Mol Biol 297:553–567
Lindow S (1983) The role of bacterial ice nucleation in frost injury to plants. Annu Rev Phytopathol 21:363–384
Lindow S (1984) Methods of preventing frost injury caused by epiphytic ice nucleation active bacteria. Plant Dis 67:327–333
Lindow S (1987) Competitive exclusion of epiphytic bacteria by competitive exclusion of epiphytic bacteria by ice-Pseudomonas syringae mutants. Appl Environ Microbiol 53:2520–2527
Lindow S, Brandl M (2003) Microbiology of the phyllosphere. Appl Environ Microbiol 69:1875–1883
Lindow S, Arny D, Upper C (1982) Bacterial ice nucleation: a factor in frost injury to plants. Plant Physiol 70:1084–1089
Liou Y, Tocilj A, Davies P et al (2000) Mimicry of ice structure by surface hydroxyls and water of a beta-helix antefreeze protein. Nature 406:322–324
Lohmann U, Feichter J (2005) Global indirect aerosol effects: a review. Atmos Chem Phys 5:715–737
Lopatina A, Krylenkov V, Severinov K (2013) Activity and bacterial diversity of snow around Russian Antarctic stations. Res Microbiol 164:e1
Lopatina A, Medvedeva S, Shmakov S, Logacheva MD, Krylenkov V, Severinov K (2016) Metagenomic analysis of bacterial communities of Antarctic surface snow. Front Microbiol 7:398
Lopez-Bueno A, Tamames J, Velazquez D, Moya A, Quesada A, Alcami A (2009) High diversity of the viral community from an Antarctic lake. Science 326:858–861
Lorv JS, Rose DR, Glick BR (2014) Bacterial ice crystal controlling proteins. Scientifica 2014:976895
Lutz S, Anesio AM, Edwards A, Benning LG (2015) Microbial diversity on Icelandic glaciers and ice caps. Front Microbiol 6:307
Lutz S, Ziolkowski LA, Benning LG (2019) The biodiversity and geochemistry of Cryoconite holes in Queen Maud Land, East Antarctica. Microorganisms 7:160
MacDonell S, Fitzsimons S (2008) The formation and hydrological significance of cryoconite holes. Prog Phys Geogr 32:595–610
Margaritis A, Bassi A (1991) Principles and biotechnological applications of bacterial ice nucleation. Crit Rev Biotechnol 11:277–295
Margesin R, Miteva V (2011) Diversity and ecology of psychrophilic microorganisms. Res Microbiol 162:346–361
Margesin R, Zacker G, Schinner F (2002) Characterization of heterotrophic microorganisms in alpine glacier cryoconite. Arct Antarct Alp Res 34:88–93
Maruyama K, Sato N, Ohta N (1999) Conservation of structure and cold-regulation of RNA-binding proteins in cyanobacteria: probable convergent evolution with eukaryotic glycine-rich RNA-binding proteins. Nucleic Acids Res 27:2029–2036
Middleton A, Brown A, Davies P et al (2009) Identification of the ice-binding face of a plant antifreeze protein. FEBS Lett 583:815–819
Middleton A, Marshall C, Faucher F et al (2012) Antifreeze protein from freeze-tolerant grass has a beta-roll fold with an irregularly structured ice-binding site. J Mol Biol 416:713–724
Miteva V (2008) Bacteria in snow and glacier ice. In: Margesin R, Schinner F, Marx JC, Gerday C (eds) Psychrophiles: from biodiversity to biotechnology. Springer-Verlag, Berlin, pp 31–50

Mori S, Castoreno A, Mulligan ME, Lammers PJ (2003) Nitrogen status modulates the expression of RNA-binding proteins in cyanobacteria. FEMS Microbiol Lett 227:203–210

Morris CE, Sands DC, Bardin M, Jaenicke R, Vogel B, Leyronas C, Ariya PA, Psenner R (2008a) Microbiology and atmospheric processes: an upcoming era of research on bio-meteorology. Biogeosci Discuss 5:191–212

Morris C, Sands D, Vinatzer B et al (2008b) The life history of the plant pathogen Pseudomonas syringae is linked to the water cycle. ISME J 2:321–334

Muldrew K, Rewcastle J, Donnelly B et al (2001) Flounder antifreeze peptides increase the efficacy of cryosurgery. Cryobiology 42:182–189

Nielsen PH (2005) Life cycle assessment supports cold-wash enzymes. SOFW J 10:1–4

Nogi Y (2008) Bacteria in the deep sea: psychropiezophiles. In: Margesin R, Schinner F, Marx JC, Gerday C (eds) Psychrophiles: from biodiversity to biotechnology. Springer Verlag, Berlin, pp 73–82

Parody-Morreale A, Murphy K (1988) Inhibition of bacterial ice nucleators by FSH antifreeze glycoproteins. Nature 333:782–783

Pearce R (2001) Plant freezing and damage. Ann Bot 87:417–424

Pearce DA (2009) Antarctic subglacial lake exploration: a new frontier in microbial ecology. ISME J 3:877–880

Pearce DA, Bridge PD, Hughes KA, Sattler B, Psenner R, Russell NJ (2009) Microorganisms in the atmosphere over Antarctica. FEMS Microbiol Ecol 69:143–157

Pearce DA, Hughes KA, Lachlan-Cope T, Harangozo SA, Jones AE (2010) Biodiversity of air-borne microorganisms at Halley station, Antarctica. Extremophiles 14:145–159

Pertaya N, Marshall C, Celik Y et al (2008) Direct visualization of spruce budworm antifreeze protein interacting with ice crystals: basal plane affinity confers hyperactivity. Biophys J 95:333–341

Phadtare S (2004) Recent developments in bacterial cold-shock response. Curr Issues Mol Biol 6:125–136

Porazinska DL, Fountain AG, Nylen TH, Tranter M, Virginia RA, Wall DH (2004) The biodiversity and biogeochemistry of Cryoconite holes from McMurdo Dry Valley glaciers, Antarctica. Arct Antarct Alp Res 36:84–91

Priscu JC, Christner BC (2004) Earth's icy biosphere. In: Bull AT (ed) Microbial diversity and bioprospecting. ASM Press, Washington DC, pp 130–145

Priscu JC, Christner BC, Foreman CM, Royston-Bishop G (2007) Biological material in ice cores. In: Encyclopedia of quaternary sciences. Elsevier, Amsterdam, pp 1156–1167

Raymond J (2016) Dependence on epiphytic bacteria for freezing protection in an Antarctic moss, Bryum argenteum. Environ Microbiol Rep 8:14–19

Raymond J, DeVries A (1977) Adsorption inhibition as a mechanism of freezing resistance in polar fshes. Proc Natl Acad Sci U S A 74:2589–2593

Raymond J, Janech M, Fritsen C (2009) Novel ice-binding proteins from a psychrophilic antarctic alga (Chlamydomonadaceae, Chlorophyceae). J Phycol 45:130–136

Reid IN, Sparks WB, Lubnow S, McGrath M, Livio M et al (2006) Terrestrial models for extraterrestrial life: methanogens and halophiles at Martian temperatures. Int J Astrobiol 5:89–97

Rivkina E, Gilichinsky D, Wagener S, Tiedje J, McGrath J (1998) Biogeochemical activity of anaerobic microorganisms from buried permafrost sediments. Geomicrobiol J 15:187–193

Rivkina R, Shcherbakova V, Laurinavichius K, Petrovskaya L, Krivushin K, Kraev G, Pecheritsina S, Gilichinsky D (2007) Biogeochemistry of methane and methanogenic archaea in permafrost. FEMS Microbiol Ecol 61:1–15

Russell NJ (2008) Membrane components and cold sensing. In: Margesin R, Schinner F, Marx J-C, Gerday C (eds) Psychrophiles: from biodiversity to biotechnology. Springer, Berlin, pp 177–190

Sattler B, Puxbaum H, Psenner R (2001) Bacterial growth in super cooled cloud droplets. Geophys Res Lett 28:239–242

Saunders NF, Thomas T, Curmi PM, Mattick JS, Kuczek E, Slade R, Davis J, Franzmann PD, Boone D, Rusterholtz K, Feldman R, Gates C, Bench S, Sowers K, Kadner K, Aerts A, Dehal P, Detter C, Glavina T, Lucas S, Richardson P, Larimer F, Hauser L, Land M, Cavicchioli R (2003) Mechanisms of thermal adaptation revealed from the genomes of the Antarctic archaea Methanogenium frigidum and Methanococcoides burtonii. Genome Res 13:1580–1588

Schmid D, Pridmore D, Capitani G et al (1997) Molecular organisation of the ice nucleation protein InaV from Pseudomonas syringae. FEBS Lett 414:590–594

Scotter A, Marshall C, Graham L et al (2006) The basis for hyperactivity of antifreeze proteins. Cryobiology 53:229–239

Siddiqui KS, Cavicchioli R (2006) Cold-adapted enzymes. Annu Rev Biochem 75:403–433

Siddiqui KS, Feller G, D'Amico S, Gerday C, Giaquinto L, Cavicchioli R (2005) The active site is the least stable structure in the unfolding pathway of a multidomain cold-adapted α-amylase. J Bacteriol 187:6197–6205

Siddiqui KS, Poljak A, Guilhaus M, De Francisci D, Curmi PM, Feller G, D'amico S, Gerday C, Uversky VN, Cavicchioli R (2006) Role of lysine versus arginine in enzyme cold-adaptation: modifying lysine to homo-arginine stabilizes the cold-adapted α-amylase from *Pseudoalteramonas haloplanktis*. Proteins 64:486–501

Simon C, Wiezer A, Strittmatter AW, Daniel R (2009) Phylogenetic diversity and metabolic potential revealed in a glacier ice metagenome. Appl Environ Microbiol 75:7519–7526

Singh SM, Puja G, Bhat DJ (2006) Psychrophilic fungi from Schirmacher oasis, East Antarctica. Curr Sci 90:1388–1392

Sogin ML, Morrison HG, Huber JA, Welch DM, Huse SM, Neal PR, Arrieta JM, Herndl GJ (2006) Microbial diversity in the deep sea and the underexplored "rare biosphere". Proc Natl Acad Sci U S A 103:12115–12120

Somero GN (1995) Proteins and temperature. Annu Rev Physiol 57:43–68

Steven B, Leveille R, Pollard WH, Whyte LG (2006) Microbial ecology and biodiversity in permafrost. Extremophiles 10:259–267

Sun X, Griffith M, Pasternak J et al (1995) Low temperature growth, freezing survival, and production of antifreeze protein by the plant growth promoting rhizobacterium Pseudomonas putida GR12-2. Can J Microbiol 41:776–784

Takai K, Nakamura K, Toki T, Tsunogai U, Miyazaki M et al (2008) Cell proliferation at 122°C and isotopically heavy CH4 production by a hyperthermophilic methanogen under high-pressure cultivation. Proc Natl Acad Sci U S A 105:10949–10954

Tanghe A, Van Dijck P, Thevelein JM (2003) Determinants of freeze tolerance in microorganisms, physiological importance, and biotechnological applications. Adv Appl Microbiol 53:129–175

Thomas DN, Dieckmann GS (2002) Antarctic Sea ice: a habitat for extremophiles. Science 295:641–644

Thomashow M (1998) Role of cold-responsive genes in plant freezing tolerance. Plant Physiol 118:1–7

Tomalty HE, Walker VK (2014) Perturbation of bacterial ice nucleation activity by a grass antifreeze protein. Biochem Biophys Res Commun 452:636–641

Trotsenko YA, Khmelenina VN (2005) Aerobic methanotrophic bacteria of cold ecosystems. FEMS Microbiol Ecol 53:15–26

Tsumuki H, Konno H, Maeda T et al (1992) An ice-nucleating active fungus isolated from the gut of the rice stem borer, Chilo supressalis Walker, (lepidoptera: pyralidae). J Insect Physiol 38:119–125

Uhlig C, Kilpert F, Frickenhaus S et al (2015) In situ expression of eukaryotic ice-binding proteins in microbial communities of Arctic and Antarctic Sea ice. ISME J 9:1–4

Vaïtilingom M, Attard E, Gaiani N, Sancelme M, Deguillaume L, Flossmann AI et al (2012) Long-term 466 features of cloud microbiology at the puy de Dôme (France). Atmos Environ 56:88–100

Valluru R, Van Den Ende W (2008) Plant fructans in stress environments: emerging concepts and future prospects. J Exp Bot 59:2905–2916

Valluru R, Lemmens W, Claupein W, Van den Ende W (2008) Freezing tolerance by vesicle-mediated fructan transport. Trends Plant Sci 13:4

Varin T, Lovejoy C, Jungblut AD, Vincent WF, Corbeil J (2012) Metagenomic analysis of stress genes in microbial mat communities from Antarctica and the high artic. Appl Environ Microbiol 78:549–559

Vincent WF (2007) Cold tolerance in cyanobacteria and life in the cryosphere. In: Seckbach J (ed) Algae and cyanobacteria in extreme environments. Springer-Verlag, Berlin, pp 287–301

Wainwright M, Wickramasinghe N, Narlikar J, Rajaratnam P (2004) Microorganisms cultured from stratospheric air samples obtained at 41 km. FEMS Microbiol Lett 218:161–165

Warren G, Mueller G, McKown R (1992) Ice crystal growth suppression polypeptides and method of making. US Pat 5118792

Weedon JT, Aerts R, Kowalchuk GA, van Bodegom PM (2011) Enzymology under global change: organic nitrogen turnover in alpine and sub-arctic soils. Biochem Soc Trans 39:309–314

Whitman WB, Coleman DC, Wiebe WJ (1998) Prokaryotes e the unseen majority. Proc Natl Acad Sci U S A 95:6578–6583

Widehem P, Cochet N (2003) *Pseudomonas syringae* as an ice nucleator—application to freeze-concentration. Process Biochem 39:405–410

Williams TJ, Burg DW, Ertan H, Raftery MJ, Poljak A et al (2010a) Global proteomic analysis of the insoluble, soluble, and supernatant fractions of the psychrophilic archaeon Methanococcoides burtonii. Part II: the effect of different methylated growth substrates. J Proteome Res 9:653–663

Williams TJ, Burg DW, Raftery MJ, Poljak A, Guilhaus M et al (2010b) Global proteomic analysis of the insoluble, soluble, and supernatant fractions of the psychrophilic archaeon Methanococcoides burtonii. Part I: the effect of growth temperature. J Proteome Res 9:640–652

Williams TJ, Lauro FM, Ertan H, Burg DW, Poljak A et al (2011) Defining the response of a microorganism to temperatures that span its complete growth temperature range (−2°C to 28°C) using multiplex quantitative proteomics. Environ Microbiol 13:2186–2203

Xiao N, Suzuki K, Nishimiya Y et al (2010) Comparison of functional properties of two fungal antifreeze proteins from Antarctomyces psychrotrophicus and Typhula ishikariensis. FEBS J 277:394–403

Xu Y, Feller G, Gerday C, Glansdorff N (2003) Moritella cold-active dihydrofolate reductase: are there natural limits to optimization of catalytic efficiency at low temperature? J Bacteriol 185:5519–5526

Yang DQ, Wang JH, Bai Y, Xu S, An LZ (2008) Diversity and distribution of the prokaryotic community in near-surface permafrost sediments in the Tianshan Mountains, China. Can J Microbiol 54:270–280

Yeh CM, Kao BY, Peng HJ (2009) Production of a recombinant type 1 antifreeze protein analogue by L. lactis and its applications on frozen meat and frozen dough. J Agric Food Chem 57:6216–6223

Yu SO, Brown A, Middleton AJ, Tomczak MM, Walker VK, Davies PL (2010) Ice restructuring inhibition activities in antifreeze proteinswith distinct differences in thermal hysteresis. Cryobiology 61:327–334

Zhakia F, Jungblut AD, Taton A, Vincent WF, Wilmotte A (2008) Cyanobacteria in cold ecosystems. In: Psychrophiles: from biodiversity to biotechnology. Springer Verlag, Berlin, pp 121–135

Zhang S, Wang H, Chen G (2010) Addition of ice-nucleation active bacteria: Pseudomonas syringae pv. Panici on freezing of solidmodel food. LWT Food Sci Technol 43:1414–1418

Zhao JS, Deng Y, Manno D, Hawari J (2010) Shewanella spp. genomic evolution for a cold marine lifestyle and in-situ explosive biodegradation. PLoS One 5:e9109

Chapter 12
Microbial Stress Response to Heavy Metals

Lyudmila Ignatova, Aida Kistaubayeva, Irina Savitskaya, Akzhigit Mashzhan, and Dilfuza Egamberdieva

Abstract Contamination by heavy metals is a global environmental problem as it poses a significant danger to all living organisms including microbes. The study of the physiological mechanisms of plant resistance to the damaging effects of abiotic factors is one of the most important biotechnology problems. Its solution is of fundamental importance for understanding the strategy of plant survival in extreme conditions, as well as for developing a technology for protecting plants from the damaging effects of adverse environmental factors. This chapter provides information about studies on the effect of heavy metal ions at concentrations of 5 and 15 mM on the growth of micromycetes, *B. bassiana* T7 strain, which has multiple resistance to zinc, copper, manganese, and lead. The results of this study will help to efficiently use the potential of microorganisms in their ability to increase the resistance of plants to stress factors, which can become the basis for the development of plant cultivation technology.

Keywords Heavy metals · Heavy metals · Zinc · Resistance · Micromycetes

L. Ignatova · A. Kistaubayeva (✉) · I. Savitskaya
Department of Biotechnology, al-Farabi Kazakh National University, Almaty, Kazakhstan
e-mail: lyudmila.Ignatova@kaznu.kz; irina.savickaya@kaznu.kz

A. Mashzhan
Department of Biotechnology, al-Farabi Kazakh National University, Almaty, Kazakhstan

Department of Biological Sciences, University of Bergen, Bergen, Norway

D. Egamberdieva
Faculty of Biology, National University of Uzbekistan, Tashkent, Uzbekistan

Leibniz Centre for Agricultural Landscape Research (ZALF), Müncheberg, Germany

D. Egamberdieva et al. (eds.), *Microbial Communities and their Interactions in the Extreme Environment*, Microorganisms for Sustainability 32,
https://doi.org/10.1007/978-981-16-3731-5_12

12.1 Introduction

Currently, the anthropogenic load on ecosystems is significantly increasing. In particular, the amount of heavy metals released into the environment is growing. Pollution of agricultural land and groundwater by heavy metals is directly related to human activities. At an increased concentration in the environment, heavy metal ions are actively absorbed by the plant roots and enter their aboveground organs, causing metabolic disturbances and growth retardation (Bingham et al. 2001). The increased content of heavy metals in contaminated soils leads to a decrease in fertility and, ultimately, to a decrease in crop yields.

In this regard, the attention of many researchers is attracted by microorganisms that are capable of increasing the productivity of agricultural plants and resistance to various environmental stressors, which is one of their most important properties and the main factor of using them to create biofertilizers (Gadd 1993).

The study of the functioning of micromycete complex with plants under stressful conditions allowed us to obtain information about the stability and growth indicators of plant-microbial systems. The ability of micromycete strains, from the collection of the Department of Biotechnology of Al-Farabi Kazakh National University, to increase the resistance of inoculated plants to the negative effects of heavy metals is experimentally proved.

Research is of great practical importance for the development of environmentally safe and resource-saving agricultural production, and the search for solutions to a number of environmental problems. The results of the research will help to effectively use the potential of microorganisms in their ability to increase the resistance of plants to stress factors, which can become the basis for the development of plant cultivation technology.

The aim of the research is to study the effect of micromycetes on resistance to heavy metals.

12.1.1 Environmental Features of Heavy Metals

The manifestation of adaptability to environmental conditions in symbioses of plants and microorganisms is a complex biological system based on the interaction of partners. It is known that symbiotrophic microorganisms are highly resistant to heavy metals, which are involved in the processes of their transformation in the rhizosphere and accumulation by plants. However, the specific mechanisms of symbiose adaptation to toxic concentrations of metals and the component integration of plant-microbial systems under stressful conditions have not been sufficiently studied. HMs are necessary biologically active elements for plants and microorganisms in microconcentrations, they are toxic for plants and rhizosphere microorganisms in higher concentrations (over 0.1%). As a result of various environmental studies of heavy metals, a group was identified that includes elements that are particularly toxic to living organisms, namely such elements as cadmium, copper, arsenic, nickel, mercury, lead, zinc, and chromium. Mercury, lead, and cadmium are considered the most harmful (Gadd 1993). The increase in their number in the environment as a result of natural or anthropogenic ingress can be a global problem. The following natural sources of HMs are distinguished: rocks, volcanoes, cosmic dust, soil erosion, evaporation from the surface of the seas and oceans, and their release by plants (Gadd 1993; Szczerba et al. 2009; Nies 1999).

The leading anthropogenic sources of heavy metals include a variety of vehicles, emissions from industrial enterprises of the coal mining, metallurgical, chemical industry, and energy complex (Shree and Rudra 2007; Ghorbani et al. 2002). Also, the sources of HMs are agrotechnical measures, namely the inclusion of pesticides, chemical ameliorants, mineral and organic fertilizers. The most toxic in terms of the set and the concentration of HMs are phosphoric fertilizers, as well as fertilizers with extraction orthophosphoric acid. From organic fertilizers, active sludge from treatment plants is dangerous for plants. During the formation of both, conditions are formed for increasing the concentration of HMs relative to the original substrate (wastewater and green mass, respectively) (Shree and Rudra 2007). Moreover, fertilizers are used systematically and in large quantities, which contribute to the accumulation of HMs in the soil (Ghorbani et al. 2002).

Heavy metals can bind to various soil components, both organic (organic and humic acids) and inorganic (clay particles and aluminum and iron hydroxides). This greatly complicates their immobilization and removal and makes their influence chronic (Ghorbani et al. 2002; Gadd 2001; Hiroki 1992).

12.1.2 *Effects of Heavy Metals on Plants*

Heavy metal ions are characterized by significant biological activity. This is explained by the fact that they belong to *d*-elements (the valence electrons of their atoms are located on *d*-orbital), which form complex compounds and compounds of variable composition and valence (Vassilev et al. 2004). These properties allow them to interact differently with organic molecules, namely with protein molecules. At the same time, HMs perform a dual role in biochemical processes—in small quantities they are needed to maintain the vital activity of the body, but in large doses they have a toxic effect.

Heavy metals are trace elements necessary for the body in very small quantities. HMs are functional elements, since they act as cofactors and activators of enzymes, and are an integral part of vitamins and other BAS. According to the quantitative content in plants, HMs are divided into four types: elements of high concentration—Mn, Zn, Fe; medium—Cu, Ni, Pb, Cr; low—Co, Mo, Cd, Sn; very low—Hg, Ag (Yadav 2010; Tran and Popova 2013).

At the same time, heavy metals in high concentrations have a negative effect on almost all living organisms, including plants. The main reason for this effect is that heavy metal ions are able to very actively combine with the sulfhydryl groups (-S-H) of proteins (primarily enzymes), blocking the active centers, modifying the tertiary structure of molecules and thereby disrupting their normal functioning (Cera 2006). At the moment, more than a hundred enzymes inactivated by HMs have been established. Heavy metals inhibit a variety of enzymes that are involved in all the main processes of plant cell life—photosynthesis, membrane transport, division, and respiration.

In addition, an excessive number of HMs is one of the reasons for the development of oxidative stress—a violation of the balance between the production and quenching of free radicals in the cell. It appears as a result of inhibition of enzymes by HMs involved in the neutralization of free radicals (catalase, glutathione peroxidase), and synthesizing antioxidants. In addition, some antioxidants (glutathione) are consumed during the operation of mechanisms that neutralize HMs in the cell (Tran et al. 2013). At the same time, free radicals, accumulating as a result of oxidative stress, can enhance and complement the adverse effects of HM ions, for example, on biological membranes (Chugh and Sawhney 1999).

In general terms, the effects of HMs on plant organisms are based on the following processes:

1. disruption of enzyme systems: it is known that HM chemical properties are similar, so they are able to replace some trace elements in the composition of enzymes, contributing to the disruption of their functions (for example, Cd can take the place of Zn in zinc-containing enzymes);
2. changing the balance of nutrients in the plant body: HMs can interact with some elements useful for vital activity (for example, with phosphate ions), turning them into an insoluble compound;
3. competition between HMs and elements necessary for regular life activity for ingestion into the plant;
4. membrane modification leading to transport disruption (Chugh and Sawhney 1999; El-Helow et al. 2000).

Heavy metals in large quantities inhibit the growth of plants, reduce the germination of seeds. The effect of HMs on the plant organism is characterized by abnormal root development, chlorosis, drying, and death. Growth retardation is the most visible sign of the effect of excessive amounts of HMs on the plant. The sources of growth retardation are different—violation of photosynthesis, energy metabolism, membrane permeability, transpiration and water balance, cell division, and development. Let us discuss certain mechanisms of such impact in more detail (Nies 2003).

First of all, HMs affect the photosynthetic apparatus of the plant organism. The number of photosynthetic pigments, both chlorophylls and carotenoids, decreases (this is a consequence of impaired synthesis and accelerated degradation of existing molecules). The photosynthetic system is disrupted at the molecular level, namely the operation of photosystem II (Cd, Zn) and to a lesser extent photosystem I (Cu), photophosphorylation is suppressed, and the speed of electron transport decreases. The work of the Calvin cycle enzymes is inhibited (by blocking SH-groups and pushing out lighter coenzymes), in addition, there may be a failure in their synthesis.

HMs also change the functioning of the stomatal apparatus, which is the reason for a decrease in transpiration and CO_2 deficiency (Bakker 1993; Burzynski and Klobus 2004).

High concentrations of HMs can also cause modification of the ultrastructure of chloroplasts. For example, thylakoid membranes are damaged, which is caused by accelerated hydrolysis of triglycerides (Burzynski and Klobus 2004). The concentration of plastoglobules increases, the number of grains and thylakoids decreases in them.

HMs affect the respiration of the plant organism in a completely different way. With a small number of heavy metals, respiration is not suppressed, but rather intensifies (Prasad and Hagemeyer 1999). The reason for this physiological response is the excitation of HM ions of certain enzymes of the respiratory cycle, an increase in the need for organic acids, which are necessary for the destruction of metal ions, and also the reason is the general increase in energy consumption for the regeneration of damage caused by poisoning. At high concentrations of heavy metals, cellular respiration is suppressed by blocking the work of glycolysis enzymes, the pentose phosphate pathway, and the Krebs cycle (Prasad and Hagemeyer 1999). HMs also lead to disruption of the normal functioning of mitochondria, inhibiting enzymes located in the matrix and the electron and proton transport system, as well as changing the functions of membranes. For example, the permeability of the outer membrane to large molecules decreases, and the permeability of the inner one to protons increases, which reduces the productivity of oxidative phosphorylation (Benavides et al. 2005).

The increased number of heavy metals negatively affects the water exchange of the plant organism, which is very adversely reflected in the plant life. The main result of noticeable pollution of the environment by HMs is characterized by a decrease in the amount of water in the tissues of the plant organism. First of all, this is due to the degradation of the conducting system, both phloem and xylem (Benavides et al. 2005). This reduces the number of vessels and their radius. In addition, a decrease in the elasticity of the cell walls and a change in the permeability of the membranes, which causes a decrease in water content, are mentioned.

It should be noted that high doses of HMs have a negative effect on the root system, as a result of which the ingress of water into the plant organism worsens. Heavy metals delay root development, inhibit branching, and reduce the number of root hairs. All this makes it difficult for the plant to absorb moisture intensively. HMs also inhibit transpiration. Thus, under the influence of metals, stomatal conductivity (transpiration) under the influence of metal ions decreases (perhaps the reason is a violation of ion transport through the membrane and an increase in the amount of ABA). Under the influence of HMs, there is an increase in the number of stomata per unit area of the leaf due to non-rhythmic delay in the growth of the leaf blade and the formation of new stomata. In addition, the ions of certain HMs are able to induce thickening of the cuticle, which is the reason for the violation of transpiration. The study of the effect of HM ions on root meristems has demonstrated that the number of dividing cells decreases with a large amount of heavy metals, and the duration of the mitotic cycle phases increases (Burzynski and Klobus 2004). The length of interphase d and G2 periods also increases (Burzynski and Klobus 2004; Prasad and Hagemeyer 1999; Benavides et al. 2005). Negative effects occur as a result of blocking SH-group of proteins, by heavy metal ions, involved in DNA replication, "packing" it into a chromosome and diverging homologous chromosomes, changing their functioning, as well as as a result of interaction with DNA molecule itself (Rodríguez-Serrano et al. 2009).

The effect of HM ions on the stem meristems is negative. When exposed to heavy metals, the size of the meristem decreases, the intensity of the laying of vegetative organs decreases. At high concentrations of HMs, growth processes are delayed (Burzynski and Klobus 2004; Prasad and Hagemeyer 1999; Benavides et al. 2005). HMs can stop the growth of cells due to stretching. The main reason is the weakening of the elasticity of the cell walls due to the connection of heavy metal ions with sulfhydryl groups of proteins located in the walls, as well as a decrease in turgor due to changes in water balance. In addition, HMs lead to the formation of free radicals that oxidize the components of the cell wall, which contributes to a decrease in its elasticity (Tran and Popova 2013).

12.1.3 Mechanisms of Protection of Microorganisms from Toxic Effects of Heavy Metals

In accordance with the localization of the metal biosorption process, the mechanisms of biosorption can be classified as intracellular interaction, on the cell surface, and extracellular interaction. Active transport of metal through cell membranes leads to its intracellular accumulation, which depends on the metabolism of the bacterium (Nies 1999).

Essential metals are actively absorbed by specialized absorption systems, since they are necessary, but other, secondary, metals can also be absorbed due to the fact that they are mistaken for a trace element (Gadd 2001).

At high concentrations of toxic metals, MO actively absorbs metal ions to detoxify their environment.

Interaction on the cell surface. Sorption on the cell surface is caused by the presence in the cell walls of compounds with functional groups (phosphate, carboxyl, sulfhydryl, hydroxyl, etc.) capable of binding positively charged HM ions (Vecchio et al. 1998).

The mechanisms, by which the metal binds on the cell surface, probably include electrostatic interactions, Van der Waals forces, covalent interaction and a combination of these processes (Vecchio et al. 1998; Saxena et al. 2002). In the case of physical-chemical interaction based on physical adsorption, ion exchange and complexation between metal and functional groups of the cell surface, the absorption of metals does not depend on metabolism (Vecchio et al. 1998).

In addition, the mechanisms that ensure the resistance of microorganisms to the toxic effects of HM ions include binding in the capsule, cell wall and cytoplasm.

Binding in the capsule and cell wall. The substances that make up the capsule have polarized groups that can bind HM ions (carboxyl, hydroxyl, and in some cases amine) (Kazy et al. 2002; Teitzel and Parsek 2003). In particular, the ability to bind HM ions has been described for *Enterobacter chloaceae, Marinobacter* sp., *Klebsiella aerogenes, Acinetobacter* sp. and others (Kazy et al. 2002). There is evidence that bacteria that form biofilms are more resistant to HMs, in particular, due to the fact that polymers, that bind cells to the film, absorb HMs (Teitzel and Parsek 2003). In some cases, inhibition of exopolysaccharide synthesis not only does not reduce the resistance to HMs, but also increases it (Kazy et al. 2002). This may be due to the significant energy consumption of the synthesis of a large number of polysaccharides. Thus, energy savings by suspending their production increases the overall viability of cells.

The main components of the cell wall of bacteria (especially gram-positive ones) are peptidoglycan murein and teichoic acid, which contain a large number of polar groups and are good absorbers of heavy metals (Kazy et al. 2002).

Some bacteria are also able to release substances into the environment that form insoluble complexes with HM ions. A classic example is sulfate-reducing bacteria that secrete a large amount of hydrogen sulfide, which forms insoluble sulfides (Gadd 2001; Higham et al. 1999; Chardin et al. 2003; Slawson et al. 1992; Li et al. 2000). *Klebsiella planticola* and *Pseudomonas aeruginosa* are able to release hydrogen sulfide under anaerobic conditions, thus protecting themselves from cadmium (Higham et al. 1999; Sharma et al. 2000). It is suggested that some silver-resistant bacteria are also capable of releasing hydrogen sulfide, which forms insoluble Ag_2S with it (Slawson et al. 1992; Li et al. 2000). In addition to sulfides, bacteria can form other insoluble complexes, such as phosphates (Chardin et al. 2003).

The cytoplasmic membrane of bacteria is the most important system that provides resistance to HMs. There is evidence that *E. coli* bacteria with a genetically determined absence of porins in the membrane are more resistant to the toxic effects of Ag ions than conventional ones (Slawson et al. 1992). Thus, the cytoplasmic membrane of bacteria is a serious barrier to the path of HMs into the cell.

Many toxic ions can enter the cell through systems designed to absorb the substances it needs. For example, Mn^{2+}, Fe^{2+}, Co^{2+}, Ni^{2+}, Cu^{2+}, Zn^{2+}, and Mg^{2} ions have similar ionic radii (130–160 pm) and the same charge, which is why, in particular, the system that transports magnesium in *Ralstonia metallidurans* can also transport cadmium, zinc, nickel, cobalt, and manganese into the cell (Cupta and Keegan 2000). Another example is that oxygen-containing HM anions (chromate and arsenate) are similar in structure to sulfate and phosphate, respectively, and can enter the cell using systems that ensure the absorption of these important nutrients (Meharg 2001). In this regard, one of the protective reactions of the plasmalemma is a change in the operation of transport systems in order to limit the penetration of HMs into the cell. Normally, a low-specific transport system that uses the energy of the membrane potential functions in the microbial cell. However, under the influence of HMs, other transport systems are launched, which are much more selective, slower and often consume ATP energy (Meharg 2001; Wallace 2004).

Another function of the membrane in protecting against HMs is their active volatile release from the cytoplasm. The proteins that ensure its implementation are very diverse. They can function both by consuming the energy of ATP or other macroergic compounds, and working on the energy of the electrochemical gradient of ions (various symports and antiports). Active ion release is a very important factor in ensuring stability. For example, the minimum suppressive concentration (MSC) of zinc for *R. metallidurans* in the presence of pMOL30 plasmid in the cell, containing genes of the transport membrane protein CnrA, increases more than 50 times, MSC of cobalt—33 times, and cadmium—7 times. With the help of this group of carrier proteins, many microorganisms are protected from the effects of Zn, Co, Cd, and Ni ions (Howe and Merchart 2000).

Gram-negative bacteria have an intermediate version of this mechanism—metal ions are actively released through the cytoplasmic membrane into the periplasmic space and bind there (Nies 2003).

Binding of heavy metals in the cytoplasm**.** As in the eukaryote, cysteine-rich peptides play a significant role in the neutralization of HM ions in the bacterial cell. At the same time, metallothioneins were found only in some cyanobacteria and have less cysteine in their composition than in eukaryotes. The variety of peptides-chelators in bacteria is quite large, there can be both low-molecular and high-molecular among them, including analogues of phytochelatins. Their synthesis is induced by heavy metal ions. In addition to peptides, HM binding with glutathione was also observed. It is also logical to assume that organic acids and amino acids also play a certain role in the neutralization of HMs (Howe and Merchart 2000).

One of the signs that the cytoplasm of a bacterial cell has a noticeable potential for HM binding is the fact that the ability to absorb them in bacterial cells killed by heating not only does not decrease, but sometimes increases. It is assumed that this is due to the fact that the destroyed plasmalemma gives HM ions access to the cytoplasm, where they actively bind to various chelators (Chugh and Sawhney 1999).

Some species of the genus *Pseudomonas* are capable of accumulating copper ions in the form of complexes with proteins in the periplasmic space or outer membrane. This protective system functions in parallel with active transport through the cytoplasmic membrane—Si ions are released into the periplasmic space and bind there (Higham et al. 1999).

Reduction of HM ions. Many bacteria are able to reduce metal ions either to ions with a lower degree of oxidation, or to a metallic state. Reduced forms are less toxic or more easily immobilized. Metals are embedded in the metabolism of some microorganisms, and can serve as final electron acceptors in energy exchange. Others have special enzymatic systems that reduce HM ions (Brown et al. 2001). These forms are less toxic.

A well-known and well-studied example is mercury reduction. Highly toxic Hg^{2+} ions that enter the cell are reduced by a special enzyme to metallic mercury Hg, which is practically non-toxic (Chardin et al. 2003). Another example is sulfate-reducing bacteria that reduce the highly toxic, including for humans, Cr^{6+} ion to a much less dangerous and mobile Cr^{3+}. Currently, bacteria capable of reducing chromium are being actively studied for use in soil purification (Abskharon et al. 2010).

The arsenic protection system involves both reduction and active membrane transport. Many microorganisms have been found to have gene complexes responsible for the synthesis of reductase, which reduces AsO_4^{3-} ions to AsO_3^{3-}, as well as the membrane transport system, with the expenditure of ATP, releasing AsO_3^{3-} from the cell (Mukhopadhyay et al. 2002).

A significant part of these protective mechanisms is genetically determined, and genes can be found both in the main genome of a bacterial cell and in plasmids, which makes it possible to efficiently laterally transfer traits important for survival (Nies 1999).

Yeast and micellar fungi can adapt to toxic metals quite easily, as can many bacteria.

Fungi are able to accumulate heavy metals in significant quantities, and the concentration in them is often higher than in the environment (Gadd 1993; Szczerba et al. 2009; Nies 1999; Shree and Rudra 2007). The metal content can reach 10–20% or more per unit of dry mycelium mass (Gadd 1993). The mechanisms of accumulation of metals in the mycelium of fungi and the physiological role of this phenomenon are of great interest. Metal biosorption by fungi is an important biogeochemical process in soil and on other substrates (Kis-Papo et al. 2001; Santhiya and Ting 2006; Furimsky 1996).

It is established that the manifestation of toxic effects of heavy metals on microscopic fungi is diverse. Ions of such metals as Cu, Co, Zn, Mg, for example, disrupt the morphogenesis of fungi of the genus *Fusarium*; it is also reported that Cu and Zn ions in concentrations of more than 50 and 100μg/mL, respectively, have a lethal effect on species of the genus *Penicillium*. Cd ions completely and Cu ions partially inhibit zinc transport in the mycelium of the fungus *P. notatum*. There is evidence that a heavy metal concentration of about 10 mg/kg is an essential element for the synthesis of *A. parasiticus* aflatoxins, and its addition to the growth medium of *A. flavus* stimulates toxin formation. In fungi, heavy metals, in particular, Cu, affect the genetic apparatus, and, in addition, protein synthesis and the organization of the cell membrane. It is assumed that the content of Zn and Cu in the soil is a factor that enhances the genetically determined formation of mycotoxins. It is noted that high concentrations of copper sharply suppress the development of saprophytic fungi, enhance the phytotoxic properties of the fungi *Penicillium nigricans, P. lanosum, Trichoderma koningii*. The spectrum of secondary metabolites of fungi of the genus *Penicillium* spp. isolated from soils subjected to anthropogenic stress was studied. It has been established that some representatives of the genus *Penicillium* spp. synthesize other sets of secondary metabolites than previously known (Gadd 1993; Yang et al. 2009; Bellion et al. 2006; Burgstaller and Schinner 1993).

Changes in the structure of biocenoses may occur with an increase in the influence of anthropogenic factors on them in modern ecological conditions. For example, it was found that adaptation of saprophytic fungi of the genus *Fusarium* spp. to a parasitic lifestyle increases under anthropogenic impacts.

12.2 Methods

12.2.1 Determination of the Resistance of Micromycetes to Heavy Metals

Micromycetes were tested for resistance to heavy metals—in concentrations of 0, 5, and 15 mM. For this purpose, salts of $CuSO_4$, $ZnSO_4$, $MnSO_4$, Pb $(CH_3CO_2)_2 \times 3H_2O$, containing metal ions of a given concentration were added to the main fond of Saburo liquid medium. To stabilize the solutions, the experiments were carried out at pH 5. $KNSO_4$ was used as the acidifying factor. The studied microorganisms were cultured in semi-submerged conditions on a shaker at a rotation frequency of 10,000 G and 21–22 °C for 5 days. Using Motic BA300 microscope, the grown cultures were microscopized at 100-fold magnification. The criterion for the stability of cultures to the content of HMs in the nutrient medium was the increase in the biomass of these cultures in comparison with the control variant. The biomass was determined by gravimetric method. The centrifugation of yeast cultures was carried out for 15 min at 5–15 thousand rpm.

The weight of dry biomass was determined by the formula (12.1):

$$M = \frac{(\mathrm{A} - \mathrm{B}) \times 100}{V} \tag{12.1}$$

where *M*—weight of the dry biomass in g/mL, *A*—weight of the filter paper with the precipitate in g, *B*—weight of the filter paper without the precipitate in g, *V*—volume of culture liquid taken for centrifugation or filtration in mL (Hassen et al. 1998).

Study of the activity of strains in soil containing zinc ions. The activity of isolated resistant isolates was studied in model experiments using soil contaminated with zinc. For the study, the soil was selected on the territory of Almaty region from the upper horizons at a depth of 0–10 cm. Soil samples were cleaned from plant residues and living parts of the plants and sterilized. The soil was placed in cups of 150 g. In the experiment, water solutions of zinc salt corresponding to the maximum permissible concentrations (MPC) were used for soil treatment: 1 MPC, 2 MPC, 4 MPC, 8 MPC. MPC of Zn is calculated per 1 kg of soil and is 300 mg. 10 mL filtrate was added to the soil containing $ZnSO_4$. After 30 days, the content of mobile zinc in the soil inoculated with micromycetes was determined. Mobile forms of zinc in the soil were determined according to the GOST R 50686–94.

Effect of active micromycetes on plant development in soil contaminated with zinc ions. The object of research was the seeds of a forage plant: barley (*Hordeum vulgare*) Arna variety and active isolate. The plant seeds were washed and sterilized for 15 min by $KMnO_4$, then thoroughly rinsed. Control samples of seeds were sterilized and germinated in sterile soil. The experimental seeds were soaked for 12 h in a solution with an association of microorganisms. Zinc sulfate salt ($ZnSO_4$) was used as an inhibitor. The maximum permissible concentration of pollutants (MPC) for agricultural crops is calculated per 1 kg of soil and is 300 mg of zinc per

1 kg of soil (Howlett and Avery 2004). In the experiment, water solutions of zinc salt corresponding to MPC (1 MPC, 2 MPC) were used for soil treatment at the rate of 1 L of water. On day 14, the length and biomass of stems and roots were measured, and the content of chlorophyll and proline in plants in experimental and control variants was determined.

Spectrophotometric analysis of photosynthetic pigments without their preliminary separation. A sample of leaves (0.15 g) was ground in a porcelain mortar, adding a little calcium dioxide, washed quartz sand and 3 mL of 96% ethanol solution. 4–5 mL of ethanol was added to the ground mass and rubbed again for several minutes. After settling the solution, the resulting extract was filtered through a paper filter, the resulting filtrate was brought to 25 mL and the density of the solution was determined at the following wavelengths: 649 and 665 nm. The concentration (*C*) of pigments was calculated according to the equation made up of the basis of the experimentally obtained specific absorption coefficients. Below is the formula used to determine the pigment content (12.2):

$$\begin{aligned} C_{\text{chl.}}a(\text{mg/L}) &= 13.70\,D_{665} - 5.76\,D_{649} \\ C_{\text{chl.}}\,b(\text{mg/L}) &= 25.80\,D_{649} - 3.87\,D_{665} \\ C_{\text{chl.}}\,a - C_{\text{chl.}}b &= 6.10\,D_{665} + 20.04\,D_{649} \end{aligned} \qquad (12.2)$$

where $C_{\text{chl.}}\ a$, $C_{\text{chl.}}\ b$ and $C_{\text{chl.}}\ a + C_{\text{chl.}}\ b$—respectively, the concentration of chlorophylls *a*, *b*, their sum, mg/L; *D*—experimentally obtained values of the optical density at the corresponding wavelengths. Having calculated the concentration of pigment in the ethanol extract, the content in the test material was determined taking into account the extract and the weighed portion according to the formula (12.3):

$$A = (C_x V)/P * 1000 \qquad (12.3)$$

where *A*—the pigment content, mg/g dry (or raw weight); *C*—the concentration of the pigment, mg/L; *V*—extracts, mL; *P*—the dry (or raw) weight. Determination of proline content in the studied plants. The content of free proline was determined using an acidic ninhydrin reagent prepared without heating (1.25 g of ninhydrate +30 mL of glacial acetic acid +20 mL of 6 M H_3PO_4). A sample of fresh plant tissue of the plant leaf plate (200 mg) was poured with 5–20 mL of boiling distilled water and kept for 10 min in a water bath at a temperature of 100 °C. Then, 2 mL of glacial acetic acid, 2 mL of ninhydrin reagent were poured into a clean test tube and 2 mL of the prepared extract was added. The samples were incubated for 20 min in a water bath at 100 °C, after which they were quickly cooled to room temperature. After artificial cooling (with cold water or ice), the optical density of the reaction products was measured at a wavelength of 520 nm on a spectrophotometer. The values of proline content were calculated with the use of a calibration curve, using chemically pure proline for its construction.

Collection and statistical processing of data. The quantitative data were statistically processed using generally accepted mathematical methods to calculate the arithmetic mean, standard deviation and root mean square error. Analysis of variance (ANOVA) Microsoft Office Excel 2009.

12.3 Results and Discussion

12.3.1 Screening of Micromycetes Resistant to High Concentrations of Heavy Metals

At present, in order to increase the productivity of agricultural plants and their resistance to HMs, priority is given to ecological methods of farming. As a result, special attention is drawn to the ability of microorganisms to stimulate growth and increase plant resistance to heavy metals, which is considered one of their main properties and an important factor in the use of promising microorganisms for the creation of biofertilizers.

For the initial screening of micromycetes that are resistant to heavy metals, a culture medium with different concentrations of zinc, copper, manganese, and lead metals was used.

From the results obtained from the experiments, it was found that out of 10 studied strains of micromycetes, the fungus *Beauveria bassiana* T7 is resistant to heavy metals.

The nature of the influence of heavy metals on microorganisms is determined, as is known, by their concentration in the environment, the level of toxicity and biological properties of microbial cells (Gadd 2001). Tables 12.1 and 12.2 present data on the growth of microscopic fungi and yeasts at various concentrations of metal ions.

By studying the growth of microscopic fungi in an environment with different concentrations of HMs, an increase in biomass was observed, which varied depending on the strain and concentration of HMs.

The toxic effect of zinc on mycelial fungi was a sharp decrease in biomass compared to the control. The exception was the *B. bassiana* T7 strain, which showed a slight decrease in biomass: by 16% at a zinc concentration of 5 mM and by 26% at 15 mM.

The most resistant strains to copper were *B. bassiana* T7, *Penicillium* sp. EZ, and *Aspergillus* sp. AC4, capable of growth at copper concentration in the medium up to 15 mM, while the amount of biomass at zinc content of 15 mM was 6.23, 2.37, 2.42 mg/mL, respectively. Thus, conducting an experiment to study the microbiological and environmental indicators of copper contamination of soils in Almaty, microscopic fungi *Penicillium* and *Aspergillus* showed a restrained reaction to copper, they were noted to be resistant to copper sulfate (Santhiya and Ting 2006; Yang et al. 2009).

Of the studied micromycetes, resistance to manganese was shown by the cultures of *Trichoderma* sp. D1 and *B. bassiana* T7. For example, the biomass of

Table 12.1 The amount of biomass of microscopic fungi cultivated at different concentrations of HMs

Name of the strain	Experiment variants	Zn^{+}	Cu^{+}	Mn^{+}	Pb^{+}
		Biomass mg/mL			
1	2	3	4	5	6
Trichoderma sp. D1	Control	5.69 ± 0.22	5.04 ± 0.25	6.33 ± 0.31	5.03 ± 0.25
	5 mM	2.79 ± 0.13	2.56 ± 0.12	3.98 ± 0.19	3.23 ± 0.16
	15 mM	1.26 ± 0.06	0.87 ± 0.04	2.97 ± 0.14	1.09 ± 0.05
Mortierella alpina EB	Control	4.76 ± 0.23	3.92 ± 0.19	3.92 ± 0.19	4.00 ± 0.16
	5 mM	1.17 ± 0.05	1.31 ± 0.06	1.33 ± 0.06	2.37 ± 0.11
	15 mM	0.02 ± 0.01	0.64 ± 0.03	0.02 ± 0.01	0.03 ± 0.00
Metarhizium sp. An1	Control	8.61 ± 0.43	9.00 ± 0.36	8.43 ± 0.33	7.99 ± 0.31
	5 mM	5.08 ± 0.25	4.68 ± 0.18	3.32 ± 0.13	3.27 ± 0.13
	15 mM	1.78 ± 0.08	0.87 ± 0.03	0.74 ± 0.03	2.53 ± 0.10
Aspergillus sp. AC4	Control	6.65 ± 0.33	7.06 ± 0.28	6.69 ± 0.26	7.94 ± 0.31
	5 mM	1.86 ± 0.09	4.19 ± 0.16	2.91 ± 0.11	3.14 ± 0.12
	15 mM	0.38 ± 0.01	2.42 ± 0.09	1.83 ± 0.07	1.34 ± 0.05
Penicillium sp. EZ	Control	5.44 ± 0.21	7.35 ± 0.29	7.15 ± 0.28	7.37 ± 0.29
	5 mM	2.35 ± 0.09	4.69 ± 0.23	4.14 ± 0.16	5.03 ± 0.20
	15 mM	1.69 ± 0.06	2.37 ± 0.09	1.42 ± 0.20	2.81 ± 0.11
Beauveria bassiana T7	Control	15.90 ± 0.63	12.78 ± 0.51	11.89 ± 0.47	11.70 ± 0.46
	5 mM	13.28 ± 0.53	8.47 ± 0.33	9.36 ± 0.46	10.50 ± 0.42
	15 mM	11.84 ± 0.47	6.23 ± 0.24	8.56 ± 0.34	8.38 ± 0.35
Penicillium bilaiae Pb14	Control	7.76 ± 0.31	6.16 ± 0.24	6.22 ± 0.24	7.01 ± 0.28
	5 mM	3.75 ± 0.15	2.38 ± 0.09	1.15 ± 0.04	3.48 ± 0.13
	15 mM	0.57 ± 0.02	0.31 ± 0.01	0.18 ± 0.01	1.34 ± 0.07

Average statistical value: $p \leq 0.05$

Table 12.2 Amount of yeast biomass at different HM concentrations

Strain	Experiment variants	Zn^{+}	Cu^{+}	Mn^{+}	Pb^{+}
		Biomass mg/mL			
Metschnicowia pulcherrima MP2	Control	4.55 ± 0.18	5.68 ± 0.22	5.79 ± 0.23	5.98 ± 0.24
	5 mM	1.17 ± 0.04	2.03 ± 0.08	3.45 ± 0.14	2.32 ± 0.11
	15 mM	0.83 ± 0.03	0.35 ± 0.01	0.32 ± 0.01	0.36 ± 0.03
Rhodotorula mucilaginosa Rh	Control	9.42 ± 0.37	9.65 ± 0.38	10.28 ± 0.41	10.33 ± 0.43
	5 mM	6.34 ± 0.25	6.09 ± 0.24	7.06 ± 0.28	8.47 ± 0.34
	15 mM	3.22 ± 0.13	2.41 ± 0.09	5.94 ± 0.24	5.37 ± 0.23
Rodotorula mucilaginosa MK	Control	6.36 ± 0.25	7.89 ± 0.31	8.57 ± 0.34	7.91 ± 0.32
	5 mM	5.19 ± 0.20	3.79 ± 0.15	5.83 ± 0.23	5.27 ± 0.21
	15 mM	3.30 ± 0.13	2.15 ± 0.08	4.88 ± 0.19	3.75 ± 0.15

Average statistical value: $p \leq 0.05$

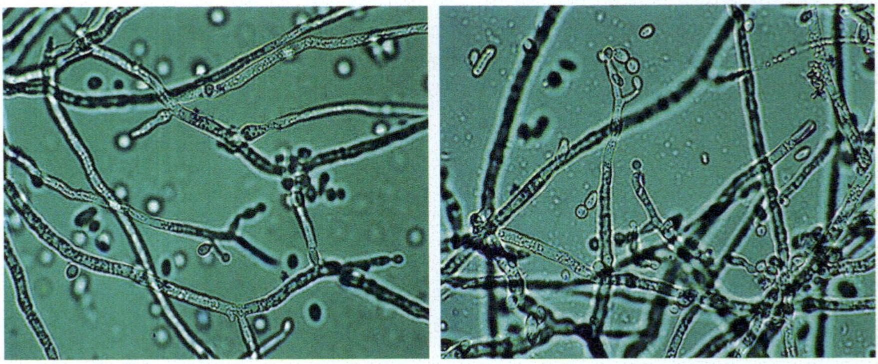

Fig. 12.1 *B. bassiana* T7 strain cultivated on a metal-free medium

Fig. 12.2 *B. bassiana* T7 strain cultured in a medium with Pb^{+} at a concentration of 5 mM

Trichoderma sp. D1 decreased by 37% at a manganese concentration of 5 mM, at 15 mM—by 53%. The biomass of *B. bassiana* T7 in a medium of 5 mM manganese decreased by 21%, at 15 mM by 28%.

Penicillium sp. EZ, *Metarhizium sp.* An1 and *B. bassiana* T7 were the most tolerant to lead. The amount of biomass of *Penicillium* sp. EZ at lead concentration of 5 mM was 5.03 mg/mL, at 15 mM of lead—2.81 mg/mL. Biomass of *Metarhizium* sp. An1 was 3.27 mg/mL at lead concentrations of 5 mM and at 15 mM—2.63 mg/mL. Thus, a number of studies have shown the use of *Penicillium chrysogenum* for the effective removal of heavy metal ions. At pH of 4.5, *P. chrysogenum* not only removes lead ions (116 mg/g) from aqueous solutions, but it is also selective for Pb (II) compared to other metal ions Cd (II), Cu (II), Zn (II).

Biomass of *B. bassiana* T7 was 10.50 mg/mL and 8.88 mg/mL at lead concentrations of 5 and 15 mM, respectively. Apparently, the stability of *B. bassiana* T7 to lead ions is explained by the ability of the fungus to accumulate and precipitate metal intracellularly and adsorb it on the cell wall (Figs. 12.1, 12.2 and 12.3).

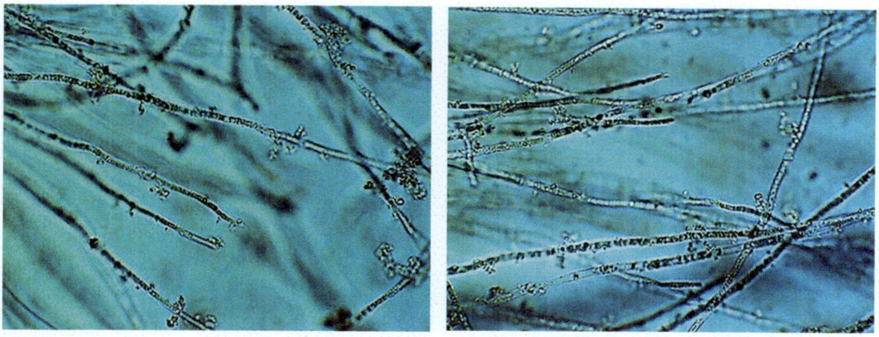

Fig. 12.3 *B. bassiana* T7 strain cultivated in a medium with Pb^{+} at a concentration of 15 mM

The presence of precipitated lead in the environment surrounding the hyphae and on the hyphae themselves may indicate that the release of certain metabolites of the fungus leads to the precipitation of lead. *B. bassiana* is known to form citric and oxalic acids, which, as a number of researchers believe, contribute to the solubilization of cadmium, copper, lead, and zinc, which are then precipitated as oxalates (Yang et al. 2009).

Studying the influence of heavy metal ions on yeast growth, it was shown that the most stable strains are *Rh. mucilaginosa* Rh and *Rh. mucilaginosa* MK (Table 12.2). They are able to grow at a concentration of HMs in the medium up to 15 mM, while depending on the species and metal concentration, the amount of biomass for *Rh. mucilaginosa* Rh strain varied in the range of 2.41–8.47 mg/mL, for *Rh. mucilaginosa* MK 2.15–5.83 mg/mL.

The results obtained are consistent with other studies (Yang et al. 2009), in which the study of the resistance of *Rhodotorula mucilaginosa, Rhodotorula aurantiaca, Rhodotorula* sp. yeasts to different concentrations of lead showed that the most resistant strains were pigmented *R. mucilaginosa* strains. They are able to grow in the medium at a concentration of lead up to 750 mg/L, while the amount of biomass was 0.96 and 0.76 g/L, respectively.

It is known that the resistance of yeast to the toxic effects of heavy metals depends on both the morphological and physiological characteristics of the cell (Alybaeva 2007).

Microscopy of yeast showed the absence of crystalline metal inclusions on the cell surface of *Rh. mucilaginosa* Rh strain (Figs. 12.4, 12.5, 12.6, 12.7, 12.8 and 12.9), while *Rh. mucilaginosa* MK strain probably converts them to an insoluble form for protection against HMs.

According to the literature (Gadd 1993), it is known that there are several mechanisms for protecting microorganisms from the toxic effects of heavy metals. The main mechanisms are the ability of microorganisms to release substances into the environment that form insoluble complexes with HM ions and the binding of heavy metals in the cell cytoplasm. It can be assumed that the resistance of micromycetes, selected as a result of screening, to heavy metals is conditioned by

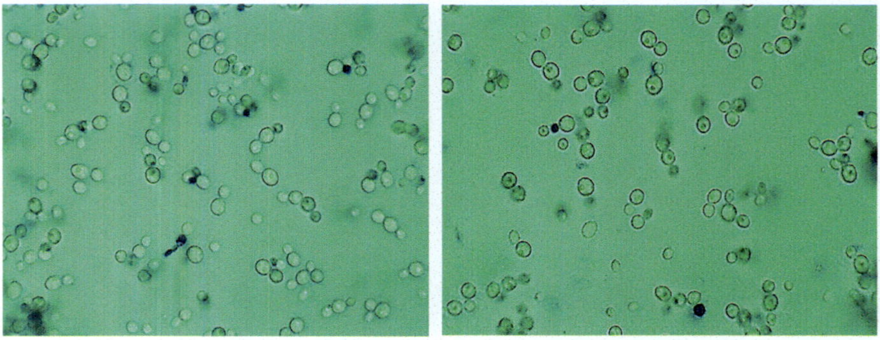

Fig. 12.4 *Rh. mucilaginosa* Rh culture on a metal-free medium

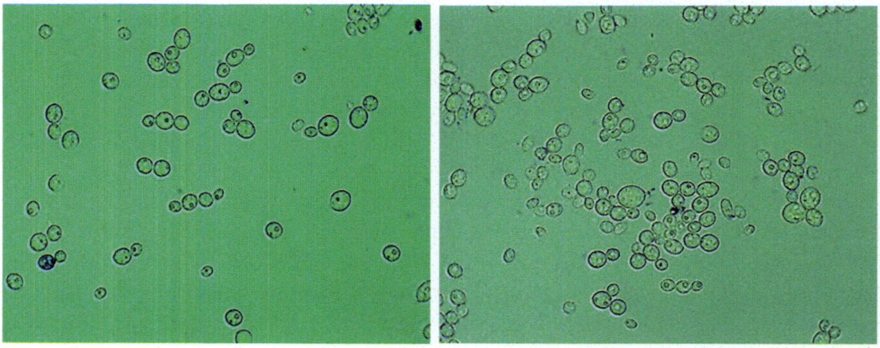

Fig. 12.5 *Rh. mucilaginosa* Rh culture on medium with Mn^{+} at a concentration of 5 mM

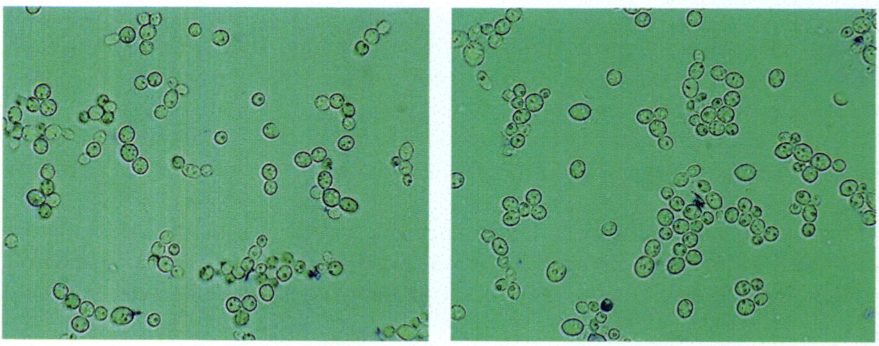

Fig. 12.6 *Rh. mucilaginosa* Rh culture in a medium with Mn^{+} at a concentration of 15 mM

various mechanisms. Thus, the microscopic picture showed that *Rh. mucilaginosa* MK and *B. bassiana* T7 strains forms insoluble metal compounds on the cell surface whereas no similar structures were found in *Rh. mucilaginosa* Rh strain.

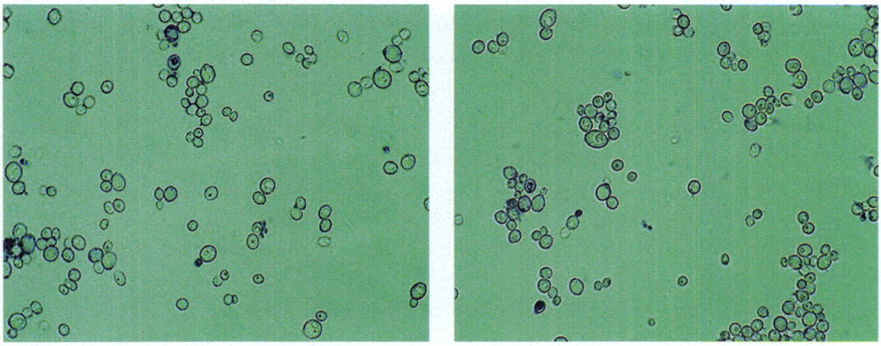

Fig. 12.7 *Rh. mucilaginosa* MK strain on a medium without metal addition

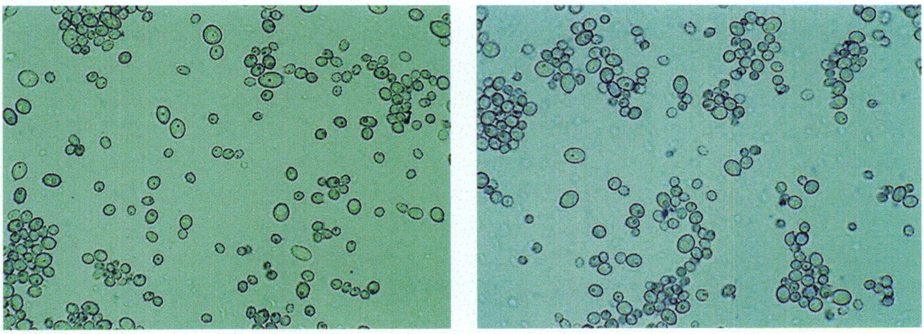

Fig. 12.8 *Rh. mucilaginosa* MK strain in a medium with Mn^+ at a concentration of 5 □M

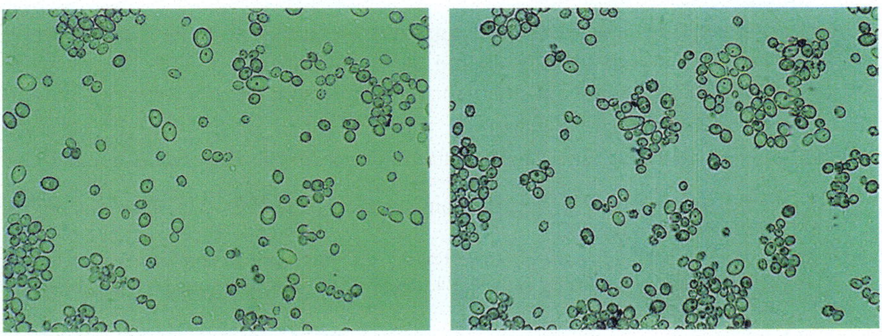

Fig. 12.9 *Rh. mucilaginosa* MK strain on Mn^+ medium at a concentration of 15 mM

Thus, the greatest interest was aroused by *B. bassiana* T7 strain, which had the maximum growth of biomass at the concentration of all the studied metals in the concentration from 5 to 15 mM. This micromycete was used to research the resistance of plants to zinc.

12.3.2 Changes in the Content of Mobile Zinc in Soil Inoculated with Micromycetes

Developing measures for detoxification of soils contaminated with HMs, the elaboration of measures to reduce the availability of toxic substances for plants is of great interest. It is obvious that the damage caused by pollution will depend more on the properties of the soil, mainly on those that affect the mobility of heavy metals and, as a result, their availability to plants and their ability to migrate (Gadd 2001). During cultivation, soil properties change under the influence of anthropogenic factors, which affects the content and availability of HMs in these soils.

According to a number of authors (Shree and Rudra 2007; Ghorbani et al. 2002; Hiroki 1992), microorganisms inserted into the soil have a positive effect not only on the physical, chemical and biological properties of the soil, but also contribute to the detoxification of contaminated soil cover.

Further study of the activity of the resistant strain *B. bassiana* T7 was carried out in model experiments using soil contaminated with zinc. In the experiment, water solutions of zinc salt corresponding to the maximum permissible concentrations (MPC) were used for soil treatment: 1 MPC, 2 MPC, 4 MPC, 8 MPC. Filtrate *B. bassiana* T7 of a volume of 10 mL was added to the soil containing $ZnSO_4$.

As a result of studying the effect of micromycetes on the formation of free zinc in the soil, it was shown that at zinc concentration corresponding to 1 MPC in the soil inoculated with micromycetes, the content of free zinc increased by 67%, 2 MPC—by 31%, 4 MPC—by 16%, 8 MPC—by 7% (Table 12.3).

The difference between the control and experimental variants can be explained by the fact that with an increase in the concentration of zinc in the soil, its toxic effect on the micromycete increases and the transfer of zinc into mobile forms by a microscopic fungus becomes difficult.

Microorganisms are capable of reducing heavy metal ions or, conversely, forming insoluble compounds with them, reducing their mobility and toxicity, or releasing chelators into the environment that form soluble complexes with HMs and

Table 12.3 Content of mobile forms of zinc in the soil under the influence of *B. bassiana* T7

Experiment no.	Experiment variants	Amount of mobile zinc (mg/kg)
1	Soil (control)	21.0 ± 0.8
2	1 MPC of zinc	264.0 ± 10.5
3	2 MPC of zinc	576.0 ± 23.0
4	4 MPC of zinc	1788.0 ± 71.5
5	8 MPC of zinc	3950.0 ± 158.0
6	Soil + *B. bassiana* T7	21.0 ± 0.8
7	1 MPC of zinc + *B. bassiana* T7	442.0 ± 17.6
8	2 MPC of zinc + *B. bassiana* T7	755.0 ± 30.2
9	4 MPC of zinc + *B. bassiana* T7	2076.0 ± 83.0
10	8 MPC of zinc + *B. bassiana* T7	4230.0 ± 169.2

Average statistical value: $p \leq 0.05$

significantly increase their mobility and bioavailability (Cupta and Keegan 2000). However, in most of these techniques, the role of microorganisms is comparable to that of plants. For example, they are extremely important in the degradation of xenobiotics. In fact, plants simply create optimal conditions for microorganisms—they provide nutrients, vitamins and oxygen. Thus, the most promising way to develop bioremediation is the creation of symbiotic systems of a microorganism (or rather several types of microorganisms)—a plant.

12.3.3 Study of the Effect of Micromycetes on Plant Resistance to Heavy Metals

Many properties of microorganisms and mechanisms of their positive action on plants can play an important role in protecting the latter from adverse environmental conditions, since the positive effects of microorganisms are oriented, among other things, against the negative influence of stressors on plants. For example, the negative effect of heavy metals is manifested in a deterioration in the consumption of nitrogen and other nutrients by plants, which in certain conditions can be the main cause of reduced resistance and inhibition of growth (Tran and Popova 2013). Inoculation with microorganisms that improve the consumption of mineral substances can increase the resistance of plants to stress (Tran and Popova 2013). In the plant, HMs induce oxidative stress, whereas microorganisms are able to activate protective reactions during inoculation, increasing the activity of the antioxidant superoxide dismutase, peroxidase, and catalase enzymes (Yadav 2010).

To study the effect of micromycetes on the resistance of plants to heavy metals, the seeds of a forage plant were used as objects of study: barley (*Hordeum vulgare*), Arna variety, and the active *B. bassiana* T7 strain. Sterile seeds soaked in the filtrate of a microscopic fungus were germinated in sterile soil treated with an aqueous solution of zinc salt of the corresponding MPC (1 MPC and 2 MPC) at the rate of 1 L of water. On day 14, morphometric parameters were measured and the content of chlorophyll and proline in barley was determined.

The effect of seed inoculation with micromycete filtrate on plants cultivated on soil contaminated with zinc at concentrations of 1 and 2 MPC was very pronounced and affected almost all the main morphological characteristics of the plant (Table 12.4).

As can be seen from Table 12.4 inoculation of seeds with micromycetes had a positive effect on their morphometric parameters. With the content of zinc in the soil at a concentration corresponding to 2 MPC, the length of the stem in the experimental version exceeded the control by 1.4 times, the length of the root—by 1.3 times. Seed germination was comparable in control and experimental variants.

Zinc, unlike cadmium, mercury, and lead, is not a toxic heavy metal. It belongs to trace elements and plays an important role in the life processes of plants, animals and humans. However, increased zinc content in the environment can have a significant

Table 12.4 Effect of seed inoculation with *B. bassiana* T7 strain on morphometric parameters of barley seedlings grown at different zinc concentrations

Variant	MPC of zinc	Sprouted seeds, %	Morphometric parameters of the plant			
			Root length (cm)	Stem length (cm)	Root biomass (g)	Shoot biomass (g)
Control	0	100	11.9 ± 0.4	26.2 ± 1.0	0.027 ± 0.002	0.201 ± 0.008
	1	95	9.2 ± 0.3	22.7 ± 0.9	0.019 ± 0.001	0.167 ± 0.007
	2	90	7.9 ± 0.3	18.3 ± 0.7	0.013 ± 0.001	0.111 ± 0.005
B. bassiana T7	0	100	13.0 ± 0.5	27.4 ± 1.1	0.031 ± 0.200	0.215 ± 0.008
	1	98	12.8 ± 0.5	25.7 ± 1.0	0.029 ± 0.002	0.205 ± 0.010
	2	92	10.4 ± 0.4	24.8 ± 1.0	0.020 ± 0.001	0.164 ± 0.008

Average statistical value: $p \leq 0.05$

negative impact on living organisms. It is known that soil contamination with zinc at the level of 5 and 25 MPC affects its agrochemical properties: acidity indicators increase, the content of mobile phosphates decreases. Therefore, zinc should be considered from the point of view of solving two problems:—insufficiency of the element for plants and animals, optimization of their nutrition;—danger of soil and vegetation contamination (Shree and Rudra 2007). The average zinc content in soils is 50 mg/kg and varies widely from 10 to 300 mg/kg. The arable layer (0–20 cm) of the main types of soils in Kazakhstan contains 5.57–80.1 mg/kg of zinc (Alybaeva 2007).

The results of experiments indicate the ability of inoculated plants to more effectively counteract the negative effect of heavy metals, increase the absorption of nutrients, which may be an important mechanism for the positive effect of micromycetes on plant resistance.

An indicator of plant resistance to adverse conditions is the content of chlorophyll *a* and *b* in the leaves of plants. The ratio of chlorophyll *a* to chlorophyll *b* characterizes the stability of the photosynthetic system, and a decrease of this indicator and a decline in the total content of chlorophyll indicate a violation of the photosynthetic activity of plants (Vassilev et al. 2004).

The next stage of research was to study the content of chlorophyll in barley seedlings grown at different concentrations of zinc—1 and 2 MPC. The quantitative content of chlorophylls *a* and *b* was determined by the change in the optical density of the pigment extract on a spectrophotometer at wavelengths corresponding to the absorption maxima of chlorophylls *a* (665 nm) and *b* (649 nm), followed by the calculation of the pigment concentration according to the equations (Vassilev et al. 2004).

Table 12.5 shows that the control variants have a decrease in the content of chlorophyll *a* by 29% and chlorophyll *b* by 19%, which is probably due to the toxic effect of zinc. Whereas, inoculation of seeds with *B. bassiana* T7 culture liquid has a protective effect on the plant, which affects a slight decrease in the content of chlorophyll *a* (by 20%) and a constant level of chlorophyll *b* (0.486–0.417 mg/g).

Table 12.5 Impact of *B. bassiana* T7 on the content of chlorophylls *a* and *b* in barley seedlings grown at different concentrations of zinc

Experiment variants	MPC of Zn	Chlorophyll *a*, mg/g fresh mass	Chlorophyll *b*, mg/g fresh mass	Total chlorophyll, mg/g fresh mass	Chlorophyll *a*/*b* ratio, %
Control	0	0.882 ± 0.035	0.438 ± 0.018	1.320 ± 0.052	2.01
	1	0.701 ± 0.028	0.385 ± 0.015	1.086 ± 0.043	1.82
	2	0.622 ± 0.024	0.357 ± 0.014	0.979 ± 0.039	1.74
B. bassiana T7	0	0.994 ± 0.039	0.476 ± 0.019	1.470 ± 0.059	2.08
	1	0.893 ± 0.035	0.444 ± 0.018	1.337 ± 0.053	2.01
	2	0.799 ± 0.031	0.417 ± 0.017	1.216 ± 0.048	1.91

Average statistical value: $p \leq 0.05$

In addition, the effect of the microscopic fungus on the stability of barley is characterized by a not significant decrease in the ratio of chlorophyll *a* to *b* forms (by 24%).

One of the main biochemical indicators by which the processes occurring in plants are judged in terms of their response to stress, including from an ecological point of view, is the content of proline amino acid in the plant mass. It was found that in response to the action of heavy metal, which is a stressor for plants, the protective system is activated. It is assumed that proline is one of the components of the stress response, the first stage of adaptation (Tran and Popova 2013).

In this regard, the content of proline in barley seedlings grown at zinc concentrations of 1 and 2 MPC was studied. The quantitative content of proline was determined by the change in the optical density of the reaction products on the spectrophotometer at a wavelength of 520 nm. The values of the proline content were calculated with the use of a calibration curve, using chemically pure proline for its construction.

As can be seen from the results of Table 12.6 proline content was only 0.216μm/g of raw mass in the barley leaves in the variant without zinc. Inoculation of seeds with fungus filtrate practically did not affect its content (0.219μm/g). In control plants, the level of proline increased by 1.5 times during 14 days under the influence of zinc. Treatment of barley seeds with *B. bassiana* T7 filtrate reduced zinc-induced proline accumulation by 7.4%.

An increase in the content of chlorophyll and proline may indicate a protective effect of inoculation by micromycetes, in particular their functioning as a biological barrier. Our data suggest that the combination of micromycetes and host plants may be a determining factor for the result of their interaction during stress caused by the action of HMs, and therefore for their possible use in phytoremediation of contaminated soils.

Table 12.6 Influence of *B. bassiana* T7 on proline content in barley seedlings grown at different zinc concentrations

Experiment variants	MPC of Zn	Proline content μM/g of raw mass
Control	0	0.216 ± 0.008
	1	0.294 ± 0.012
	2	0.325 ± 0.013
Beauveria bassiana T7	0	0.219 ± 0.008
	1	0.279 ± 0.008
	2	0.301 ± 0.012

Average statistical value: $p \leq 0.05$

12.4 Conclusion

Despite the large amount of material devoted to the adaptation mechanisms of microorganisms under stress conditions caused by heavy metals, information about filamentous fungi and yeasts is not enough. There is a definite need to study microscopic fungi in contaminated soils and the search for new species with true tolerance should continue.

Moreover, we are still at the beginning of elucidating the molecular mechanisms that are involved in metal homeostasis, detoxification, and tolerance of filamentous fungi in general. A better understanding of metal transport mechanisms including their regulation and the underlying biochemical and physiological mechanisms of tolerance are of key interest.

As a result of our own studies of the effect of heavy metal ions at concentrations of 5 and 15 mM on the growth of micromycetes, *B. bassiana* T7 strain was selected, which has multiple resistance to zinc, copper, manganese and lead.

The level of biomass accumulation of this strain at the concentration of HMs in the medium of 15 mM varied in the range of 6230–11,848 mg/mL. Under conditions of increased zinc concentration in the soil (2 MPC), by inoculating seeds with *B. bassiana* T7 filtrate, the stem length increased by 1.4 times, the root length by 1.3 times, and the biomass of the aboveground and underground parts by 1.5 times. A slight decrease in the content of chlorophyll *a* (by 20%) was shown with an unchanged level of chlorophyll *b* (0.486–0.417 mg/g), as well as a decrease in the level of proline accumulation induced by zinc.

References

Abskharon RNN, Gaad El-Rab SMF, Hassan SHA, Shoreit AAM (2010) Reduction of toxic hexavalent chromium by *E. coli*. GJBBR 5(2):129–135

Alybaeva RA (2007) Assessment of the ecological state of soils of the city of Ust-Kamenogorsk. In: Bulletin of KazNU, ecological series, vol 2, pp 40–44

Bakker EP (ed) (1993) Cell K+ and K+ transport systems in prokaryotes. Alkali cation transport systems in prokaryotes. CRC, Boca Raton, FL

Bellion M, Courbot M, Jacob C, Blaudez D, Chalot M (2006) Extracellular and cellular mechanisms sustaining metal tolerance in ectomycorrhizal fungi. FEMS Microbiol Lett 254:173–181
Benavides MP, Gallego SM, Tomaro ML (2005) Cadmium toxicity in plants. Braz J Plant Physiol 17(1):21–34
Bingham FT, Pereyea J, Jarrell WM (2001) Metal toxicity to agricultural crop. Met Ions Biol Syst 20:119–156
Brown NL, Shih YC, Leang C et al (2001) Mercury transport and resistance. Biochem Soc Trans 4:715–718
Burgstaller W, Schinner FJ (1993) Leaching of metals with fungi. J Biotechnol 27(2):91–116
Burzynski M, Klobus G (2004) Changes of photosynthetic parameters in cucumber leaves under Cu, Cd and Pb stress. Photosynthetica 42(4):505–510
Cera EA (2006) Structural perspective on enzymes activated by monovalent cations. J Biol Chem 2 (6):1305–1308
Chugh LK, Sawhney SK (1999) Effect of cadmium on activities of some enzymes of glycolysis and pentose phosphate pathway in pea. Biol Plant 42(3):401–407
Chardin B, Giudici-Orticoni MT, De Luca G, Guigliarelli B, Bruschi M (2003) Hydrogenases in sulfate-reducing bacteria function as chromium reductase. Appl Microbiol Biotechnol 63:315–321
Cupta G, Keegan B (2000) Bioaccumulation and biosorption of lead by poultry litter microorganisms. Poult Sci 77(3):400–404
El-Helow ER, Sabry SA, Amer RM (2000) Cadmium biosorption by a cadmium resistant strain of *Bacillus thuriengiensis*: regulation and optimization of cell surface afınity for metal cations. BioMetals 4:273–280
Furimsky E (1996) Spent refinery catalysts: environment, safety and utilization. Catal Today 30:223–286
Gadd GM (1993) Interactions of fungi with toxic metals. New Phytol 124:25–60
Ghorbani NR, Salehrastin N, Moenia A (2002) Heavy metals affect the microbial populations and their activities. In: 17th WCSS, Thailand. Symposium, 14–21 August 2002
Gadd GM (2001) Heavy metal accumulation by bacteria and other microorganisms. Cell Mol Life Sci 8:834–840
Hassen A, Saidi N, Cherif M, Boudabous A (1998) Resistance of environmental bacteria to heavy metals. Bioresour Technol 64:7–44
Higham DP, Sadler PJ, Scawen MD (1999) Cadmium-binding proteins in *Pseudomonas putida*: pseudothioneins. Environ Health Perspect 3:5–11
Hiroki M (1992) Effects of heavy metal contamination on soil microbial population. Soil Sci Plant Nutr 38:141–147
Howe G, Merchart S (2000) Heavy metal-actived synthesis of peptides in Chlamydomonas reinhardtii. Plant Physiol 98(1):127–136
Howlett NG, Avery SV (2004) Induction of lipid peroxidation during heavy metal stress in Saccharomyces cerevisiae and inuence of plasma membrane fatty acid unsaturation. Appl Environ Microbiol 63(8):2971–2976
Kis-Papo T, Grishkan I, Oren A, Wasser SP, Nevo E (2001) Spatiotemporal diversity of filamentous fungi in the hypersaline Dead Sea. Mycol Res 105(6):749–756
Kazy SK, Sar P, Sen AK, D'Souza SF (2002) Extracellular polysaccharides of a copper-sensitive and a copperresistant *Pseudomonas aeruginosa* strain: synthesis, chemical nature and copper binding. World J Microbiol Biotechnol 18(6):583–588
Li X-Z, Nikaido H, Williams KE (2000) Silver-resistant mutants of *Escherichia coli* display active efflux of Ag^{+} and are deficient in porins. J Bacteriol 179(19):6127–6132
Meharg AA (2001) The role of the plasmalemma in metal tolerance in angiosperms. Physiol Plantarum 88:191–198
Mukhopadhyay R, Rosen BP, Phung LT, Silver S (2002) Microbial arsenic: from geocycles to genes and enzymes. FEMS Microbiol Rev 26(3):311–325
Nies DH (1999) Microbial heavy-metal resistance. Appl Microbiol Biotechnol 51:730–750

Nies DH (2003) Efflux-mediated heavy metal resistance in prokaryotes. FEMS Microbiol Rev 27:313–339

Prasad MNV, Hagemeyer J (eds) (1999) Heavy metal stress in plants: from molecules to ecosystems. Springer-Verlag, Berlin

Rodríguez-Serrano M, Romero-Puertas MC, Pazmiño DM, Testillano PS, Risueño MC, Del Río LA, Sandalio LM (2009) Cellular response of pea plants to cadmium toxicity: cross talk between reactive oxygen species, nitric oxide, and calcium. Plant Physiol 150(1):229–243

Santhiya D, Ting YP (2006) Use of adapted *Aspergillus Niger* in the bioleaching of spent refinery processing catalyst. J Biotechnol 121:62–74

Shree NS, Rudra DT (2007) Environmental bioremediation technologies. Springer-Verlag, Berlin

Sharma PK, Balkwill DL, Frenkel A, Vairavamurthy MA (2000) A new Klebsiella planticola strain (Cd-1) grows anaerobically at high cadmium concentrations and precipitates cadmium sulide. Appl Environ Microbiol 66(7):3083–3087

Slawson RM, Trevors JT, Lee H (1992) Silver accumulation and resistance in *Pseudomonas stutzeri*. Arch Microbiol 158(6):398–404

Szczerba MW, Britto DT, Kronzucker HJ (2009) K^+ transport in plants: physiology and molecular biology. J Plant Physiol 166:447–466

Saxena D, Joshi N, Srivastava S (2002) Mechanism of copper resistance in a copper mine isolate *Pseudomonas putida* strain S4. Curr Microbiol 6:410–414

Teitzel GM, Parsek MR (2003) Heavy metal resistance of bioilm and planctonic *Pseudomonas aeruginosa*. Appl Environ Microbiol 69(4):2313–2320

Tran TA, Popova LP (2013) Functions and toxicity of cadmium in plants: recent 16 advances and future prospects. Turk J Bot 37(1):1–13

Tran L, Nunan L, Redman RM, Mohney LL, Pantoja CR, Fitzsimmons K, Lightner DV (2013) Determination of the infectious nature of the agent of acute hepatopancreatic necrosis syndrome affecting penaeid shrimp. Dis Aquat Org 105(1):45–55

Vecchio A, Finoli C, Di Simine D, Andreoni V (1998) Heavy metal biosorption by bacterial cells. Fresenius J Anal Chem 361(4):338–342

Vassilev A, Lidon F, Scotti P, Da Graca M, Iordanov I (2004) Cadmium-induced changes in chloroplast lipids and photosystem activities in barley plants. Biol Plant 48:153–156

Wallace A (2004) Excess trace metal effects on calcium absorpion in plants. Commun Soil Sci Plant Anal 13:146–151

Yadav SK (2010) Heavy metals toxicity in plants: an overview on the role of glutathione and phytochelatins in heavy metal stress tolerance of plants. S Afr J Bot 76(2):167–179

Yang J, Wang Q, Wang Q, Wu T (2009) Heavy metals extraction from municipal solid waste incineration fly ash using adapted metal tolerant *Aspergillus Niger*. Bioresour Technol 100 (1):254–260

Chapter 13
Heavy Metal Resistance in Prokaryotes: Mechanism and Application

Armine Margaryan, Hovik Panosyan, and Nils-Kåre Birkeland

Abstract Metal-rich natural and artificial habitats are extreme environments for the development and evolution of unique microbial communities, which have adapted to the toxic levels of the metals. Diverse bacterial groups have developed abilities to deal with the toxic metals by bioaccumulation of the metal ions inside the cell actively or passively, extracellular precipitation, efflux of heavy metals outside to the microbial cell surface, biotransformation of toxic metals to less toxic forms, and metal adsorption on the cell wall. Metalophilic microbes are found in all bacterial and archaeal groups studied, but mostly appear among aerobic and facultative anaerobic chemoheterotrophic and chemolithoautotrophic microorganisms of the *Bacillus, Pseudomonas, Staphylococcus, Actinobacteria, Cuprividus, Acidobacterium, Acidithiobacillus, Thiobacillus, Ferroplasma,* and *Sulfolobus* genera. The phenomenon of microbial heavy metal resistance has fundamental importance and is particularly relevant in microbial ecology, especially in connection with the roles of microbes in biogeochemical cycling of heavy metals and in the bioremediation of metal-contaminated environments. The heavy metal resistance mechanisms and different applications of metal resistant/metalophilic bacteria and archaea have been expounded deeply in this chapter.

Keywords Heavy metals · Metalophilic microbes · Heavy metal resistance · Bioremediation · Bioleaching

A. Margaryan (✉) · H. Panosyan
Department of Biochemistry, Microbiology and Biotechnology, Faculty of Biology, Yerevan State University, Yerevan, Armenia
e-mail: arminemargaryan@ysu.am; hpanosyan@ysu.am

N.-K. Birkeland
Department of Biological Sciences, University of Bergen, Bergen, Norway
e-mail: nils.birkeland@uib.no

D. Egamberdieva et al. (eds.), *Microbial Communities and their Interactions in the Extreme Environment*, Microorganisms for Sustainability 32,
https://doi.org/10.1007/978-981-16-3731-5_13

13.1 Heavy Metals and Its Toxicity on Microbes

There is no widely agreed criterion-based definition of a heavy metal. In metallurgy, a heavy metal may be defined on the basis of density, in physics the differentiating criterion might be atomic number, and in chemistry or biology the distinguishing criteria could be atomic mass (Hawkes 1997; Ali and Khan 2018; Meija et al. 2016). Based on density definition, the heavy metals are those elements that have a density above 5 g/cm^3 (Nies 1999). Based on atomic number definition, heavy metals are those elements which atomic number greater than 20 (Ca), sometimes this is capped at 92 (U). Definitions based on atomic number have been criticized for including metals with low densities. Atomic mass definitions can range: it reserved those elements with an atomic mass greater than Na (atomic mass 22.98), greater than 50 (Ni (58.69), Cu (63.54), Mo (95.95), etc.) or more than 200 (e.g., Hg (200), TI (204), Pb (207), Bi (209), and the Th series) (Baldwin and Marshall 1999; Ali and Khan 2018; Pourret and Hursthouse 2019).

Correspondingly, the list of heavy metals according to different definitions will include different elements. Of the 90 natural elements, 21 are non-metals, 16 are light metals, and the remaining 53 (including As) are heavy metals (Ali and Khan 2018).

Most heavy metals are transition elements with incompletely filled *d* orbitals. These *d* orbitals provide heavy metal cations with the ability to form complex compounds which may or may not be redox-active. Thus, the heavy metal cations which play an important role as micronutrients in the vital processes of microorganisms or other living organisms are essential metals. For example, Mo(II), Fe(II), Cu (II), Mn(II), Zn(II), Ni(II), and Co(II) are involved in the catalytic acceleration of biochemical processes. They can serve as cofactors or be part of enzymes such as nitrogenases, superoxide dismutases, dehydrogenases, cytochrome oxidases, ureases, etc. (Ehrlich 1997a; Nies 1999). Cu(II) and Ni(II) are involved in bacterial cell's redox processes (Nies 1999). Zn(II) ions stabilize the structure of DNA and proteins of the bacterial cell wall, since they have redox stability at certain pH and Eh values of biological media (Nies 1999). A significant number of bacteria and archaea are able to use ions of certain metals (Fe(III), Mn(II), Cr(VI), etc.) and metalloids as donors or acceptors of electrons in energy metabolism (Ehrlich 1997a). Thus, many archaeal and bacterial species have the ability to derive energy from the reduction of a variety of metals. Archaeal species *Archaeoglobus fulgidus*, *Pyrococcus furiosus,* and bacterial species *Desulforomonas, Desulfovibrio* are capable of reducing Fe(III), and two *Pyrobaculum* sp. can effectively grow respiring Fe(III) (Vargas et al. 1998; Feinberg et al. 2008; Kashefi et al. 2008). At least one archaeal species, *Pyrobaculum arsenaticum*, can use arsenate as a terminal electron acceptor for growth (Oremland and Stolz 2005).

Nickel is another important requirement for methanogens: it is required for methanogenesis in *Methanobacterium* strains (Hartzell et al. 1988) and in the methanogenic archaea *Methanobrevibacter smithii* and *M. barkeri* for incorporation

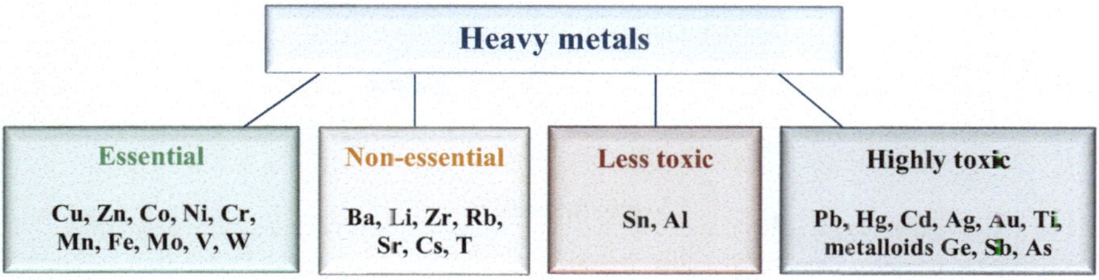

Fig. 13.1 Classification of heavy metals based on their biological role and effects

into cofactor, a yellow chromophore found in the methylreductase of *Methanobacterium* (Diekert et al. 1981; Ellefson et al. 1982).

Tungsten and molybdenum have similar chemical properties. Molybdenum is a trace metal required by virtually every species, and tungsten can replace molybdenum in some instances (Kletzin and Adams 1996). Tungsten is an essential trace metal for hyperthermophile archaea *P. furiosus*, as it involves in aldehyde oxidoreductases activity. *Thermococcus litoralis* uses another tungsten-containing enzyme, FOR (Dhawan et al. 2000). Several bacterial species, including strains of *Pseudomonas, Chloroflexus, Thiobacillus, Alcaligenes,* and *Thermus* genera and archaea *Pyrobaculum arsenaticum, P. aerophilum* can generate energy either by oxidation or reduction of specific arsenic oxyanions (Ben Fekih et al. 2018).

Some heavy metal ions, for example Cd(II), Pb(II), Sn(II), Hg(II), and Ag(I), do not have vital biological significance for microorganisms, besides form strong toxic complexes, which makes them too dangerous for any physiological function (Bruins et al. 2000). These heavy metals can also show more specific forms of chemical attack through mimicry. In this regard the toxic metals may act as mimics of essential metals, binding to physiological sites that normally are reserved for an essential element. Through mimicry, the toxic metals may gain access to, and potentially disrupt, a variety of important or even critical metal-mediated cellular functions (Cousins et al. 2006; Kasprzak 2002). In the Fig. 13.1 is presented the diagram showing the heavy metal's classification based on their toxicity.

At high concentrations, all heavy metals (both those that are essential and those that do not have biological significance) are toxic to microbes and other organisms (Nies 1999). Toxicity of heavy metals is manifested in detrimental effects on microorganisms, such as changes in the conformational structures of nucleic acids and proteins, in violation of redox processes and in maintaining the osmotic balance (Ehrlich 1997a; Nies 1999; Igiri et al. 2018). Cd, Hg, Ag ions tend to connect within the cell with sulfhydryl groups, inhibiting the activity of sensitive enzymes. The cations of some metals can replace physiologically significant ions in biomolecules, thereby violating their functions. Ni and Co ions can displace Fe, Zn—Mg ions, Cd and Zn ions—Ca ions (Ehrlich 1997a; Nies 1999). Heavy metal cations can combine with glutathione groups of gram-negative bacteria, forming a bisglutathione complex, which tends to interact with molecular oxygen to form oxidized glutathione (GS-SG) (Kachur et al. 1998). The latter can be reduced in NADPH-dependent reactions, and as a result, the formed metal cations bind other glutathione molecules,

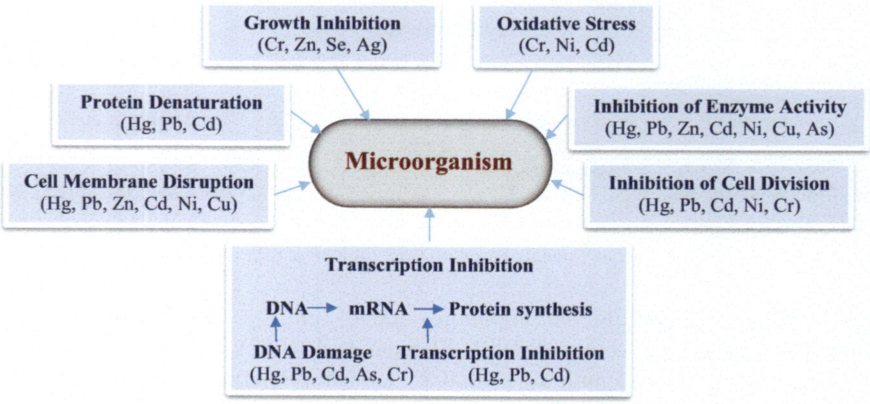

Fig. 13.2 Heavy metal toxicity mechanisms to microorganisms

thereby causing oxidative stress. Oxygen-containing anions of some heavy metals and metalloids can be involved in the metabolism of structurally similar anions of vital elements, such as S and P. For example, a chromate ion can affect the metabolism of a sulfate ion, arsenate—a metabolism of phosphate (Nies 1999; White and Gadd 2000). Cd and Pb pose deleterious effect on microbes, damage cell membranes, and destroy the structure of DNA. This harmfulness is generated by the displacement of metals from their native binding sites or ligand interactions.

Arsenic is a metalloid that occurs naturally in the environments mainly in tow forms: the trivalent species (As(III)), commonly as the oxyanion arsenite (AsO_2^-), and the pentavalent species (As(V)), or arsenate (AsO_4^{3-}). Arsenite is more toxic than arsenate as it is able to bind strongly to sulfhydryl groups in proteins and weakly to thiol groups, such as those in glutathione, lipoic acid, and cysteine. The primary toxic effects of arsenate arise from its transformation to arsenite, besides arsenate has ability to compete with phosphate oxyanions for both transport and energetics functions (Ben Fekih et al. 2018).

The morphology, metabolism, and growth of microbes are affected by changing the nucleic acid structure, causing functional disturbance, disrupting cell membranes, inhibiting enzyme activity, and oxidative phosphorylation (Fig. 13.2) (Ahemad 2012; Igiri et al. 2018).

Cr(VI) is usually present as the oxyanion chromate and based on its high oxidizing potential, considered as the most toxic form of chromium. Toxic effects of chromate for bacteria are associated with its structural similarity to sulfate (SO_4^{2-}). The CrO_4^{2-} crosses the cell membrane in some species via the sulfate transport system and cases an oxidative damage to biomolecules. Cr(VI) does not interact directly with DNA, hence its genotoxicity is attributed to its intracellular reduction to Cr(III) via reactive intermediates. The resulting types of DNA damage that are produced can be grouped into two categories: (1) oxidative DNA damage and (2) Cr(III)-DNA interactions (Cervantes and Campos-García 2007; Díaz-Magaña et al. 2009; Luo et al. 2019).

13.2 Microbial Heavy Metal Transporters

To have any physiological or toxic effect, most heavy metal ions should enter the microbial cell. Microorganisms have two main types of transport systems for heavy metal ions. The first type of transport system is fast, nonspecific, which is expressed constitutively and is controlled through the cytoplasmic membrane of bacteria by the proton gradient (pmf—proton motive force) (Silver 1996; Sar et al. 1998; Nies 1999). The second type is a substrate-specific slow transport, often requiring ATP as an energy source in addition to the proton gradient (Table 13.1). This "energetically expensive" type of transport system is inducible and is used by the cell in certain metabolic states, for example, in a state of hunger (Nies 2003, 2007; Nies and Silver 1995).

ATP-binding cassette (ABC) transporters are a major category of membrane-associated bacterial protein structures involved in the transport of a wide range of substrates including heavy metals. For example, Ni can be absorbed by the NikA-E transport system (ABC family transporter), which consists of five components (NikA periplasmic Ni-binding protein, NikB and NikC transmembrane pores for passage of Ni, NikD and NikE ions hydrolyze ATP and use energy to ion transport Ni(II)). The NikA protein can also bind Co, Cu, and Fe ions, but with a tenfold low affinity (Eitinger and Mandrand-Berthelot 2000; Mulrooney and Hausinger 2003). In different microbes, the Znu transport system of the ABC family absorbs Zn ions and has a similar structure to the Nik transporter.

Heavy metal ions like Ni, Co, Zn, and Mn can be accumulated also in gram-negative bacteria and archaea by the fast and nonspecific CorA system (metal inorganic transporter of the MIT family) (Smith and Maguire 1995; Hynninen 2010). In *B. subtilis*, Mg, Ni, Mn, Co, and Zn ions can be absorbed by the metal citrate transport protein CitM and CitH (Hantke 2001; Krom et al. 2000).

The fast ion transport along the concentration gradient is an important factor contributing to the toxicity of heavy metals. When cells are exposed to high concentrations of heavy metals, which can accumulate through nonspecific transport systems, the "passage" into the cytoplasm can remain open, even at "toxicologically dangerous" concentrations of metals in the cytoplasm, since this process is constitutive (Nies 1999). Despite heavy metal toxicity, microbes possessing different metal resistance strategies, such as detoxification, metal absorption, uptake and accumulation, extracellular precipitation, efflux of heavy metals from the cells.

Table 13.1 Heavy metal transporters in prokaryotic cells

Transporter type	Member	Organism	Function	Energy	Metal ions	Comments	Reference
ABC	NikA-E, ZnuABC, SitABCD, PsaABC, TroABC, MtsABC, FimA, Pzp 1, FepCDG, FecECD, FhuBBC, Sfu, Tfe, NiCoT, WtpA	*E. coli, Salmonella enterica, Streptococcus pneumonia, S. pyogenes, Treponema pallidum, Hemophilus influenzae, Methanosarcina acetivorans, M. barkeri, M. mazei, Sulfolobus solfataricus, Thermoplasma acidophilum, Methanomicrobium* sp., *M. acetivorans, Pyrococcus furiosus*	Uptake	ATP	Mn(II), Zn(II), Ni(II), Cu(II), Fe(III), Fe(II), W(VI), Mo(II)	2 membrane-integral partsa +2 ATPase parts = ABC core + periplasmic binding protein	Hohle and O'Brian (2009), Nies (2003), Porcheron et al. (2013), Zhang et al. (2009), Bini (2010) and Margaryan et al. (2013)
P-type	ZntA, CadA, PbrA, CopAB, MgtA, KdpB, Mt., Mba, Fa, Af	*Cupriavidus metallidurans, B. subtilis, Enterobacter hirae, Pseudomonas syringae, P. aeruginosa, X. campestris, E. coli, S. aureus, Stenotrophomonas maltophilia, Deinococcus*	Both	ATP	Mg(II), Mn(II), Ca(II), K(I), Cu(II), Zn(II), Cd(II), Pb(II), Ag(I)	1 membrane-bound protein as core	Rademacher and Masepohl (2012), Hantke (2001), Mills et al. (1994), Nies (2003), Aguilar-Barajas et al. (2010), Moraleda-Muñoz et al. (2010), Margaryan et al. (2013), De Hertogh et al.

		radiodurans, Methanocaldococcus jannaschii, Thermoplasma acidophilum, T. volcanium, Myxococcus xanthus, Methanobacterium thermoautotrophicum, Methanosarcina barkeri, Ferroplasma acidarmanus, Archaeoglobus fulgidus					(2004) and Adekanmbi et al. (2019)
MIT	CorA		Uptake	Chemiosmotic	Most cations	Membrane-integral protein (CorA)	Nies (2003)
MntH	–	*E. coli, Bradyrhizobium japonicum*	Uptake	–	Mn(II), Fe(II)	MntH family proteins are proton-driven metal ion transporters with homology to eukaryotic NRAMP proteins	Peng et al. (2018) and Hohle and O'Brian (2009)
HoxN			Uptake	Chemiosmotic	Co(II), Ni(II)	Membrane-integral protein	Nies (2003)
CBA family /RND	CzcABC, CusABC, CnrABC/NccABC	*R. metallidurans, Mesorhizobium loti, Alcaligenes eutrophus, Myxococcus xanthus, Pseudomonas aeruginosa*	Efflux	Proton gradient	Co(II), Zn(II), Cd(II), Ni(II), Cu(II), Ag(I)	1 CPM proton/cation antiporter + membrane fusion protein (dimer) + outer membrane factor:	Silver (1996), Collard et al. (1994), Diels et al. (1995), Hantke (2001), Moraleda-Muñoz et al.

(continued)

Table 13.1 (continued)

Transporter type	Member	Organism	Function	Energy	Metal ions	Comments	Reference
						CBA transport systems	(2010), Aguilar-Barajas et al. (2010) and Adekanmbi et al. (2019)
CDF	CzcD, ZitB, ZneA, CusA	*E. coli; B. subtilis; Staphylococcus aureus, Thermus thermophiles, B. subtilis, C. metallidurans*	Efflux	Chemiosmotic	Zn(II), Cd(II), Co(II)	Membrane-integral protein	Grass et al. (2001), Hantke (2001), Guffanti et al. (2002), Nies (2003), Spada et al. (2002), Anton et al. (1999) and Nikaido (2018)
CHR	ChrA, Orf1, Orf2, SrpC	*Pseudomonas aeruginosa, Cupriavidus metallidurans, Ralstonia metallidurans, Acinetobacter calcoaceticus, B. subtilis, Synechococcus* sp., *Synechocystis* sp., *Methanococcus jannaschii, Proteus mirabilis*	Antiport	Chemiosmotic	Chromate	ChrA is a membrane-integral protein that confers resistance to the toxic ion chromate through the energy-dependent chromate efflux from the cytoplasm. In P. aeruginosa and A. eutrophus, chromate is accumulated by sulfate uptake systems, and expression of ChrA leads to reduced	Nies (2003), Díaz-Pérez et al. (2007) and Díaz-Magaña et al. (2009)

						accumulation of chromium. ChrA protein of *A. eutrophus* may be a chromate uptake system when expressed alone. CHR family proteins can also catalyze chromate/sulfate antiport	
MerR	MerA, ZntA, PbrA	*E. coli, B. subtilis, R. metallidurans, Hydrogenobaculum* sp. *Hydrogenivirga* sp., *Thermus thermophilus*	Reduction/efflux			MerA is the protein subunit of the homodimeric mercuric reductase (MR) enzyme, the central function of the mer system	Brown et al. (2003), Freedman et al. (2012) and Boyd and Barkay (2012)
Arsenical pump membrane protein	ArsB	*E. coli, Campylobacter jejuni*	Efflux	ATP	As(III), Sb(III)	ArsB is an antiporter that extrudes As(III) or Sb(III) from cells by $H^+/As(OH)_3$ exchange coupled to the electrochemical proton gradient. It has been proposed that ArsB transport a polymeric ring composed of three $As(OH)_3$ molecule	Garbinski et al. (2019) and Mourão et al. (2020)
RART (bile/arsenite/	Acr3	*Alkaliphilus metalliredigens*	Efflux	ATP	As(III)	mAcr3–1 has 10 transmembrane-	

(continued)

Table 13.1 (continued)

Transporter type	Member	Organism	Function	Energy	Metal ions	Comments	Reference
riboflavin transporter superfamily)						spanning segments, with the N-and C-termini localized in the cytosol. Acr3 compared with ArsB, has two additional transmembrane-spanning segments. Two residues may be involved in metalloid translocation	
MFS	ArsK	*Agrobacterium tumefaciens, Bacillus* sp.	Efflux	–	As(III), Sb(III), MAs(III), Rox (III)	ArsK is a 411 amino acid residue transmembrane protein that can be predicted to have 12 transmembrane segment (TMs). ArsK has the broadest substrate specificity of any known trivalent arsenic efflux permease, including both inorganic and organic species	
	ArsJ	*Pseudomonas aeruginosa, E. coli*	Efflux	–	As(V)	ArsJ is a 410-residue membrane protein with	

						10 predicted TMs. It has a large putative extracellular loop between TMs 3 and 4, and a smaller intracellular loop between TMs 4 and 5	
MAs(III)-selective permeas	ArsP	*Campylobacter jejuni*	Efflux	–	As(V), methylarsenite (MAs(III)), roxarsone (Rox(III))	It is smaller than MFS transporters and is predicted to have only 8 TM. ArsP is an organic arsenic transporter for the pentavalent forms of roxarsone and nitarsone	
Feo	FeoB	*B. subtilis*, *E. coli*, *S. thermophiles*, *Pyrococcus furiosus*, *Klebsiella pneumoniae*, *L. pneumophila*, *Thermotoga maritima*, *Gallionella capsiferriformans*	Uptake	–	Fe(II)	Starting from the N-terminal end, the G-protein domain is the first protein domain of FeoB. The G-protein domain resides in the cytoplasm and is covalently tethered to the polytopic transmembrane region of FeoB through the so-called GDI-domain	Lau et al. (2016)

13.3 Heavy Metal Resistance in Prokaryotes

Heavy metal ions cannot undergo degradation or significant modification like toxic organic compounds in the environment. Microbes, leaving in the heavy metal-polluted environment, develop different mechanisms to tolerate toxic concentrations of the metals (Nies 2007). Microbes can have one or a combination of several different strategies of metal resistance (Bruins et al. 2000; Nies and Silver 1995).

In bacteria, all existing mechanisms that allow surviving in the presence of toxic concentrations of heavy metals in the medium can be attributed to several main types. This is an active release of metal from the cell, restriction of metal intake due to changes in cell permeability, intracellular metal binding and detoxification, extracellular binding, enzymatic metal detoxification into a less toxic form and a decrease in the metal sensitivity of cellular components (Fig. 13.3) (Nies and Silver 1995; Nies 2007; Bruins et al. 2000; Ahemad 2015).

Among the Archaea, thermophiles and hyperthermophiles of the Crenarchaeota and the methanogens and thermophiles of Euryarchaeota utilize P-type ATPases and

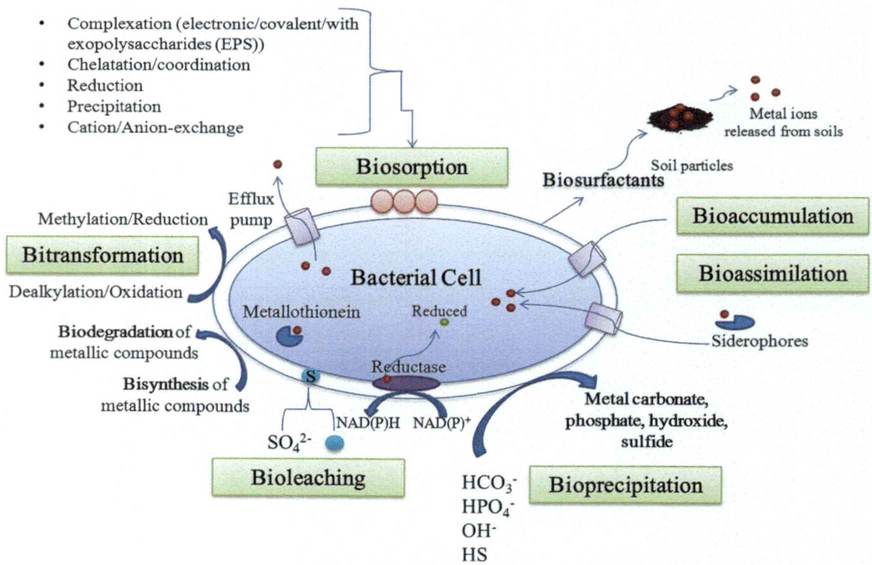

Fig. 13.3 Various bacterial interactions with heavy metals in metal-polluted soil. *Biosorption*: Precipitation/crystallization of metals occurs due to bacteria-mediated reactions or as a result of the production of specific metabolites. *Bioaccumulation:* Plasmid-DNA-encoded efflux transporters (e.g., ATPase pumps or chemiosmotic ion/proton pumps) expel the accumulated metals outside the cell. *Bioprecipitation:* Metals bind to the anionic functional groups (e.g., sulfhydryl, carboxyl, hydroxyl, sulfonate, amine, and amide groups) of extracellular materials present on cell surfaces. *Bioleaching:* Organic acids secreted by bacteria solubilize the insoluble metal minerals. *Biotransformation:* Some bacteria utilize methylation as an alternative for metal resistance/detoxification mechanism, which involves the transfer of methyl groups to metals and metalloids

ABC transporters for metal transport and homeostasis (Coombs and Barkay 2005; Bartolucci et al. 2013).

13.3.1 Active Transport of Heavy Metals

Microbes use active transport mechanisms to efflux toxic metals from the cytoplasm. Metals that do not have physiological significance usually enter into the cell through transport systems designed for the necessary cations, but then quickly get out of the cell by efflux pumps (Ehrlich 1997a). It was found that active ion efflux systems can be either ATP-independent or using ATP energy (see Table 13.1). All of them are highly specific for cations or anions that are exported from the cell (Nies and Silver 1995; Hynninen 2010). A large number of varieties of this mechanism of metal resistance in bacteria and archaea are described (Table 13.1). Three families of transport systems are mainly involved in the export of heavy metal ions from the cell: a three-component transmembrane transporter in Gram-negative bacteria is Capsule biogenesis/assembly (CBA) family transporter, which acts as a chemosmotic antiport; cation diffusion facilitator (CDF), which acts as a chemosmotic ion-proton exchanger and P-type ATPase located in the inner membrane and using ATP energy to export metal ions from the cytoplasm to periplasm (Fig. 13.4) (Hynninen 2010; Nies 2003, 2007; Grass et al. 2001).

13.3.2 CBA Family Transporters

CBA family transporters are a three-component protein complex that span the whole cell wall of Gram-negative bacteria and expel ions from cyto- and periplasm to

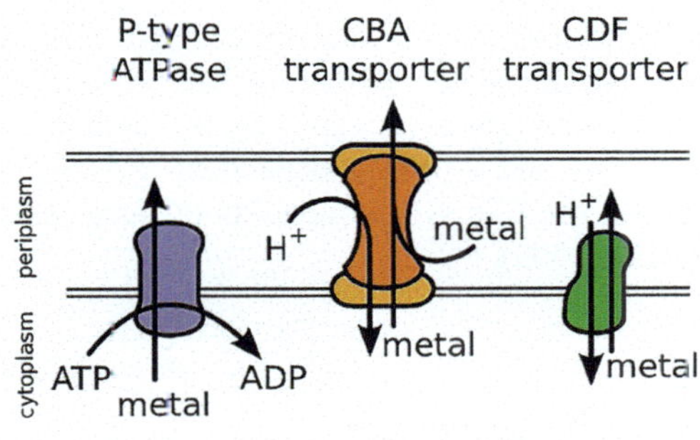

Fig. 13.4 The main transporter families that determine bacterial heavy metal resistance. *P-type ATPases* pump their substrates from cytoplasm to periplasm using energy provided by ATP hydrolysis. *CBA transporters* are three-component complexes in Gram-negative bacteria that efflux ions from cyto- and periplasm to outside using a chemiosmotic gradient. *CDF transporters* are driven by a proton motive force and they export ions from cytoplasm to periplasm (Hynninen 2010)

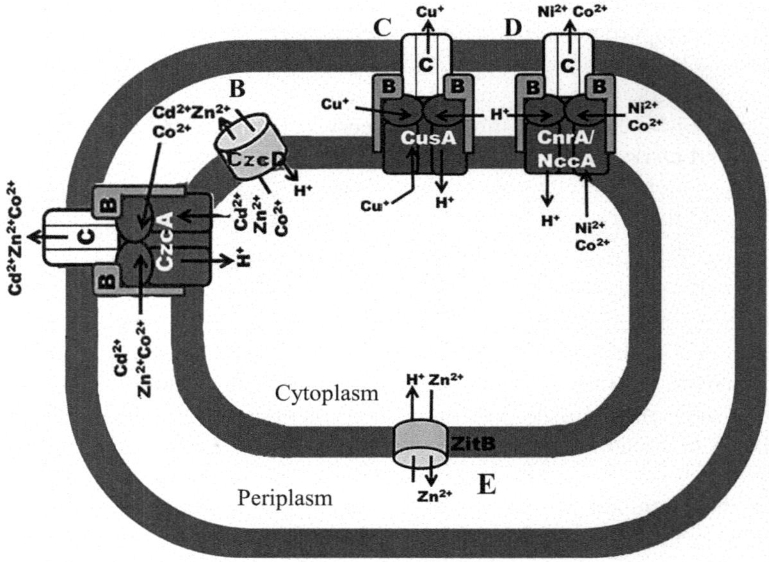

Fig. 13.5 Structural models of CBA and CDF families pumps. (**a**) Czc, functioning as a proton/cation antiport, consisting of intramembrane (CzcA), extramembrane (CzcC) and integral (CzcB) proteins, (**b**) CzcD, transporting Cd, Zn and Co ions in *B. subtilis*, (**c**) CusABC, and (**d**) CnrABC / NccABC are similar in structure and function to the CzcABC system, (**e**) ZitB is similar to the CzcD system (modified from Aguilar-Barajas et al. 2010)

outside using a chemiosmotic gradient. The most important component of the transporter is the intramembrane protein RND (resistance, nodulation, and division), which was first described as a bacterial transport protein involved in the resistance processes of heavy metals in *R. metallidurans*, nodulation of *Mesorhizobium loti*, and cell division of *E. coli* (Nies 2003).

An example of the RND family transporter is the Czc system for the active export of Cd(II), Zn(II), Co(II) cations from a bacterial cell. The Czc system is described and studied in detail in the facultative chemolithoautotrophic bacteria *Alcaligenes eutrophus* CH34. The Czc system is regulated by a proton concentration gradient across the inner membrane and is ATP-independent (Silver 1996; Collard et al. 1994; Diels et al. 1995). The Czc system consists of three main parts (Fig. 13.5) (Rosen 2002; Anton et al. 1999).

13.3.3 CDF Family Transporters

The cation diffusion facilitators (CDFs) are a family of membrane-bound proteins that maintain cellular homeostasis of essential metal ions. Proteins of the secondary cationic CDF transporter catalyzing the efflux of heavy metals and were found in both prokaryotes and eukaryotes. All proteins of the CDF family are substrate-specific. The main substrate for CDF transporters is Zn(II) ions, but Co(II), Ni(II), Cd(II), and Fe(II) can also initiate transporter. The CDF system is regulated by a proton concentration gradient, ΔΨ, ΔpH, or K(I) concentration gradient (Nies 2007; Guffanti et al. 2002; Paulsen and Saier 1997).

CDF coding genes were found in the chromosomes of a number of microorganisms, but protein functionality has been characterized only in few microbes. In *B. subtilis*, *czcD* genes are located in the operon along with *trikA* dehydrogenase gene (Nies 2003). The *czcD-trkA* operon is complementary to the K(I) transport system in *E. coli* (Guffanti et al. 2002). CzcD was first described in bacteria *Ralstonia metallidurans* CH34 as a regulator of *czcABC* gene expression, but CzcD (Fig. 13.5) can also participate in the transport of Cd(II), Zn(II), Co(II) in the absence of the CzcABC system (Anton et al. 1999; Nies 2003; Scherer and Nies 2009; von Rozycki and Nies 2009).

In *B. subtilis*, CzcD is regulated by a K(I) concentration gradient and leads to the emission of Cd(II), Zn(II) and Co(II) (Guffanti et al. 2002). In *E. coli* cells, the CzcD system is regulated by a proton concentration gradient and leads to the emission of Zn(II) and Cd(II) ions, but not Co(II) (Nies 2003; Paulsen and Saier 1997).

In *E. coli*, the ZitB protein (product of the *ybgR* gene) of the CDF family has also been described, which determines resistance to Zn ions, reducing ion accumulation (Fig. 13.5) (Grass et al. 2001).

In *Staphylococcus aureus*, CzcD determines resistance to Zn(II) and Co(II), in *Thermus thermophilus* determines resistance to Zn(II) and Cd(II). CDF proteins can also export Pb ions (Spada et al. 2002; Xiong and Jayaswal 1998).

13.3.4 P-Type ATPase Family Transporters

P-type ATPase is a family of transport protein that exports ions against a concentration gradient using ATP. It is highly substrate-specific. The substrates are Na, K, Mg, Ca, Cu, Ag, Zn, Cd, Co, and Pb cations. Heavy metal-transporting ATPases have a metal-binding domain (MBD) and are described in both gram-positive and gram-negative bacteria. Prototype of P-type ATPase is ZntA system for active efflux of Zn(II), Cd(II), and Pb(II) from *E. coli* cell (Fig. 13.6) and CadA for active efflux of Cd(II) from *S. aureus* cell (Fig. 13.6).

CadA consists with six domains located in the membrane, four of which are involved in translocation of cations, and a conservative Cys-Pro-Cys tripeptide. Two intracellular domains common to all P-type ATPases are aspartyl kinase and phosphatase domain. During metal transport, ATP phosphorylates the protein, probably at the location of the invariant aspartic acid (Asp 415). Phosphorylation occurs only in the presence of Cd ions (Tsai et al. 1992). The transport system CadA was also found in the bacteria *Bacillus subtilis, Pseudomonas metallidurans, Cupriavidus metallidurans, Synechocystis* sp. etc. (Lee et al. 2001; Scherer and Nies 2009).

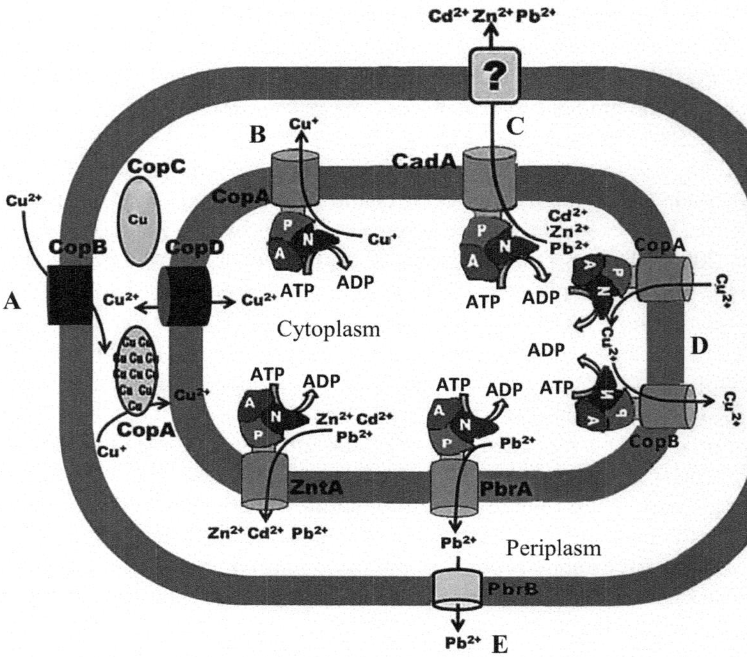

Fig. 13.6 Transport systems of metals—P-type ATPases: (**a**) CopABCD copper transport system; (**b**) CopA ATPase P-type transformation of Cu(I) into Cu(II); (**c**) CadA ATPase of the P-type, removal of Cd, Zn, and Pb ions from the cytoplasm; (**d**) CopA system of absorption of Cu(II) and CopB of Cu(II) export in *E. hirae* cells; (**e**) PbrA ATPase P-type removal of Pb ions from the cytoplasm; ZntA ATPase P-type removal of Zn, Cd and Pb ions from the cytoplasm (modified from Aguilar-Barajas et al. 2010)

PbrA system, the member of P-type ATPase, actively removes Pb ions from the cytoplasm of the bacterium *Cupriavidus metallidurans* (Fig. 13.6). The structure and function of the protein PbrA is similar to the CadA and ZntA. PbrB lipoprotein is located on the outer membrane, which probably transports Pb ions from periplasm to the environment (Aguilar-Barajas et al. 2010).

In *Enterobacter hirae* have been found the CopA and CopB system of the P-type ATPase family, which are for Cu(II) transport. CopA determines the absorption of Cu ions, and CopB efflux of Cu ions from the cytoplasm. The synthesis of the both proteins is regulated by operon genes (Fig. 13.6) (Argüello et al. 2013). The promoter region of the operon is controlled by the CopY repressor, regulated by Cu ions. CopZ protein, together with Cu ions, activates the promoter. The binding of copper to CopZ leads to the formation of the complex, which attached to CopY, as a result the operon, is activated (Rademacher and Masepohl 2012).

Homologous systems have been described in *Pseudomonas syringae, Xanthomonas campestris,* and *E. coli* (Cooksey 1994). In the copper metabolism of *P. syringae*, two regulatory *copRS* genes and four structural *copABCD* genes were found, while in *X. campestris* and *E. coli,* the corresponding genes are called *pcoRS* and *pcoABCD*. The *copR* and *copS* genes are located immediately after the copper tolerance operon (*copABCD*) on the pPT23D plasmid and transcribed as an operon from two genes of the same constitutive promoter (Mills et al. 1994; Rademacher and Masepohl 2012) (Fig. 13.6).

The product of *copS* gene is the copper-sensitive CopS protein, which located in the inner membrane. The product of *copR* gene is the regulatory protein CopR, which located in the cytoplasm. With an increased periplasmic concentration of Cu (II), CopR transphosphorylates the CopS protein and activates transcription of the *cop* operon (Rademacher and Masepohl 2012; Mills et al. 1994).

The plasmid operon *copABCD* in the bacterium *Pseudomonas syringae* is one of the first described copper resistance systems in bacteria. The *copABCD* operon encodes a system that prevents the penetration of copper into the cell cytoplasm. CopA and CopC are periplasmic proteins that bind copper. The proteins CopA and CopC able to bind 11 and 1 copper atoms, respectively, on the same polypeptide (Aguilar-Barajas et al. 2010). The activation of transporters leads to the accumulation of copper in the periplasmic space, which protects the cell from the toxic effect of the ion. CopA also exhibits oxidase activity, transforming Cu(I) into Cu(II), thereby protecting the periplasmic enzymes from the toxic effect of copper (Argüello et al. 2013).

CopC is probably a chaperone protein that transports Cu ions to the integral CopD protein. CopD consists of eight transmembrane segments and transports copper both into the cytoplasm and from the cytoplasm to the periplasm. CopB is an outer membrane protein that absorbs copper (Aguilar-Barajas et al. 2010).

P-type ATPase has been found in 17 archaea species, by screening the databases from TIGR, NCBI, DOE, and TCDB. In all analyzed archaea species contained 1–3 metal ATPases, which belong to six different phylogenetic TC (Transport Classification) clusters. The proteins belonging to these clusters export (more rarely import), a variety of monovalent or divalent metals (copper, zinc, lead, cadmium, or silver) (De Hertogh et al. 2004). Only three transmembrane motifs for metal-transporting ATPases identified in archaea, which correspond to the group IB-1 (Cu(I)/Ag(I)), group IB-2 (Zn(II)/Cd(II)/Pb(II)), and group IB-3 (Cu(II)/Cu(I)/Ag(I)) motifs (Argüello et al. 2003).

Two metal-transporting ATPase genes *CopA* and *CopB* from the thermophilic archae *Archaeoglobus fulgidus* were cloned in *E. coli*, purified, and their ATPase activity were biochemically characterized (Mana-Capelli et al. 2003; Mandal and Argüello 2003). The thermophilic ATPase activity of *CopA* was best activated by the monovalent metals Ag(I) and Cu(I) while *CopB* was activated by the divalent Cu(II).

ATPases along with the ABC transporters, transcriptional regulators, and certain metallochaperones were found to be involved in metal resistance and homeostasis in the haloarchaeon *Halobacterium* sp. strain NRC-1 (Kaur et al. 2006). The list of archaea P-type ATPases are shown in the Table 13.2.

Table 13.2 The list of archaeal P-Type ATPases (De Hertogh et al. 2004)

Microorganism	Substrate	Function	TC typical organism
Aeropyrum pernix	Zn(II), Cd(II), Pb(II)	Efflux	Bacteria; plants; fungi; protozoa
Archaeoglobus fulgidus	Cu(I)/Ag(I)	Efflux	Archaea (CopA), Bacteria
Ferroplasma acidarmanus	Cu(II)	Uptake	Bacteria
Halobacterium sp.	Cu(II)	Uptake	Bacteria
	Zn(II), Cd(II), Pb(II), Cu(I)/Ag(I)	Efflux	Bacteria; plants; fungi; protozoa, archaea (CopA)
Methanosarcina acetivorans	Cu(I), Ag(I), Zn (II) Cd2C-, Pb2C	Efflux	Bacteria; plants; fungi; protozoa
Methanosarcina barkeri	Mg(II)/Ni(II), Cu(I), Ag(I), Zn(II), Cd(II), Pb(II)	Efflux	Archaea, eukaryotes (Wilson's disease), Bacteria; plants; fungi; protozoa
Methanosarcina mazei	Cu(I), Ag(I)	Efflux	Eukaryotes (Wilson's disease), Bacteria
Methanobact. thermoautotrophicum	Cu(I), Ag(I), Zn(II), Cd(II), Pb(II)	Efflux	Archaea (CopA), Bacteria; plants; fungi; protozoa
Pyrobaculum aerophilum, Pyrococcus furiosus, Sulfolobus solfataricus, Sulfolobus tokodaii, Thermoplasma acidophilum, Thermotoga maritima, Thermoplasma volcanium	Cu(I)/Ag(I)	Efflux	Archaea (CopA)

13.3.5 Limitation of Metal Intake Due to Changes in Cell Permeability

When a cell is exposed to concentrations of heavy metals in the environment, the microbe may undergo structural changes in the cell wall, membrane, and cytoplasmic membrane. These processes are not always the result of the toxic effects of the metals. They can be a manifestation of induced defense mechanisms that limit the flow of toxic ions into the cell cytoplasm (Bruins et al. 2000; Ehrlich 1997a).

The first sites of cell and heavy metal interaction are at the cell surface. The bacterial cytoplasmic membrane, and to a lesser extent the outer membrane in Gram-negative bacteria, are a major barrier to the entry of hydrophilic substances, including metal ions, into the interior of the cell. In Gram-negative bacteria, like *E. coli*, the outer membrane contains protein channels called porins, that allow low-molecular-weight substances such as metal ions to diffuse across the membrane into the periplasmic space. In *E. coli* synthesis of the major porin can be prevented by mutations in a single gene resulting in increased metal resistance. The outer membrane can also act as a limited (i.e., saturable) trap for heavy metals by nonspecifically binding them, therefore contributing to the natural metal tolerance of cells (Rouch et al. 1995).

An unusual mechanism of metal resistance is found in *Pseudomonas syringae,* which accumulate blue Cu(II) ions in the periplasmic space and outer membrane. At least part of this copper sequestering activity is determined by copper-binding periplasmic CopA protein products of the copper resistance operon (*cop*). Copper resistance operons related to *cop* have been described in the related plant pathogen *Xanthomonas campestris* and in *E. coli*, but these resistance systems may differ functionally from the *P. syringae* system (Cooksey 1994).

A significant advantage for survival in environments contaminated with heavy metals is reducing bioavailability or mobility of heavy metal ions by the released exopolysaccharide (EPS). Anionic property of EPS allows the biopolymer to effectively sequester positively charged heavy metal ions and restricts the entry of metal ions into the cell. The anionic property of EPS imparts by abundant active and ionisable functional groups and non-carbohydrate substituents like phosphodiester (techoic acid), phosphate, hydroxyl groups, or acetamido group of chitin, structural polysaccharides of fungi. On contrary to homopolysaccharides, extracellular heteropolysaccharides are often polyanionic due to association of some of such functional groups with polysaccharide backbone. The sorption and immobilization again occurs via different mechanisms like ion exchange, complexation, precipitation, etc. (Gupta and Diwan 2017). As an example can be serve *Ochrobactrum anthropi*, isolated from activated sludge. This bacteria producing the most EPS for the removal of Cr(VI), Cd(II) and Cu(II) (Ozdemir et al. 2003).

Staphylococcus xylosus and *Staphylococcus carnosus* strains were characterized by production of surface-exposed chimeric two different polyhistidyl peptides, His_3-Glu-His_3 and His_6 due to the expression of recombinant plasmid genes designed for binding to divalent metal ions. As a result, the entry of Cd ions and other toxic metals into the cell is limited, which suggests that such bacteria could find use in bioremediation of heavy metals (Samuelson et al. 2000).

It has been shown, that EPS synthesized by *Arthrobacter viscosus* accumulate 2.3 times more Cd(II) than an equivalent weight of intact cells and have 13.7 times the sorptive capacity of *Arthrobacter globiformis* cells, which do not produce EPS (Hrynkiewicz et al. 2015).

Numerous halophilic bacteria and archaea can also tolerate high concentrations of heavy metals by secrete EPS. *Halomonas* strains can tolerate high concentrations of Pb(II) and Cd(II) (5 mM) by the EPS-mediated adsorption of the metallic ions (Voica et al. 2016). Dry biomass of the haloarchaeon *Halobacterium* sp. GUSF was an effective adsorbent for Mn(II) from saline solutions, the process of adsorption involving cell surface carboxyl, amino, phosphate and hydroxyl groups (Naik and Furtado 2014). The high EPS-producing halotolerant cyanobacterium *Aphanothece halophytica* grown at 6% NaCl (w/v) was capable of accelerated Zn (II) adsorption up to a critical cell density that may result in aggregation, reducing the matrix surface available for metal binding (Incharoensakdi and Kitjaharn 2002).

13.3.6 Intracellular Binding of Toxic Metals and Their Detoxification

The accumulation of metal in the cytoplasm and its detoxification can occur due to the binding of toxic ions to specific proteins, like low-weight cysteine-rich proteins and peptides. A variety of metal-binding peptides like glutathione (GSH) and proteins like metallothioneins and phytochelatins produced by certain microbes like *Cyanobacterium synococcus, Synechococcus* sp., *E. coli, P. putida* (Gupta and Diwan 2017; Bruins et al. 2000; Silver 1996).

Citrobacter sp., isolated from metal-polluted soil can resist Cd(II) toxicity by forming insoluble complexes of Cd-phosphate ($CdHPO_4$); this transformation is mediated by a cell-bound phosphatase that precipitates inorganic phosphate with heavy metals. A strain of *Pseudomonas putida* isolated from sewage can sequester intracellular Cd(II) by producing three low-molecular-weight cysteine-rich proteins related to eukaryotic metallothioneins, while *K. aerogenes* excretes sulfur into the surrounding environment to immobilize Cd(II) ions as insoluble Cd-sulfide (Hrynkiewicz et al. 2015). The ability to intracellularly accumulate lead phosphate in the form of granules was exhibited by the *P. aeruginosa* (Naik et al. 2012). For Mycobacterium scrofulaceum, the ability to intracellular accumulation of Cu(II) in the form of sulfide was found (Bruins et al. 2000).

Cyanobacteria at toxic concentrations of free Cu ions in the medium produce extracellular chelating ligands that bind to Cu(II) ions, reducing their bioavailability. In *Synechococcus* spp., a ubiquitous and important group of phytoplankton, synthesis of chelating ligands is regulated by the concentration of free Cu(II) ions in the medium according to the feedback mechanism (Moffett and Brand 1996).

In some cases, the formation of metal precipitating anions may result from normal cellular metabolism, such as the formation of sulfides under anaerobic conditions by sulfate-reducing bacteria of the genus *Desulphovibrio*. In other cases, the process is inducible under certain environmental conditions, for example, the formation of sulfides by bacteria of the genus *Clostridium* (Karnachuk et al. 2003).

Metallothioneins and phytochelatins are not represented in archaeal genomes, however members of the CutA family of metal-binding proteins are found in archaea, bacteria, and eukaryotes. The crystal structure of the *Pyrococcus horikoshii* CutA has been determined with and without copper, contributing to the clarification of the protein's function. In fact, binding of heavy metals induced the reversible multimerization of CutA. Thus, a role has been proposed for CutA in the capture and precipitation of metal ions. Interestingly, while the metal-binding site of the *E. coli* homolog contains Cys and His residues, these amino acids are absent in the *Pyrococcus* protein (Bini 2010).

13.3.7 Reduction of Heavy Metal Ions and Enzymatic Detoxification

Bacteria and Archaea are reducing a broad spectrum of heavy metal ions: chromate, molybdate, vanadate, iron, etc. (Table 13.3). Some bacteria and archaea can use metals and metalloids as electron donors or acceptors for energy generation. Metals in the oxidized form could serve as terminal acceptors of electrons during anaerobic respiration.

The most studied example of the manifestation of the metal resistance mechanism in bacteria associated with the process of intracellular enzymatic metal detoxification is the Hg ion resistance system (Nies 1999). Stability is due to the functioning of the operon and was revealed both in gram-positive (*S. aureus, Bacillus* sp.) and gram-negative bacteria (*Escherichia coli, Pseudomonas aeruginosa, Serratia marcescens, Thiobacillus ferrooxidans*) (Bruins et al. 2000). As a result of the expression of the genes that make up the *mer* operon, Hg(II) in two stages is reduced to metallic mercury, which then diffuses through the cell membrane and is released into the environment (Fig. 13.7). Due to the volatility of metallic mercury, its content in the medium can rapidly decrease (Silver 1996).

In *Alcaligenes faecalis* bacteria, the mechanisms of enzymatic oxidation of As (III) compounds present in the form of AsO_2 to As(V) compounds in the form of AsO_4, which are less toxic, have been studied and described (Anderson et al. 2003).

Microorganisms have developed, or acquired, various genetic systems to cope with arsenic toxicity. These systems include the *ars* operons, groups of genes widely

Table 13.3 Reduction of metals and metalloids by different microorganisms (modified from Ianeva 2009)

Reduction process	Microorganism
Hg(II)/Hg(0)	*Bacillus cereus, Klebsiella pneumonia, P. stutzeri*
Fe(III)/Fe(II)	*Geobacter* sp., *G. metallireducens, Bacillus thermoamylovorans, Ferroplasma* spp., *Thermoplasma* spp.
Cr(VI)/Cr(III)	*Desulfomicrobium norvegicum, Microbacterium* sp., *Ochrobacterium intermedium, Brevibacterium* sp., *Pseudomonas* spp.
As(V)/As(III)	*S. aureus*
U(VI)/U(IV)	*Desulfovibrio desulfuricans, Shewanella putrefaciens, Thermoterrabacterium ferrireducens, Metallosphaera prunae, M. sedula*
Mn(IV)/Mn(II)	*Shewanella putrefaciens*
Se(VI)/Se(IV)/Se(0) Se(IV)/Se(0)	*R. metallidurans, B. thermoamylovorans, Shewanella oneidensis*
V(V)/V(IV)	*S. oneidensis, G. metallireducens*
Tc(VII)/Tc(IV)	*Geobacter sulfurreducens, S. putrefaciens*
Mo(VI)/Mo(V)	*Thiobacillus ferrooxidans*
Au(III)/Au(0)	*Stenotrophomonas* sp.
Te(IV)/Te(0)	*B. thermoamylovorans, S. oneidensis*

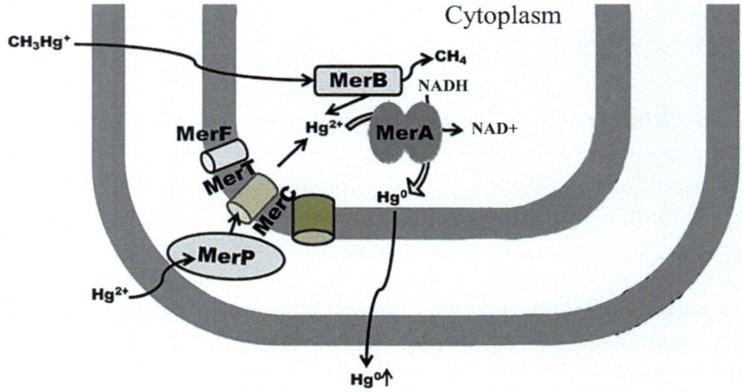

Fig. 13.7 Mechanism of detoxification by the Hg(II) Mer system (Aguilar-Barajas et al. 2010)

distributed in bacterial and archaeal species. *ars* operons frequently occur in most prokaryotic genomes, and it has been stressed that they are more common than genes for tryptophan biosynthesis. This operon first has been found in the plasmid pI258 in the clinical bacteria *Staphylococcus aureus*. The plasmid pI258 was found to encode multiple resistances to antibiotics, arsenate, arsenite and other heavy metal derivatives (Ben Fekih et al. 2018; Novick and Roth 1968). Arsenic resistance genes have identified in R773 plasmid in *Escherichia coli* strain isolated from a patient with a urinary tract infection (Hedges and Baumberg 1973). The nucleotide sequence of the determinants from the *E. coli* R773 plasmid identified the *arsRDABC* operon involved in the arsenic resistance phenotype, and staphylococcal plasmids pI258 and pSX267 both contained similar, but simpler *arsRBC* operons encoding proteins with homology to those encoded by R773 (Ben Fekih et al. 2018). The distribution of *ars* operon genes in bacteria and archaea are presented in the Table 13.4.

Nearly every organism has resistance pathways for inorganic arsenic. The minimal constituents are usually an As(III)-responsive repressor (ArsR), and an As(III) efflux permease (ArsB or ACR3) that functions to extrude trivalent As(III) from cells. The As(III)-stimulated ATPase (ArsA), and the As(III) metallochaperone (ArsD), which are always associated in ars operons, appears to be later adaptations that enhances the ability of ArsB to extrude As(III) and increase resistance. ArsC and other arsenate reductases are required for resistance to arsenate (Yang and Rosen 2016; Ben Fekih et al. 2018). Recently, a parallel pathway for organic arsenicals has been identified. The ars genes responsible for the organo-arsenical detoxification include *arsM*, which encodes an As(III) S-adenosylmethionine methyltransferase, *arsI*, which encodes a CeAs bond lyase, and *arsH*, which encodes a methylarsenite oxidase (Fig. 13.8).

Pentavalent inorganic arsenate (As(V)) is reduced by the ArsC arsenate reductase to trivalent arsenite (As(III)). Some microbes encode As(III) S-adenosylmethionine methyltransferases ArsM protein, that transform As(III) into the considerably more toxic (for humans, carcinogenic) organo-arsenical MAs(III). Other microbes can

Table 13.4 Distribution of *ars* genes in arsenic-resistant bacteria and archaea

Microbe	Plasmid/ Chr	Operon/gene	Comments	Reference
E. coli	R773	*arsRDABC*	The ArsB is an integral membrane protein which acts as an anion channel. The ArsA protein is the energy-transducing ATPase subunit, with specific-binding sites for ATP and arsenite. Binding of ArsC to the complex either changes the specificity to arsenate or increases the range of substrates to allow recognition of both arsenate and arsenite. The ArsR and ArsD are regulatory proteins	Chen et al. (1986) and Carlin et al. (1995)
Acidiphilum multivorum	pKW301			Suzuki et al. (1998)
Staphylococcus aureus	pI258	*arsRBC*	*ars* operon induced by arsenate [As(V)], arsenite [As(III)], and antimonite Sb(III)	Ji and Silver (1992)
S. xylosus	pSX267			Rosenstein et al. (1992)
E. coli	Chr			Carlin et al. (1995)
Pseudomonas aeruginosa	pKW301			Cai et al. (1998)
P. fluorescens				Prithivirajsingh et al. (2001)
Rhodopseudomonas palustris	Chr	*arsRCBH* *arsRM* *arsRC*	ArsM catalyzes the formation of a number of methylated intermediates from [As(III)], with trimethylarsine as the end product	Qin et al. (2006)
Bacillus subtilis	Tn	*arsRBC*	ArsR, ArsB, and ArsC function as a negative regulator, a membrane-associated protein need for extrusion of arsenite, and arsenate reductase, respectively	Sato and Kobayashi (1998)
Serratia marcescens	IncHI2	*arsR, arsB, arsC, and arsH*	272-kb plasmid encoding a variety of antibiotic and heavy metal resistances including resistance to arsenate, arsenite, antimony, mercury, tellurite, tetracycline, chloramphenicol, and kanamycin	Ryan and Colleran (2002) and Whelan and Colleran (1992)
P. putida	Chr	*arsRBCH*	The operon encodes self-repressed transcriptional regulator (ArsR), a membrane-bound transporter that extrudes AsIII out of the cell (ArsB), an arsenate reductase (ArsC) for transformation of AsV to AsIII and an ArsH of unknown function but also important for arsenic resistance	Páez-Espino et al. (2015)

(continued)

Table 13.4 (continued)

Microbe	Plasmid/ Chr	Operon/gene	Comments	Reference
Streptomyces sp.	pHZ227	*arsRBOCT*	*arsO* encodes putative flavin-binding monooxygenase) and *arsT* encodes a putative thioredoxin reductase)	Wang et al. (2006)
Corynebacterium glutamicum	Chr	*arsRBC, ars1, ars2*	ArsB a regulatory protein, ArsB an arsenite permease, and ArsC an arsenate reductase, the operon *ars1* contains an additional arsenate reductase gene (*arsC1*). Additional arsenite permease and arsenate reductase genes (*arsB3* and *arsC4*) scattered on the chromosome were also identified	Ordóñez et al. (2005)
Shewanella oneidensis	Chr	*arsDABC*	ArsB and ArsC may be useful for As(V)-respiring bacteria in environments where As concentrations are high	Saltikov et al. (2003)
Leptospirillum ferriphilum	Chr TnLfArs	*arsRCB* *arsRCDA* *arsB*	ArsR in a negative regulator, ArsC is a arsenate reductase, ArsD is a second repressor and ArsA is an ATPase that associates with ArsB and links arsenite export to ATP hydrolysis. These genes are followed by genes encoding ORF7 (an NADH-like oxidoreductase), ORF8 (a cystathione-β-synthase (Tuffin et al.) domain-containing protein), and ArsB, the arsenite-efflux pump	Tuffin et al. (2006)
Acidithiobacillus caldus	TnAtcArs	*arsRCDADA* *arsB*	A series of genes consisting of *arsR*, two tandem copies of *arsA* and *arsD*, two *ORFs* and *arsB* is situated between the resolvase and transposase genes	Tuffin et al. (2005)
A. ferrooxidans	Chr	*arsCR, arsBH*	ArsR is promoter in response to arsenic and antimonite	Butcher and Rawlings (2002)
Thiobacillus ferrooxidans	Chr	*arsB, arsC, arsH, and* a putative *arsR*	Genes encoding for ArsB (arsenite export) and ArsC (arsenate reductase)	Butcher et al. (2000)
Ferroplasma acidarmanus	Chr	*arsR, arsB*	Genes encoding for ArsR (arsenite-sensitive regulator) and ArsB (arsenite-efflux pump)	Gihring et al. (2003)
Halobacterium sp.	Chr pNRC100	*arsB* *arsADRC, arsR2M*	It is suggesting *arsM* gene produced a second novel mechanism of arsenic resistance involving a putative arsenite(III)-methyltransferase	Wang et al. (2004)

Herminiimonas arsenicoxydans	Chr	*aoxABCD, aoxRS, arsRCBCH, arsM*	Expression of the *aoxAB* operon promotes [As(III)] oxidation. *arsRCBCH* operone code an ArsR regulator, an [As(III)] extrusion pump, one or two arsenate reductases (ArsC), an ArsH putative flavoprotein, *arsM* gene expose arsenic methylation activity	Muller et al. (2007)
Desulfovibrio desulfuricans	Chr	*arsRBCC, arsC1*	*arsR* operon gene is encoded the repressor protein, which control operon. *arsC1* gene constitutively expressed and allows a rapid response to an influx of arsenate into the cell, as the arsenate is reduced by ArsC1	Li and Krumholz (2007)
Sinorhizobium sp.	pSinA	*arsRC, arsRCB, arsHR*	Removal of this plasmid from cells of the host strain caused the loss of resistance to arsenic and heavy metals (Cd, Co, Zn, and Hg)	Drewniak et al. (2013)
Sinorhizobium meliloti	Chr	*aqpS, arsC, arsH*	The *ars* operon includes an aquaglyceroporin (*aqpS*) in place of *arsB*, that facilitates transport of arsenite. ArsH is involved in [As(III)] detoxification	Ye et al. (2007) and Yang et al. (2005)
Ochrobactrum tritici	Chr	*ars1, ars2*	*ars1* operon contains five genes encoding the following proteins: ArsR, ArsD, ArsA, CBS-domain-containing protein and ArsB. The *ars2* operon is composed of six genes that encode two other ArsR, two ArsC (belonging to different families of arsenate reductases), one ACR3 and one ArsH-like protein	Branco et al. (2008)
Geobacillus kaustophilus	Chr	*arsRBC, arsC*	Operon consists with *arsC1, arsC2, arsC3* genes. *arsC3* is a monocistronic locus, sequencing of the regions flanking *arsC1* and *arsC2* revealed the presence of additional genes encoding a putative arsenite transporter and an ArsR-like regulator upstream of each arsenate reductase, indicating the presence	Cuebas et al. (2011)
Thermomonospora curvata	Chr	*arsI, arsM*	ArsI is a microbial non-heme, ferrous-dependent dioxygenase that transforms toxic methylarsenite (Nadar et al.) to less toxic inorganic arsenite [As(III)] by C–As bond cleavage	Nadar et al. (2016)

Chr chromosomal genes

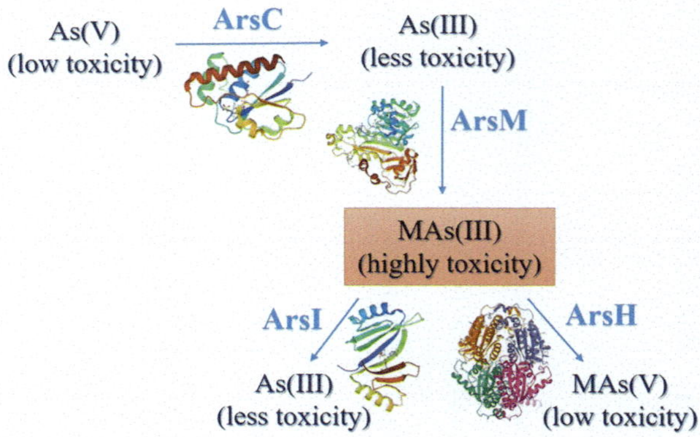

Fig. 13.8 Enzymes of organo-arsenical production and detoxification

produce the ArsI C-As lyase, a dioxygenase that cleaves off the methyl group, forming inorganic As(III). Since As(III) is less toxic than MAs(III), this reaction detoxifies the organo-arsenical product. Other bacteria have the ArsH NADPHFMN oxidoreductase that oxidizes MAs(III) to relatively nontoxic pentavalent MAs(V), also a detoxification process (Yang and Rosen 2016). The protein structure of the ArsC (from *S. aureus*), ArsM (from *Cyanidioschyzon* sp.), ArsI (from *T. curvata*) and ArsH (from *S. meliloti*), presented in the illustration, were used from Protein Data Bank (https://www.rcsb.org/).

Outer example of the heavy metal detoxification is hexavalent chromate reduction. Bacterial developed the mechanisms for reduction of Cr(VI) to the Cr(III) species and efflux of chromate from cell cytoplasm. Several chromate reductases have been identified in diverse bacterial species (Table 13.5). Most characterized enzymes belong to the NAD(P)H-dependent flavoprotein family of reductases.

Candidatus "Methanoperedens" independently utilizes chromate as electron acceptor to form Cr(III) compounds, or it can oxidizes methane to generate intermediates or electrons, which will be utilized to reduce chromate to Cr(III) compounds by unknown chromate reducers synergistically (Luo et al. 2019).

Efflux of chromate by the ChrA membrane transporter, a plasmid-encoded protein, has been demonstrated in *Pseudomonas* and *Cupriavidus* species (Fig. 13.9). Chromate efflux by ChrA consists of an energy-dependent process driven by the membrane potential. The CHR protein family, which includes putative ChrA homologs, currently contains about 135 sequences from all three domains of life. Other mechanisms of bacterial resistance to chromate involve the expression of components of the machinery for repair of DNA damage as well as free-radical scavenging enzymes (Cervantes and Campos-García 2007; Díaz-Magaña et al. 2009).

Table 13.5 The sours and properties several chromate-reducing enzymes (Pradhan et al. 2016; Singh et al. 2015)

Organism	Enzyme (function)	Substrates
P. ambigua G-l	Chr (chromate and nitroreductase)	Chromate, nitro-compounds
P. putida	Chr (chromate and quinoneductase)	Quinones, chromate, 2,6-Dichloroindo phenol, potassium Ferricyanide
E. coli	YieF, ChrA (chromate and quinone)	Quinones, chromate, 2,6-Dichloroindo phenol, potassium Ferricyanide, V(V), Mo(VI)
	NfsA (chromate and nitroreductase)	Chromate, nitro-compounds
	NemA (chromate reductase)	Chromate
E. coli K12	ChrR (quinoneductase)	Quinones
T. scotoductus SA-01	Chr (chromate reductase)	Chromate
Rhodobacter sphaeroides	Chr (chromate reductase)	Chromate
	ApcA (chromate and azoreductase)	Chromate, chromate bitrate, TNT
Vibrio harveyi	NfsB (nitroreductase)	Nitrofurazone, Trinitroluene, chromate
Gluconobacter hansenii	Gh-ChrR (chromate reductase)	Chromate uranyl
B. subtilis	YcnD (FMN reductase)	Chromate, Nitroaromatic compounds, Quinones
Desulfovibrio vulgaris	Cytochrome c_3 (periplasmic c type cytochrome)	Chromate
D. desulfuricans	Thioredoxin oxidoreductase	Chromate, Mo, U, Se, Te
D. alaskensis		
Desulfuromonas acetoxidans	Cytochrome c_7 (periplasmic c type cytochrome)	Chromate
Acidiphilium cryptum	ApcA (chromate and azoreductase)	Chromate, chromate bitrate, TNT
Methanobacterium sp.	FMN reductase	Chromate

13.4 Application and Prospects of Heavy Metal Resistant Microbes

Accumulation of high concentrations of heavy metals in environments can cause many human health risks and serious ecological problems. The ability of microorganisms to adsorb heavy metals or change the forms of their presence in the environment attracts wide attention of researchers in connection with the possibility of biotechnological use of heavy metal resistant bacteria or archaea for wastewater treatment, bioremediation of contaminated environments, as well as in biogeotechnology of metals (Volesky 1994; Gadd 2005; White and Gadd 2000).

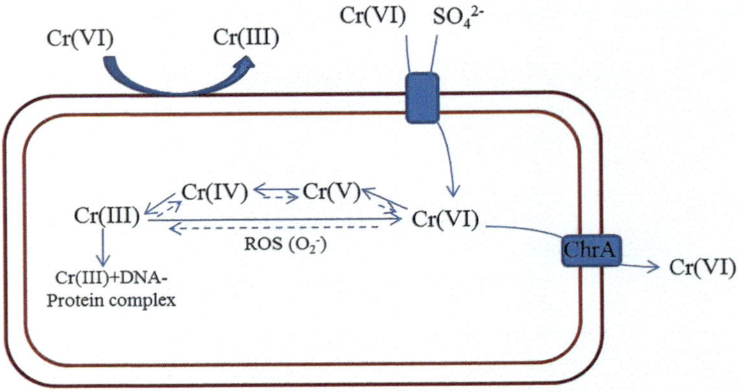

Fig. 13.9 Schematic diagram of Cr(VI) transport into bacterial cell, its reduction pathways, and efflux (modified from Pradhan et al. 2016)

Bioremediation using microorganisms is receiving much attention due to their good performance and employed in order to transform toxic heavy metals into a less harmful state (Ndeddy Aka and Babalola 2016; Akcil et al. 2015) or using microbial enzymes to clean-up polluted environment (Okoduwa et al. 2017). The technique is environmentally friendly and cost effective in the revitalization of the environment (Turpeinen et al. 2004; Ma et al. 2016). In the Table 13.6 showed a number of microbes which can be used for removing metal ions from solutions. However, bioremediation of heavy metals has limitations. Among these are production of toxic metabolites by microbes and non-biodegradability of heavy metals.

Bioremediation of the environment from toxic metal can be achieved by biosorption ability of the microbes. Biosorption is the group of all processes, during which alive or dead microbial biomass removes heavy metals or other pollutants from solutions (Gavrilescu 2004). Biosorption occurring with the participation of microorganisms may be conducted by surface adsorption concerning the gathering of metals on the cell surface and linking them with extracellular polymers, such as exopolysaccharide (EPS). EPS released out of self-defense against harsh conditions of starvation, pH and temperature, hence it displays exemplary physiological, rheological and physiochemical properties. The ionic nature of metals, its size and charge density in turn regulates its interaction with negatively charged EPS (Gupta and Diwan 2017). In the Table 13.7 is given some microbial EPS involved in heavy metal remediation.

It is often when biosorption occurs as the first phase of the following intracellular accumulation and the process of surface adsorption occurring very fast—during several minutes may have a dominant role in metal linking or may lead to high metal accumulation in the middle of the cell in a longer time (Gavrilescu 2004).

The practical application of biosorption to the removal or the recovery of heavy metals is mainly the result of the reversibility of this process. Desorption allows the recovery of metals (which is profitable in the case of more valuable heavy metals like

Table 13.6 Remediation of heavy metal by microorganisms (modified from Igiri et al. 2018)

Bioremediator	Metals	Metal ion concentration (mg/L)	Sorption efficiency (%)
Acinetobacter sp.	Cr	16	87
Sporosarcina saromensis (M52)		50	82.5
Bacillus cereus		1500	81
B. cereus (immobilized)		1500	96
B. circulans MN1		1100	71.4
B. cereus plus 0.5 glucose		1	78
B. cereus		1	72
Bacillus sp. SFC		25	80
		50	43
B. subtilis		057	99.6
Desulfovibrio desulfuricans (KCTC 5768) (immobilize on zeolite)		200 100 50	56.1 99.8 99.6
Staphylococcus sp.		4.108	45
Bacillus sp. (B2)		50–37.06 200–81.5	74.1 40.75
Bacillus sp. (B4)		50–36.57	73.14
Bacillus sp. (B9)		50–30.75 100–60 200–78.7	61.5 60 39.39
Bacillus sp. (B2)		100–42.15	42.15
Bacillus sp. (B4)		100–73.41 200–97.76	73.41 48.88
Micrococcus sp.		100	90
Acinetobacter sp. B9 (MTCC10506)		7 15 16 30	93.7 81 78 67
Streptomyces sp.		6.42	72
Immobilized *B. subtilis*		570–2	99.6
Bacillus subtilis		570–2	99.6
Immobilized *P. aeruginosa*		570–4	99.3
Pseudomonas aeruginosa		570–2	99.6
Stenotrophomonas sp.		16.59	81.27
Spirulina sp.		5	98.3
Acinetobacter sp. + *Arthrobacter* sp.		16	78
P. aeruginosa + *B. subtilis*		570–2	99/5
Pseudomonas aeruginosa	Hg	150	29.83
Vibrio parahaemolyticus (PG02)		5 10	90 80
Bacillus licheniformis		0.1	73
Vibrio fluvialis		0.25	60
Klebsiella pneumonia		100	28.65

(continued)

Table 13.6 (continued)

Bioremediator	Metals	Metal ion concentration (mg/L)	Sorption efficiency (%)
Cellulosimicrobium sp. (KX710177)	Pb	50 100 200 300	99.33 96.98 84.62 62.28
Gemella sp.		0.3	55.16 ± 0.06
Micrococcus sp.		0.3	36.55 ± 0.01
Pseudomonas sp.		1	87.9
Staphylococcus sp.		0.183	82.6
Streptomyces sp.		0.286	32.5
B. iodinium		100–1.8	87
Desulfovibrio desulfuricans (KCTC 5768) (immobilize on zeolite)	Cu	50 100 200	97.4 98.2 78.7
Staphylococcus sp.		1.536	42
Streptomyces sp.		1.129	18
Enterobacter cloacae		100	20
Desulfovibrio desulfuricans (immobilize on zeolite)		100	98.2
Flavobacterium sp.		1.194	20.3
Arthrobacter strain D9		0.05	22
Enterobacter cloaceae		100	65
Micrococcus sp.		0.3	38.64 ± 0.06
Gemella sp.		0.3	50.99 ± 0.01
Pseudomonas sp.		1	41
Flavobacterium sp.		0.161	25
A. faecalis (GP06)		100–19.2	70
Pseudomonas aeruginosa (CH07)		100–17.4	75
Desulfovibrio desulfuricans (immobilize on zeolite)	Ni	50 100 200	90.3 90.1 90.1
Micrococcus sp.		50	55
Pseudomonas sp.		1	53
Acinetobacter sp. B9		51	68.94
Enterobacter cloacae	Co	100	8
Bacillus firmus	Zn	–	61.8
Pseudomonas sp.		1	49.8
Aeratedmicrobial sediment fuel cells (A-SMFCs)	Cr Cu Ni	–	80.7 72.72 80.37
Non-aerated microbial sediment fuel cells (NA-SMFCs)	Cr Cu Ni	–	67.36 59.36 52.74

Table 13.7 Heavy metal remediation by microbial EPS

EPS-producing microbes	Metal ion removed	Remarks	Reference
Hyphomonas MHS-3, *Hyphomonas* sp.	Cu(II), Hg(II), Pb(II), Cd(II), Zn(II)	Adsorbent system was effective over wide range of pH (1–11) and temperature range (0–200 °C). The marine strains were able to remove the metal ions from an initial concentration of 50–100 ppb to US EPAa drinking water standards	Chmurny et al. (1998)
Arthrobacter viscosus	Cr(VI)	Devised for industrial applications for hexavalent chromium removal, through the retention of metal ions in the biofilms, in solutions with concentrations between 50 and 250 mg/L	Tavares and Neves (2008)
Ochrobactrum anthropi	Cr(VI), Cd (II), Cu(II)	57.8 mg Cr(VI)/g EPS at initial metal load of 280 ppm, 26 mg Cu(II)/g EPS at initial metal load of 91.6 ppm	Ozdemir et al. (2003)
Acetobacter	Fe(III), Cu (II), Mn(II), Zn(II), Co(II)	90% reduction from initial metal load of 0.1 mmol/dm^3 (Fe(III) > Cu (II) > Mn(II) > Zn(II); Co(II))	Oshima et al. (2008)
Bacillus firmus	Pb(II), Zn(II), Co(II)	1103 mg Pb(II)/g EPS (98.3%,), 860 mg Cu(II)/g EPS (74.9%)	Salehizadeh and Shojaosadati (2003)
Methylobacterium organophilum	Pb(II), Cu(II)	21% Cu(II),18% Pb(II) removal from 0.04 ppm initial metal load	Kim et al. (1996)
Herminiimonas arsenicoxydans	Arsenic	Up to 5 mmol/L metal ion uptake	Marchal et al. (2010)
Halomonas sp.	Trace metals	Metal analysis of the purified EPS revealed that it contained high levels of K, Ca, Mg and several essential trace metals, including Zn, Cu, Fe and the metalloid Si Capacity to sequester trace metals and mediate their bioavailability to eukaryotic phytoplankton	Gutierrez et al. (2012)
Shewanella oneidensis	Cd(II)	80% Cd(II) removal	Ha et al. (2010)
Azotobacter chroococcum	Pb(II), Hg(II)	40.48% Pb(II)(33.5 mg Pb(II)/g of EPS); 47.87% Hg(II) (38.9 mg of Hg (II)/g EPS)	Rasulov et al. (2013)
Cupriavidus pauculus	Cd(II), Ni(II), Cu(II), Co(II)	The tolerance levels of *C. pauculus* 1490 to Cd(II), Ni(II), Cu(II) and Co (II) were 300 mg/L, 400 mg/L, 400 mg/L and 400 mg/L, respectively. EPS yield reaching 956.12 ± 10.59 mg/g(DW) at 100 mg/L	Zeng et al. (2020)
Anabaena spiroides	Mn(II)	8.52 mg Mn(II)/g EPS	Freire-Nordi et al. (2005)

(continued)

Table 13.7 (continued)

EPS-producing microbes	Metal ion removed	Remarks	Reference
Gloeocapsa gelatinosa	Pb(II)	82.22 _ 4.82 mg Pb(II)/g CPS	Raungsomboon et al. (2006)
Calothrix marchica		65 mg Pb(II)/g CPS	Ruangsomboon et al. (2007)
Cyanospira capsulata	Cu(II)	115 mg Cu(II)/g EPS at 12.3 ppm initial metal load	De Philippis et al. (2007)
Nostoc PCC7936		85.0 ± 3.2 mg Cu(II)/g EPS at 12.3 ppm initial metal load	Sharma et al. (2008)

gold, copper, and zinc) or their removal (Wang and Chen 2009). As the biosorbents can be:

- Biomass of microorganisms is the secondary product in the sewage or pharmaceutical industry and in sewage treatment processes;
- Microorganisms from cultured and proliferated on a special base indicating the ability to efficiently metals;
- Sorbents of vegetable or animal origin (as nutshells, crust-rich tannins, sea plants, humus, moss peat, etc.).

The direct use of microorganisms with distinctive features of catabolic potential and/or their products such as enzymes and bio surfactant is a novel approach to enhance and boost their remediation efficacy (Le et al. 2017; Schenk et al. 2012). Different alternatives have also been anticipated to widen the applications of microbiological techniques toward the remediation of heavy metals. For instance, the use of microbial fuel cell to degrade recalcitrant heavy metals has been explored. Biofilm mediated bioremediation can be applied for cleaning up of heavy metal-contaminated environment.

High bioremediation potential and feasibility of the microbial detoxification of arsenic by reduction, oxidation, and methylation process, make bacteria an impending foundation for green chemistry to exterminate arsenic in the environment (Sher and Rehman 2019).

Many microorganisms are capable of precipitating metal ions. The method of precipitation of metals in the form of sulfides is based on the ability of sulfate-reducing bacteria (*Desulfovibrio, Desulfotomaculum, Desulfomonas, Desulfobacter, Desulfobulbus, Desulfococcus, Desulfosarcina, Desulfonema*) to form H_2S, which precipitates metals from solutions almost completely. Thus, from solutions containing 8.6 g/L Cu, the extraction of Cu was 98.5%. Toxic metals can also precipitate during their recovery. For example, chromium-reducing bacteria under anaerobic conditions reduce Cr(VI) to Cr(III), which is precipitated (Cervantes and Campos-García 2007).

Soil microorganisms, including plant growth promoting bacteria, through toxic metal stress evading mechanisms, can be used as bioinoculant or biofertilizers, which substantially improve the growth of plants implanted in heavy

Table 13.8 Microorganisms important for biohydrometallurgy

Microbes	Energy source	Optimum growth condition	Reference
Bacteria			
Acidithiobacillus ferrooxidans	Sulfide minerals, S, S (II), Fe(II), FeS_2	pH 1.7–2.0 (1.0–5.5); 30–35 °C (2–40 °C); O_2	Quatrini and Johnson (2019)
Leptospirillum ferrooxidans	Fe(II), FeS_2	pH 2.0–2.5 (1.0–4.0); 30–45 °C (2–50 °C); O_2	Sand et al. (1992)
A. thiooxidans	S, S(II)	pH 2.0–2.5 (0.5–6.0); 30 °C (2–40 °C); O_2	Yang et al. (2019)
A. caldus	S, S(II)	pH 2.0–2.5 (0.5–6.0); 45 °C (30–52 °C); O_2	Chen et al. (2012)
Sulfobacillus thermosulfidooxidans, S. acidophilus	Fe(II), S, S(II), sulfide minerals	pH 1.7–2.4 (1.1–5.0); 48–50 °C (20–60 °C); O_2	Norris et al. (1996)
Archaea			
Acidianus brierleyi	S, S(II), Fe(II)	pH 1.5–2.0; 70 °C (45–75 °C); O_2	Segerer et al. (1986)
Metallosphaera sedula		pH 1.0–4.5; 75 °C (50–80 °C); O_2	Huber et al. (1989)
Sulfolobus metallicus	S, sulfide minerals, Fe (II)	pH 1.0–4.5; (50–75 °C); O_2	Huber and Stetter (1991)
Ferroplasma acidiphilum	FeS_2	pH 1.7–1.8 (1.3–2.2); 35 °C (15–45 °C); O_2	Golyshina et al. (2000)

metal-contaminated soils by lowering the metal toxicity (Madhaiyan et al. 2007; Wani and Khan 2010; Khan et al. 2012). In addition, there are other mechanisms of plant growth promotion by bacteria e.g., they protect colonizing plants from the pathogens attack directly by inhibiting/killing pathogens through the production of antibiotics, hydrogen cyanide, and phenazines, etc. (Saravanakumar et al. 2007; Cazorla et al. 2007).

Metalophilic bacteria and archaea play an important role in the process of leaching of metals from ores, concentrates, rocks and solutions, thus they are widely used in biogeometallurgy. In the Table 13.8 showed chemolithotrophic bacteria that oxidize Fe(II), S(II), S, and sulfide minerals important for biohydrometallurgy (Sand et al. 1992; Ehrlich 1997b; Vardanyan and Vardanyan 2018).

Many prokaryotes, including archaea, are capable of transforming the oxidation state of metals in processes leading to either their solubilization or biomineralization. Although these phenomena have been observed in the environment and studied in cultures, there is still much to be learned about the genetic determinants of these metal transformations.

Acknowledgments This work was supported by grants from the Eurasia Programme of the Norwegian Center for International Cooperation in Education (CPEA-2011/10081, CPEA-LT-2016/10095) and partially supported by Research Grant from MESCS Science Committee of Armenia, to AM (19YR-1F065), and the Armenian National Science and Education Fund based in New York, USA, to AM (ANSEF-NS-microbio-3869 and 4619).

References

Adekanmbi AO, Adelowo OO, Okoh AI, Fagade OE (2019) Metal-resistance encoding gene-fingerprints in some bacteria isolated from wastewaters of selected printeries in Ibadan, South-Western Nigeria. J Taibah Univ Sci 13(1):266–273

Aguilar-Barajas E, Ramírez-Díaz MI, Riveros-Rosas H, Cervantes C (2010) Heavy metal resistance in pseudomonads. In: Ramos JL, Filloux A (eds) Pseudomonas: volume 6: molecular microbiology, infection and biodiversity. Springer, Dordrecht, pp 255–282

Ahemad M (2012) Implications of bacterial resistance against heavy metals in bioremediation: a review. IIOABJ 3(3):39–46

Ahemad M (2015) Phosphate-solubilizing bacteria-assisted phytoremediation of metalliferous soils: a review. 3 Biotech 5(2):111–121

Akcil A, Erust C, Ozdemiroglu S, Fonti V, Beolchini F (2015) A review of approaches and techniques used in aquatic contaminated sediments: metal removal and stabilization by chemical and biotechnological processes. J Clean Prod 86:24–36

Ali H, Khan E (2018) What are heavy metals? Long-standing controversy over the scientific use of the term 'heavy metals'—proposal of a comprehensive definition. Toxicol Environ Chem 100 (1):6–19

Anderson GL, Love M, Zeider BK (2003) Metabolic energy from arsenite oxidation in Alcaligenes faecalis. J Phys IV (Proc) 107:49–52

Anton A, Grosse C, Reissmann J, Pribyl T, Nies DH (1999) CzcD is a heavy metal ion transporter involved in regulation of heavy metal resistance in Ralstonia sp. strain CH34. J Bacteriol 181 (22):6876–6881

Argüello JM, Mandal AK, Mana-Capelli S (2003) Heavy metal transport CPx-ATPases from the thermophile Archaeoglobus fulgidus. Ann N Y Acad Sci 986:212–218

Argüello JM, Raimunda D, Padilla-Benavides T (2013) Mechanisms of copper homeostasis in bacteria. Front Cell Infect Microbiol 3:73

Baldwin DR, Marshall WJ (1999) Heavy metal poisoning and its laboratory investigation. Ann Clin Biochem 36(Pt 3):267–300

Bartolucci S, Contursi P, Fiorentino G, Limauro D, Pedone E (2013) Responding to toxic compounds: a genomic and functional overview of archaea. Front Biosci (Landmark Ed) 18:165–189

Ben Fekih I, Zhang C, Li YP, Zhao Y, Alwathnani HA, Saquib Q, Rensing C, Cervantes C (2018) Distribution of arsenic resistance genes in prokaryotes. Front Microbiol 9:2473

Bini E (2010) Archaeal transformation of metals in the environment. FEMS Microbiol Ecol 73 (1):1–16

Boyd ES, Barkay T (2012) The mercury resistance operon: from an origin in a geothermal environment to an efficient detoxification machine. Front Microbiol 3:349

Branco R, Chung A-P, Morais PV (2008) Sequencing and expression of two arsenic resistance operons with different functions in the highly arsenic-resistant strain Ochrobactrum tritici SCII24T. BMC Microbiol 8(1):95

Brown NL, Stoyanov JV, Kidd SP, Hobman JL (2003) The MerR family of transcriptional regulators. FEMS Microbiol Rev 27(2–3):145–163

Bruins MR, Kapil S, Oehme FW (2000) Microbial resistance to metals in the environment. Ecotoxicol Environ Saf 45(3):198–207

Butcher BG, Rawlings DE (2002) The divergent chromosomal ars operon of *Acidithiobacillus ferrooxidans* is regulated by an atypical ArsR protein. Microbiology (Reading) 148 (Pt 12):3983–3992

Butcher BG, Deane SM, Rawlings DE (2000) The chromosomal arsenic resistance genes of *Thiobacillus ferrooxidans* have an unusual arrangement and confer increased arsenic and antimony resistance to *Escherichia coli*. Appl Environ Microbiol 66(5):1826–1833

Cai J, Salmon K, DuBow MS (1998) A chromosomal ars operon homologue of *Pseudomonas aeruginosa* confers increased resistance to arsenic and antimony in *Escherichia coli*. Microbiology (Reading) 144(Pt 10):2705–2713
Carlin A, Shi W, Dey S, Rosen BP (1995) The ars operon of *Escherichia coli* confers arsenical and antimonial resistance. J Bacteriol 177(4):981–986
Cazorla FM, Romero D, Pérez-García A, Lugtenberg BJJ, Vicente A d, Bloemberg G (2007) Isolation and characterization of antagonistic *Bacillus subtilis* strains from the avocado rhizoplane displaying biocontrol activity. J Appl Microbiol 103(5):1950–1959
Cervantes C, Campos-García J (2007) Reduction and efflux of chromate by bacteria. In: Nies DH, Silver S (eds) Molecular microbiology of heavy metals. Springer, Berlin, pp 407–419
Chen CM, Misra TK, Silver S, Rosen BP (1986) Nucleotide sequence of the structural genes for an anion pump. The plasmid-encoded arsenical resistance operon. J Biol Chem 261 (32):15030–15038
Chen L, Ren Y, Lin J, Liu X, Pang X (2012) *Acidithiobacillus caldus* sulfur oxidation model based on transcriptome analysis between the wild type and sulfur oxygenase reductase defective mutant. PLoS One 7(9):e39470
Chmurny AB, Quintero EJ, Kneer R (1998) Novel heavy metals sorbents produced from hyphomonas and method of use. Worldwide applications
Collard JM, Corbisier P, Diels L, Dong Q, Jeanthon C, Mergeay M, Taghavi S, van der Lelie D, Wilmotte A, Wuertz S (1994) Plasmids for heavy metal resistance in *Alcaligenes eutrophus* CH34: mechanisms and applications. FEMS Microbiol Rev 14(4):405–414
Cooksey DA (1994) Molecular mechanisms of copper resistance and accumulation in bacteria. FEMS Microbiol Rev 14(4):381–386
Coombs JM, Barkay T (2005) New findings on evolution of metal homeostasis genes: evidence from comparative genome analysis of Bacteria and archaea. Appl Environ Microbiol 71 (11):7083
Cousins RJ, Liuzzi JP, Lichten LA (2006) Mammalian zinc transport, trafficking, and signals. J Biol Chem 281(34):24085–24089
Cuebas M, Villafane A, McBride M, Yee N, Bini E (2011) Arsenate reduction and expression of multiple chromosomal ars operons in *Geobacillus kaustophilus* A1. Microbiology (Reading) 157(Pt 7):2004–2011
De Hertogh B, Lantin A-C, Baret PV, Goffeau A (2004) The archaeal P-type ATPases. J Bioenerg Biomembr 36(1):135–142
De Philippis R, Paperi R, Sili C (2007) Heavy metal sorption by released polysaccharides and whole cultures of two exopolysaccharide-producing cyanobacteria. Biodegradation 18(2):181–187
Dhawan IK, Roy R, Koehler BP, Mukund S, Adams MW, Johnson MK (2000) Spectroscopic studies of the tungsten-containing formaldehyde ferredoxin oxidoreductase from the hyperthermophilic archaeon *Thermococcus litoralis*. J Biol Inorg Chem 5(3):313–327
Díaz-Magaña A, Aguilar-Barajas E, Moreno-Sánchez R, Ramírez-Díaz MI, Riveros-Rosas H, Vargas E, Cervantes C (2009) Short-chain chromate ion transporter proteins from *Bacillus subtilis* confer chromate resistance in *Escherichia coli*. J Bacteriol 191(17):5441
Díaz-Pérez C, Cervantes C, Campos-García J, Julián-Sánchez A, Riveros-Rosas H (2007) Phylogenetic analysis of the chromate ion transporter (CHR) superfamily. FEBS J 274 (23):6215–6227
Diekert G, Konheiser U, Piechulla K, Thauer RK (1981) Nickel requirement and factor F430 content of methanogenic bacteria. J Bacteriol 148(2):459–464
Diels L, Dong Q, van der Lelie D, Baeyens W, Mergeay M (1995) Theczc operon of *Alcaligenes eutrophus* CH34: from resistance mechanism to the removal of heavy metals. J Ind Microbiol 14 (2):142–153
Drewniak L, Dziewit L, Ciezkowska M, Gawor J, Gromadka R, Sklodowska A (2013) Structural and functional genomics of plasmid pSinA of *Sinorhizobium* sp. M14 encoding genes for the arsenite oxidation and arsenic resistance. J Biotechnol 164(4):479–488
Ehrlich HL (1997a) Microbes and metals. Appl Microbiol Biotechnol 48(6):687–692

Ehrlich HL (1997b) Technical potential for bioleaching and biobeneficiation of ores to recover base metals (other than iron or copper), platinum-group metals and silver. In: Rawlings DE (ed) Biomining: theory, microbes and industrial processes. Springer, Berlin, pp 129–150

Eitinger T, Mandrand-Berthelot MA (2000) Nickel transport systems in microorganisms. Arch Microbiol 173(1):1–9

Ellefson WL, Whitman WB, Wolfe RS (1982) Nickel-containing factor F430: chromophore of the methylreductase of *Methanobacterium*. Proc Natl Acad Sci U S A 79(12):3707–3710

Feinberg LF, Srikanth R, Vachet RW, Holden JF (2008) Constraints on anaerobic respiration in the hyperthermophilic archaea *Pyrobaculum islandicum* and *Pyrobaculum aerophilum*. Appl Environ Microbiol 74(2):396–402

Freedman Z, Zhu C, Barkay T (2012) Mercury resistance and mercuric reductase activities and expression among chemotrophic thermophilic Aquificae. Appl Environ Microbiol 78 (18):6568–6575

Freire-Nordi CS, Vieira AAH, Nascimento OR (2005) The metal binding capacity of *Anabaena spiroides* extracellular polysaccharide: an EPR study. Process Biochem 40(6):2215–2224

Gadd G (2005) Microorganisms in toxic metal-polluted soils. In: Varma A, Buscot F (eds) Microorganisms in soils: roles in genesis and functions soil biology. Springer, Berlin, pp 325–356

Garbinski LD, Rosen BP, Chen J (2019) Pathways of arsenic uptake and efflux. Environ Int 126:585–597

Gavrilescu M (2004) Removal of heavy metals from the environment by biosorption. Eng Life Sci 4 (3):219–232

Gihring TM, Bond PL, Peters SC, Banfield JF (2003) Arsenic resistance in the archaeon "*Ferroplasma acidarmanus*": new insights into the structure and evolution of the ars genes. Extremophiles 7(2):123–130

Golyshina OV, Pivovarova TA, Karavaiko GI, Kondratéva TF, Moore ER, Abraham WR, Lünsdorf H, Timmis KN, Yakimov MM, Golyshin PN (2000) *Ferroplasma acidiphilum* gen. Nov., sp. nov., an acidophilic, autotrophic, ferrous-iron-oxidizing, cell-wall-lacking, mesophilic member of the *Ferroplasmaceae* fam. Nov., comprising a distinct lineage of the archaea. Int J Syst Evol Microbiol 50(Pt 3):997–1006

Grass G, Fan B, Rosen BP, Franke S, Nies DH, Rensing C (2001) ZitB (YbgR), a member of the cation diffusion facilitator family, is an additional zinc transporter in *Escherichia coli*. J Bacteriol 183(15):4664

Guffanti AA, Wei Y, Rood SV, Krulwich TA (2002) An antiport mechanism for a member of the cation diffusion facilitator family: divalent cations efflux in exchange for K^+ and H^+. Mol Microbiol 45(1):145–153

Gupta P, Diwan B (2017) Bacterial exopolysaccharide mediated heavy metal removal: a review on biosynthesis, mechanism and remediation strategies. Biotechnol Rep (Amst) 13:58–71

Gutierrez T, Biller DV, Shimmield T, Green DH (2012) Metal binding properties of the EPS produced by *Halomonas* sp. TG39 and its potential in enhancing trace element bioavailability to eukaryotic phytoplankton. Biometals 25(6):1185–1194

Ha J, Gélabert A, Spormann AM, Brown GE (2010) Role of extracellular polymeric substances in metal ion complexation on Shewanella oneidensis: batch uptake, thermodynamic modeling, ATR-FTIR, and EXAFS study. Geochim Cosmochim Acta 74(1):1–15

Hantke K (2001) Bacterial zinc transporters and regulators. Biometals 14(3):239–249

Hartzell PL, Escalante-Semerena JC, Bobik TA, Wolfe RS (1988) A simplified methylcoenzyme M methylreductase assay with artificial electron donors and different preparations of component C from *Methanobacterium thermoautotrophicum* delta H. J Bacteriol 170(6):2711–2715

Hawkes SJ (1997) What is a "heavy metal"? J Chem Educ 74(11):1374

Hedges RW, Baumberg S (1973) Resistance to arsenic compounds conferred by a plasmid transmissible between strains of *Escherichia coli*. J Bacteriol 115(1):459–460

Hohle TH, O'Brian MR (2009) The mntH gene encodes the major Mn(2+) transporter in *Bradyrhizobium japonicum* and is regulated by manganese via the Fur protein. Mol Microbiol 72(2):399–409

Hrynkiewicz K, Złoch M, Kowalkowski T, Baum C, Niedojadło K, Buszewski B (2015) Strain-specific bioaccumulation and intracellular distribution of Cd^{2+} in bacteria isolated from the rhizosphere, ectomycorrhizae, and fruitbodies of ectomycorrhizal fungi. Environ Sci Pollut Res Int 22(4):3055–3067

Huber G, Stetter KO (1991) *Sulfolobus metallicus*, sp. nov., a novel strictly chemolithoautotrophic thermophilic archaeal species of metal-mobilizers. Syst Appl Microbiol 14(4):372–378

Huber G, Spinnler C, Gambacorta A, Stetter KO (1989) Metallosphaera sedula gen, and sp. nov. represents a new genus of aerobic, metal-mobilizing, thermoacidophilic archaebacteria. Syst Appl Microbiol 12(1):38–47

Hynninen A (2010) Zinc, cadmium and lead resistance mechanisms in bacteria and their contribution to biosensing. Academic Dissertation in Microbiology, University of Helsinki. pp 1–52

Ianeva OD (2009) Mechanisms of bacteria resistance to heavy metals. Mikrobiol Zh 71(6):54–65

Igiri BE, Okoduwa SIR, Idoko GO, Akabuogu EP, Adeyi AO, Ejiogu IK (2018) Toxicity and bioremediation of heavy metals contaminated ecosystem from tannery wastewater a review. Hindawi J Toxicol 2018:2568038

Incharoensakdi A, Kitjaharn P (2002) Zinc biosorption from aqueous solution by a halotolerant cyanobacterium *Aphanothece halophytica*. Curr Microbiol 45(4):261–264

Ji G, Silver S (1992) Regulation and expression of the arsenic resistance operon from *Staphylococcus aureus* plasmid pI258. J Bacteriol 174(11):3684–3694

Kachur AV, Koch CJ, Biaglow JE (1998) Mechanism of copper-catalyzed oxidation of glutathione. Free Radic Res 28(3):259–269

Karnachuk OV, Kurochkina SY, Nicomrat D, Frank YA, Ivasenko DA, Phyllipenko EA, Tuovinen OH (2003) Copper resistance in *Desulfovibrio* strain R2. Antonie Van Leeuwenhoek 83 (1):99–106

Kashefi K, Moskowitz BM, Lovley DR (2008) Characterization of extracellular minerals produced during dissimilatory Fe(III) and U(VI) reduction at 100 °C by *Pyrobaculum islandicum*. Geobiology 6(2):147–154

Kasprzak KS (2002) Oxidative DNA and protein damage in metal-induced toxicity and carcinogenesis1, 3 1 This article is part of a series of reviews on "Oxidative DNA Damage and Repair." The full list of papers may be found on the homepage of the journal. 3Guest Editor: Miral Dizdaroglu. Free Radic Biol Med 32(10):958–967

Kaur A, Pan M, Meislin M, Facciotti MT, El-Gewely R, Baliga NS (2006) A systems view of haloarchaeal strategies to withstand stress from transition metals. Genome Res 16(7):841–854

Khan MS, Zaidi A, Wani PA, Oves M (2012) Erratum to: role of plant growth promoting rhizobacteria in the remediation of metal contaminated soils. Environ Chem Lett 10(1):105–106

Kim S-Y, Kim J-H, Kim C-J, Oh D-K (1996) Metal adsorption of the polysaccharide produced from Methylobacterium organophilum. Biotechnol Lett 18(10):1161–1164

Kletzin A, Adams MW (1996) Tungsten in biological systems. FEMS Microbiol Rev 18(1):5–63

Krom BP, Warner JB, Konings WN, Lolkema JS (2000) Complementary metal ion specificity of the metal-citrate transporters CitM and CitH of *Bacillus subtilis*. J Bacteriol 182(22):6374–6381

Lau CK, Krewulak KD, Vogel HJ (2016) Bacterial ferrous iron transport: the Feo system. FEMS Microbiol Rev 40(2):273–298

Le TT, Son M-H, Nam I-H, Yoon H, Kang Y-G, Chang Y-S (2017) Transformation of hexabromocyclododecane in contaminated soil in association with microbial diversity. J Hazard Mater 325:82–89

Lee SW, Glickmann E, Cooksey DA (2001) Chromosomal locus for cadmium resistance in *Pseudomonas putida* consisting of a cadmium-transporting ATPase and a MerR family response regulator. Appl Environ Microbiol 67(4):1437–1444

Li X, Krumholz LR (2007) Regulation of arsenate resistance in *Desulfovibrio desulfuricans* G20 by an arsRBCC operon and an arsC gene. J Bacteriol 189(10):3705

Luo J-H, Wu M, Liu J, Qian G, Yuan Z, Guo J (2019) Microbial chromate reduction coupled with anaerobic oxidation of methane in a membrane biofilm reactor. Environ Int 130:104926

Ma Y, Rajkumar M, Zhang C, Freitas H (2016) Beneficial role of bacterial endophytes in heavy metal phytoremediation. J Environ Manage 174:14–25

Madhaiyan M, Poonguzhali S, Sa T (2007) Metal tolerating methylotrophic bacteria reduces nickel and cadmium toxicity and promotes plant growth of tomato (*Lycopersicon esculentum* L.). Chemosphere 69(2):220–228

Mana-Capelli S, Mandal AK, Argüello JM (2003) *Archaeoglobus fulgidus* CopB is a thermophilic Cu^{2+}-ATPase: functional role of its histidine-rich-N-terminal metal binding domain. J Biol Chem 278(42):40534–40541

Mandal AK, Argüello JM (2003) Functional roles of metal binding domains of the *Archaeoglobus fulgidus* cu(+)-ATPase CopA. Biochemistry 42(37):11040–11047

Marchal M, Briandet R, Koechler S, Kammerer B, Bertin PN (2010) Effect of arsenite on swimming motility delays surface colonization in *Herminiimonas arsenicoxydans*. Microbiology 156(8):2336–2342

Margaryan AA, Panosyan HH, Birkeland N-K, Trchounian AH (2013) Heavy metal accumulation and the expression of the *copA* and *nikA* genes in *Bacillus subtilis* AG4 isolated from the Sotk gold min in Armenia. Yerevan, Armenia: biolog. J Armenia 3(65):51–57

Meija J, Coplen TB, Berglund M, Brand WA, De Bièvre P, Gröning M, Holden NE, Irrgeher J, Loss RD, Walczyk T, Prohaska T (2016) Atomic weights of the elements 2013 (IUPAC technical report). Pure Appl Chem 88(3):265–291

Mills SD, Lim C-K, Cooksey DA (1994) Purification and characterization of CopR, a transcriptional activator protein that binds to a conserved domain (cop box) in copper- inducible promoters of *Pseudomonas syringae*. Mol Gen Genet 244(4):341–351

Moffett JW, Brand LE (1996) Production of strong, extracellular cu chelators by marine cyanobacteria in response to cu stress. Limnol Oceanogr 41(3):388–395

Moraleda-Muñoz A, Pérez J, Extremera AL, Muñoz-Dorado J (2010) Differential regulation of six heavy metal efflux Systems in the Response of *Myxococcus xanthus* to copper. Appl Environ Microbiol 76(18):6069

Mourão J, Rebelo A, Ribeiro S, Peixe L, Novais C, Antunes P (2020) Tolerance to arsenic contaminant among multidrug-resistant and copper-tolerant *salmonella* successful clones is associated with diverse ars operons and genetic contexts. Environ Microbiol 22(7):2829–2842

Muller D, Médigue C, Koechler S, Barbe V, Barakat M, Talla E, Bonnefoy V, Krin E, Arsène-Ploetze F, Carapito C, Chandler M, Cournoyer B, Cruveiller S, Dossat C, Duval S, Heymann M, Leize E, Lieutaud A, Lièvremont D, Makita Y, Mangenot S, Nitschke W, Ortet P, Perdrial N, Schoepp B, Siguier P, Simeonova DD, Rouy Z, Segurens B, Turlin E, Vallenet D, Van Dorsselaer A, Weiss S, Weissenbach J, Lett MC, Danchin A, Bertin PN (2007) A tale of two oxidation states: bacterial colonization of arsenic-rich environments. PLoS Genet 3(4):e53

Mulrooney SB, Hausinger RP (2003) Nickel uptake and utilization by microorganisms. FEMS Microbiol Rev 27(2–3):239–261

Nadar VS, Yoshinaga M, Pawitwar SS, Kandavelu P, Sankaran B, Rosen BP (2016) Structure of the ArsI C-as Lyase: insights into the mechanism of degradation of Organoarsenical herbicides and growth promoters. J Mol Biol 428(11):2462–2473

Naik S, Furtado I (2014) Equilibrium and kinetics of adsorption of Mn^{2+} by Haloarchaeon *Halobacterium* sp. GUSF (MTCC3265). Geomicrobiol J 31(8):708–715

Naik MM, Pandey A, Dubey SK (2012) *Pseudomonas aeruginosa* strain WI-1 from Mandovi estuary possesses metallothionein to alleviate lead toxicity and promotes plant growth. Ecotoxicol Environ Saf 79:129–133

Ndeddy Aka RJ, Babalola OO (2016) Effect of bacterial inoculation of strains of *Pseudomonas aeruginosa, Alcaligenes feacalis* and *Bacillus subtilis* on germination, growth and heavy metal (cd, cr, and ni) uptake of brassica juncea. Int J Phytoremediation 18(2):200–209

Nies DH (1999) Microbial heavy-metal resistance. Appl Microbiol Biotechnol 51(6):730–750

Nies DH (2003) Efflux-mediated heavy metal resistance in prokaryotes. FEMS Microbiol Rev 27 (2–3):313–339

Nies D (2007) Bacterial transition metal homeostasis. In: Nies D, Silver S (eds) Molecular microbiology of heavy metals microbiology monographs. Springer, Berlin, pp 117–142

Nies DH, Silver S (1995) Ion efflux systems involved in bacterial metal resistances. J Ind Microbiol 14(2):186–199

Nikaido H (2018) RND transporters in the living world. Res Microbiol 169(7–8):363–371

Norris PR, Clark DA, Owen JP, Waterhouse S (1996) Characteristics of *Sulfobacillus acidophilus* sp. nov. and other moderately thermophilic mineral-sulphide-oxidizing bacteria. Microbiology (Reading) 142(Pt 4):775–783

Novick RP, Roth C (1968) Plasmid-linked resistance to inorganic salts in *Staphylococcus aureus*. J Bacteriol 95(4):1335–1342

Okoduwa SIR, Igiri B, Udeh CB, Edenta C, Gauje B (2017) Tannery effluent treatment by yeast species isolates from watermelon. Toxics 5(1)

Ordóñez E, Letek M, Valbuena N, Gil JA, Mateos LM (2005) Analysis of genes involved in arsenic resistance in *Corynebacterium glutamicum* ATCC 13032. Appl Environ Microbiol 71 (10):6206–6215

Oremland RS, Stolz JF (2005) Arsenic, microbes and contaminated aquifers. Trends Microbiol 13 (2):45–49

Oshima T, Kondo K, Ohto K, Inoue K, Baba Y (2008) Preparation of phosphorylated bacterial cellulose as an adsorbent for metal ions. React Funct Polym 68(1):376–383

Ozdemir G, Ozturk T, Ceyhan N, Isler R, Cosar T (2003) Heavy metal biosorption by biomass of *Ochrobactrum anthropi* producing exopolysaccharide in activated sludge. Bioresour Technol 90(1):71–74

Páez-Espino AD, Durante-Rodríguez G, de Lorenzo V (2015) Functional coexistence of twin arsenic resistance systems in *Pseudomonas putida* KT2440. Environ Microbiol 17(1):229–238

Paulsen IT, Saier MH Jr (1997) A novel family of ubiquitous heavy metal ion transport proteins. J Membr Biol 156(2):99–103

Peng ED, Oram DM, Battistel MD, Lyman LR, Freedberg DI, Schmitt MP (2018) Iron and zinc regulate expression of a putative ABC metal transporter in *Corynebacterium diphtheriae*. J Bacteriol 200(10):e00051–e00018

Porcheron G, Garenaux A, Proulx J, Sabri M, Dozois C (2013) Iron, copper, zinc, and manganese transport and regulation in pathogenic Enterobacteria: correlations between strains, site of infection and the relative importance of the different metal transport systems for virulence. Front Cell Infect Microbiol 3:90

Pourret O, Hursthouse A (2019) It's time to replace the term "heavy metals" with "potentially toxic elements" when reporting environmental research. Int J Environ Res Public Health 16(22):4446

Pradhan SK, Singh NR, Rath BP, Thatoi H (2016) Bacterial chromate reduction: a review of important genomic, proteomic, and bioinformatic analysis. Crit Rev Environ Sci Technol 46 (21–22):1659–1703

Prithivirajsingh S, Mishra SK, Mahadevan A (2001) Detection and analysis of chromosomal arsenic resistance in *Pseudomonas fluorescens* strain MSP3. Biochem Biophys Res Commun 280 (5):1393–1401

Qin J, Rosen BP, Zhang Y, Wang G, Franke S, Rensing C (2006) Arsenic detoxification and evolution of trimethylarsine gas by a microbial arsenite S-adenosylmethionine methyltransferase. Proc Natl Acad Sci U S A 103(7):2075–2080

Quatrini R, Johnson DB (2019) Acidithiobacillus ferrooxidans. Trends Microbiol 27(3):282–283

Rademacher C, Masepohl B (2012) Copper-responsive gene regulation in bacteria. Microbiology 158:2451–2464

Rasulov BA, Yili A, Aisa HA (2013) Biosorption of metal ions by exopolysaccharide produced by *Azotobacter chroococcum* XU1. J Environ Prot 4:989–993

Raungsomboon S, Chidthaisong A, Bunnag B, Inthorn D, Harvey NW (2006) Production, composition and Pb^{2+} adsorption characteristics of capsular polysaccharides extracted from a cyanobacterium *Gloeocapsa gelatinosa*. Water Res 40(20):3759–3766

Rosen BP (2002) Transport and detoxification systems for transition metals, heavy metals and metalloids in eukaryotic and prokaryotic microbes. Comp Biochem Physiol A Mol Integr Physiol 133(3):689–693

Rosenstein R, Peschel A, Wieland B, Götz F (1992) Expression and regulation of the antimonite, arsenite, and arsenate resistance operon of *Staphylococcus xylosus* plasmid pSX267. J Bacteriol 174(11):3676–3683

Rouch DA, Lee BTO, Morby AP (1995) Understanding cellular responses to toxic agents: a model for mechanism-choice in bacterial metal resistance. J Ind Microbiol 14(2):132–141

Ruangsomboon S, Chidthaisong A, Bunnag B, Inthorn D, Harvey NW (2007) Lead (Pb^{2+}) adsorption characteristics and sugar composition of capsular polysaccharides of cyanobacterium *Calothrix marchica*. Songklanakarin J Sci Technol 29(2):529–541

Ryan D, Colleran E (2002) Arsenical resistance in the IncHI2 plasmids. Plasmid 47(3):234–240

Salehizadeh H, Shojaosadati SA (2003) Removal of metal ions from aqueous solution by polysaccharide produced from *Bacillus firmus*. Water Res 37(17):4231–4235

Saltikov CW, Cifuentes A, Venkateswaran K, Newman DK (2003) The ars detoxification system is advantageous but not required for as(V) respiration by the genetically tractable *Shewanella* species strain ANA-3. Appl Environ Microbiol 69(5):2800–2809

Samuelson P, Wernérus H, Svedberg M, Ståhl S (2000) Staphylococcal surface display of metal-binding polyhistidyl peptides. Appl Environ Microbiol 66(3):1243–1248

Sand W, Rohde K, Sobotke B, Zenneck C (1992) Evaluation of *Leptospirillum ferrooxidans* for leaching. Appl Environ Microbiol 58(1):85–92

Sar P, Kazy SK, Asthana RK, Singh SP (1998) Nickel uptake by *Pseudomonas aeruginosa*: role of modifying factors. Curr Microbiol 37(5):306–311

Saravanakumar D, Vijayakumar C, Kumar N, Samiyappan R (2007) PGPR-induced defense responses in the tea plant against blister blight disease. Crop Prot 26(4):556–565

Sato T, Kobayashi Y (1998) The ars operon in the skin element of *Bacillus subtilis* confers resistance to arsenate and arsenite. J Bacteriol 180(7):1655–1661

Schenk PM, Carvalhais LC, Kazan K (2012) Unraveling plant–microbe interactions: can multi-species transcriptomics help? Trends Biotechnol 30(3):177–184

Scherer J, Nies DH (2009) CzcP is a novel efflux system contributing to transition metal resistance in *Cupriavidus metallidurans* CH34. Mol Microbiol 73(4):601–621

Segerer A, Neuner A, Kristjansson JK, Stetter KO (1986) *Acidianus infernus* gen. Nov., sp. nov., and Acidianus brierleyi comb. nov.: facultatively aerobic, extremely acidophilic thermophilic sulfur-metabolizing archaebacteria. Int J Syst Evol Microbiol 36(4):559–564

Sharma M, Kaushik A, Somvir B, K. and Kamra, A. (2008) Sequestration of chromium by exopolysaccharides of *Nostoc* and *Gloeocapsa* from dilute aqueous solutions. J Hazard Mater 157(2):315–318

Sher S, Rehman A (2019) Use of heavy metals resistant bacteria-a strategy for arsenic bioremediation. Appl Microbiol Biotechnol 103(15):6007–6021

Silver S (1996) Bacterial resistances to toxic metal ions. Gene 179(1):9–19

Singh R, Dong H, Liu D, Zhao L, Marts AR, Farquhar E, Tierney DL, Almquist CB, Briggs BR (2015) Reduction of hexavalent chromium by the thermophilic methanogen *Methanothermobacter thermautotrophicus*. Geochim Cosmochim Acta 148:442–456

Smith RL, Maguire ME (1995) Distribution of the CorA Mg^{2+} transport system in gram-negative bacteria. J Bacteriol 177(6):1638–1640

Spada S, Pembroke JT, Wall JG (2002) Isolation of a novel *Thermus thermophilus* metal efflux protein that improves *Escherichia coli* growth under stress conditions. Extremophiles 6 (4):301–308

Suzuki K, Wakao N, Kimura T, Sakka K, Ohmiya K (1998) Expression and regulation of the arsenic resistance operon of *Acidiphilium multivorum* AIU 301 plasmid pKW301 in *Escherichia coli*. Appl Environ Microbiol 64(2):411–418

Tavares MT, Neves IC (2008) Biosorption system produced from biofilms supported on Faujasite (FAU) zeolite, process for obtaining it and its usage for removal of hexavalent chromium (Cr (VI)). International patent, W0 2007/020588 A1, EP1912905

Tsai KJ, Yoon KP, Lynn AR (1992) ATP-dependent cadmium transport by the cadA cadmium resistance determinant in everted membrane vesicles of *Bacillus subtilis*. J Bacteriol 174 (1):116–121

Tuffin IM, de Groot P, Deane SM, Rawlings DE (2005) An unusual Tn21-like transposon containing an ars operon is present in highly arsenic-resistant strains of the biomining bacterium *Acidithiobacillus caldus*. Microbiology (Reading) 151(Pt 9):3027–3039

Tuffin IM, Hector SB, Deane SM, Rawlings DE (2006) Resistance determinants of a highly arsenic-resistant strain of *Leptospirillum ferriphilum* isolated from a commercial biooxidation tank. Appl Environ Microbiol 72(3):2247–2253

Turpeinen R, Kairesalo T, Häggblom MM (2004) Microbial community structure and activity in arsenic-, chromium- and copper-contaminated soils. FEMS Microbiol Ecol 47(1):39–50

Vardanyan NS, Vardanyan AK (2018) Thermophilic chemolithotrophic bacteria in mining sites. In: Egamberdieva D, Birkeland N-K, Panosyan H, Li W-J (eds) Extremophiles in Eurasian ecosystems: ecology, diversity, and applications. Springer, Singapore, pp 187–218

Vargas M, Kashefi K, Blunt-Harris EL, Lovley DR (1998) Microbiological evidence for Fe(III) reduction on early earth. Nature 395(6697):65–67

Voica DM, Bartha L, Banciu HL, Oren A (2016) Heavy metal resistance in halophilic Bacteria and archaea. FEMS Microbiol Lett 363(14):1–9

Volesky B (1994) Advances in biosorption of metals: selection of biomass types. FEMS Microbiol Rev 14(4):291–302

von Rozycki T, Nies DH (2009) *Cupriavidus metallidurans*: evolution of a metal-resistant bacterium. Antonie Van Leeuwenhoek 96(2):115–139

Wang J, Chen C (2009) Biosorbents for heavy metals removal and their future. Biotechnol Adv 27 (2):195–226

Wang G, Kennedy SP, Fasiludeen S, Rensing C, DasSarma S (2004) Arsenic resistance in *Halobacterium* sp. strain NRC-1 examined by using an improved gene knockout system. J Bacteriol 186(10):3187–3194

Wang L, Chen S, Xiao X, Huang X, You D, Zhou X, Deng Z (2006) *arsRBOCT* arsenic resistance system encoded by linear plasmid pHZ227 in *Streptomyces* sp. strain FR-008. Appl Environ Microbiol 72(5):3738–3742

Wani PA, Khan MS (2010) *Bacillus* species enhance growth parameters of chickpea (*Cicer arietinum* L.) in chromium stressed soils. Food Chem Toxicol 48(11):3262–3267

Whelan KF, Colleran E (1992) Restriction endonuclease mapping of the HI2 incompatibility group plasmid R478. J Bacteriol 174(4):1197–1204

White C, Gadd GM (2000) Copper accumulation by sulfate-reducing bacterial biofilms. FEMS Microbiol Lett 183(2):313–318

Xiong A, Jayaswal RK (1998) Molecular characterization of a chromosomal determinant conferring resistance to zinc and cobalt ions in *Staphylococcus aureus*. J Bacteriol 180(16):4024–4029

Yang HC, Rosen BP (2016) New mechanisms of bacterial arsenic resistance. Biom J 39(1):5–13

Yang HC, Cheng J, Finan TM, Rosen BP, Bhattacharjee H (2005) Novel pathway for arsenic detoxification in the legume symbiont *Sinorhizobium meliloti*. J Bacteriol 187(20):6991–6997

Yang L, Zhao D, Yang J, Wang W, Chen P, Zhang S, Yan L (2019) *Acidithiobacillus thiooxidans* and its potential application. Appl Microbiol Biotechnol 103(19):7819–7833

Ye J, Yang H-C, Rosen BP, Bhattacharjee H (2007) Crystal structure of the flavoprotein ArsH from *Sinorhizobium meliloti*. FEBS Lett 581(21):3996–4000

Zeng W, Zhang S, Xia M, Wu X, Qiu G, Shen L (2020) Insights into the production of extracellular polymeric substances of *Cupriavidus pauculus* 1490 under the stimulation of heavy metal ions. RSC Adv 10(34):20385–20394

Zhang Y, Rodionov DA, Gelfand MS, Gladyshev VN (2009) Comparative genomic analyses of nickel, cobalt and vitamin B12 utilization. BMC Genomics 10(1):78

Chapter 14
Bioremediation Potential of Soil Bacteria of Heavy Metal Polluted Environments of Kyrgyzstan

Cholpon Omurgazieva and Makhabat Konurbaeva

Abstract It is sad to realize, but our time has become in the history of the Earth the time of the most drastic change (almost to the point of destruction) by mankind of the natural habitat. As a result of anthropogenic impact, thousands of species of saprophytic microorganisms, plants, and animals may disappear on the planet. This process must be stopped, because further reduction of biodiversity can lead to destabilization of ecosystems. Environmental pollution with heavy metals and radionuclides is one of the most important environmental problems of the late twentieth to early twenty-first centuries. As you know, many heavy metals pose a danger to living organisms due to their toxicity and mutagenicity; cleaning the environment from these compounds is now becoming increasingly important. The phenomenon of microbial heavy metal resistance has major importance and is crucial in microbial ecology, especially in processes of biogeochemical cycling of heavy metals and in the bioremediation of metal-contaminated ecological systems. A biological approach such as bioremediation offers significantly more advantages over conventional technology to reduce the content of heavy metals in the environment due to its less expensive and environmentally friendly nature (Feruke-Bello, J Bioremediation 22(1–2): 28–42, 2018). The bioremediation method is based on the activation of aboriginal soil microflora, as well as on the introduction, instead of pollution, of specially selected microorganisms that actively utilize the pollutant, which significantly accelerates the processes of soil restoration to the level of their economic, including agricultural use. This work presents an experimental study of the interaction of heavy metals and microorganisms in the soil, the possibility of using microorganisms for bioremediation of soils from pollution by xenobiotics, including heavy metals. The data on the selection of highly resistant strains of

C. Omurgazieva (✉)
Institute of Biology, National Academy of Science of the Kyrgyz Republic, Bishkek, Kyrgyzstan

M. Konurbaeva
Department of Plant Protection, Kyrgyz-Turkish University, Bishkek, Kyrgyzstan

D. Egamberdieva et al. (eds.), *Microbial Communities and their Interactions in the Extreme Environment*, Microorganisms for Sustainability 32,
https://doi.org/10.1007/978-981-16-3731-5_14

microorganisms to the effects of heavy metals in soils and their use for the detoxification of these elements for cleaning the natural environment are presented.

Keywords Bioremediation · Uranium tailings · Heavy metals · Strains · Soil microorganisms

14.1 Introduction

A significant environmental threat to the Kyrgyz Republic is posed by tailing dumps of radioactive waste and heavy metals. On the territory of the republic, the volume of stored waste exceeds 620 million tons of cubic meters. The National Plan for Environmental Protection, developed with the involvement of foreign specialists, provides a description of the most problematic objects from a geoecological point of view. These include storage facilities for radioactive waste in Min-Kush, Kazhi-Say, Kara-Balta; waste from processing of mercury and antimony in Khaidarken, Kadamzhay, Chuvay; rare earth metals in Orlovka, Ak-Tyuz, Kichi-Kemin; gold in Kazarman. Currently, the condition of these dumps and storage facilities is not at the proper level. In ecologically unfavorable such and adjacent areas, the content of toxic elements—mercury, antimony, selenium, arsenic, copper, nickel, cadmium, etc., is tens and hundreds of times higher than the maximum permissible concentration (MPC) and is contaminated with heavy metals of the first and second classes, dangerous to human health (Omurgazieva 2007).

Many heavy metals, especially in high concentrations, are highly toxic, and since they accumulate in the soil and plant organisms in significant quantities, they continuously enter the human body through food webs. The need for living organisms such elements as lead, mercury, and cadmium has not been established, and they are considered the most toxic, especially for animals and humans. Their ions, entering the human body through the trophic chain, can cause allergic diseases, bronchial asthma and pulmonary emphysema, impairment of generative functions and redox processes, and neoplasms (Sharshenalieva and Koleno 1990).

It is known that soil accumulates pollution to a greater extent than the atmosphere and natural waters. Radionuclides and heavy metals, actively sorbed on soil particles, primarily affect the living phase of soils, the functioning of soil fauna, including soil microorganisms. Among living things, microorganisms are sensitive and quickly react to various influences of environmental factors.

Bacteria are living organisms that consume the pollutant as a source of energy for life and convert it into environmentally friendly products of their own metabolism—carbon dioxide, water, and mineral salts. And enzymes and biosurfactants serve as a catalyst for the biological process: they quickly break down organic molecules, thereby significantly facilitating the assimilation of pollutants by bacteria (Korolchenko 2007).

In a review of foreign works on the purification of the environment from xenobiotics, it was shown that of the 27 types of purification technologies developed, six are based on methods using microorganisms (Vel'kov 1995). Microbial degradation of heavy metals, radionuclides and hydrocarbons forms the basis of modern technology for bioremediation of contaminated environmental objects (Feruke-Bello et al. 2018; Trindade et al. 2005; Kuznetsov et al. 2010; Egorova et al. 2010). The undoubted advantages of this method are efficiency, economy, environmental safety and the absence of harmful pollution (Alexander 1999; Feruke-Bello et al. 2018). Bioremediation is perfectly complemented by phytoremediation (Muratova et al. 2010), more precisely, chemo-phytostabilization (sowing herbs and planting trees or shrubs with the introduction of ameliorants and fertilizers). Sowing of perennial grasses after the completion of the bioremediation cycle ensures complete restoration of soil fertility as a result of the activity of the rhizosphere microflora (Turkovsvaya and Muratova 2005). The high density of rhizosphere microbial populations indicates more favorable and stable conditions in the rhizosphere compared to soil without plants. Thus, a plant under conditions of pollution is a factor that maintains and/or increases the number of soil microbial populations, providing them with a niche and additional nutrition for reproduction and performing protective functions against the effects of a pollutant, thereby intensifying the processes of self-purification of contaminated soil. The selection of plants exhibiting a pronounced resistance to soil pollution of each pollutant is an important first step toward screening promising phytoremediates. The study of the features of their germination and development in the presence of a pollutant will make it possible to determine the parameters of artificial introduction of vegetation in contaminated areas for the purpose of their phytoremediation. The advantages of bio- and phytoremediation as an economically profitable, ecologically safe and aesthetically attractive biotechnology for the restoration of contaminated areas in situ have been shown by many researchers (Zharikov et al. 2013; Wang et al. 2012).

Development and improvement of bioremediation technologies, especially soils contaminated with heavy metals (Bagaeva et al. 2013; Bekasova et al. 1999; White et al. 1997), oil and petroleum products (Baryshnikova et al. 2001; Davydova et al. 1988; Ismailov 1988; Kobzev et al. 2001; Koronelli et al. 2001; Loginov et al. 2002) is currently an area of active fundamental and applied research. The most widespread methods of bioremediation are based on the activation of aboriginal soil microflora, potentially capable of utilizing the pollutant through the use of a number of agrotechnical measures (soil loosening, moistening, the use of fertilizers, etc.), as well as on the introduction of specially selected microorganisms into the place of pollution, actively utilizing pollutant, which significantly speeds up the processes of soil restoration. The most widespread methods of bioremediation are based on the activation of aboriginal soil micro flora. Potentially capable of utilizing the pollutant with a number of agrotechnical measures (soil loosening, moistening, the use of fertilizers, etc.), as well as on the introduction of specially selected microorganisms into the place of pollution, actively utilizing pollutant, which significantly speeds up the processes of soil restoration. The schemes of such reclamation technologies are corrected and modified depending on the individual characteristics of the place of pollution and the properties of pollutants. The first approach gives rather high, stable results and is mainly used at low pollution levels (up to 5%). The process of self-healing of the contaminated environment has been going on for more than 25–30 years (Oborin et al. 1988). The application of this method in places of pollution significantly accelerates the processes of soil restoration.

The research is aimed at solving the problem of conservation, restoration and rational use of natural resources and relates to biotechnology of environmental protection in the field of mining.

The bioremediation method has not yet been applied in the territories of industrial enterprises polluted with waste—tailing dumps of the Kyrgyz Republic and remains unexplored.

In this regard, we are focused on the study of microorganisms to develop an environmentally friendly method of cleaning soil from pollution; on the search and isolation of highly resistant strains of microorganisms to high concentrations of heavy metals and radionuclides in the areas affected by industrial emissions from mining enterprises of Kyrgyzstan.

14.2 Kyrgyz Republic

The Kyrgyz Republic is located in the southeastern part of Central Asia, within the Tien Shan and Pamir-Alai mountain ranges. The lowest point (488 m above sea level) is the point where the Naryn River crosses the border with the Republic of Uzbekistan, and the highest is Pobeda Peak (7439 m). The average height of the territory above sea level is 2630 m. Kyrgyzstan borders on Kazakhstan, Tajikistan, Uzbekistan and China. All the variety of landscapes and natural and climatic conditions of Kyrgyzstan can be combined into four natural and climatic zones: valley-foothill—up to 1200 m, mid-mountain—from 1200 to 2200 m, high-mountain—from 2200 to 3500 m, and nival—above 3500 m. Less 20% of the territory of the republic belongs to areas with comfortable living conditions. Large systems of mountain ranges, oriented in different directions, led to the creation of several regions, the climate in which is quite homogeneous and noticeably different from each other. The climate of the Kyrgyz Republic is sharply continental, mostly arid, somewhat smoothed by increased cloudiness and precipitation due to the high-mountainous relief. The peculiarities of the climate are determined by the location of the republic in the Northern Hemisphere in the center of the Eurasian continent, as well as by the distance from significant water bodies and the close proximity of deserts.

The hydrographic network of Kyrgyzstan is complex and varied: it includes numerous rivers flowing from mountain ranges into valleys; a large number of lakes; powerful alpine glaciers, accumulators of moisture and sources of river power. The country's rivers belong to three main drainless basins: the Aral Sea (Naryn-Syrdarya), Lake Issyk-Kul and the basin of the river. Tarim (Saryjaz). The area between these basins is distributed accordingly: 76.5; 10.8; and 12.4%. And only 0.3% of the area in the east of Kyrgyzstan belongs to the basin of Lake Balkhash.

14.3 Uranium Tailings in Kyrgyzstan

Industrial mining and processing of radioactive ores in Kyrgyzstan began in 1907 at the Tuya-Muyun mine. As a result of the long-term activity of uranium mines, a significant amount of radioactive waste was formed, which is stored in mountain dumps and tailings in various regions of the country. The main facilities for the extraction and processing of radioactive ores in Kyrgyzstan include the enterprises of the former Leninabad Mining and Chemical Combine in Mailuu-Suu, Shekaftar, Kyzyl-Dzhar; enterprises of the Kara-Balta mining plant (KGRK) in the city of Kara-Balta, the settlement of Min-Kush, the settlement of Kadzhi-Sai, as well as the enterprises of the Kyrgyz mining and metallurgical plant in the settlement of Ak-Tyuz, the settlement of Orlovka (Fig. 14.1) country (Torgoev and Aleshin 2001).

There are 72 radioactive waste storage facilities in the Kyrgyz Republic (tailings and mining dumps). The total volume of solid radioactive waste exceeds 130 million m^3, and the area occupied by them is 650 hectares (ha). The greatest danger is posed by 35 radioactive tailing dumps with a total volume of 48.3 million m^3, including 29 tailings with uranium production waste with a total volume of up to 41 million m^3 of tailing material. Additionally, 35 facilities (waste rock dumps) with a low uranium ore content with a total volume of 83 million m^3 are also located in the country. In the past, there have been numerous accidents at tailing dumps in Mailuu-Suu, Min-Kush, Ak-Tyuz with catastrophic consequences in the form of radioactive contamination of the transboundary territories of Kyrgyzstan, Kazakhstan and Uzbekistan (http://www.caresd.net/img/docs/5586.pdf).

Radioactive waste, heavy metals and toxic substances pollute the environment (soil, surface and underground waters, atmosphere and plants).

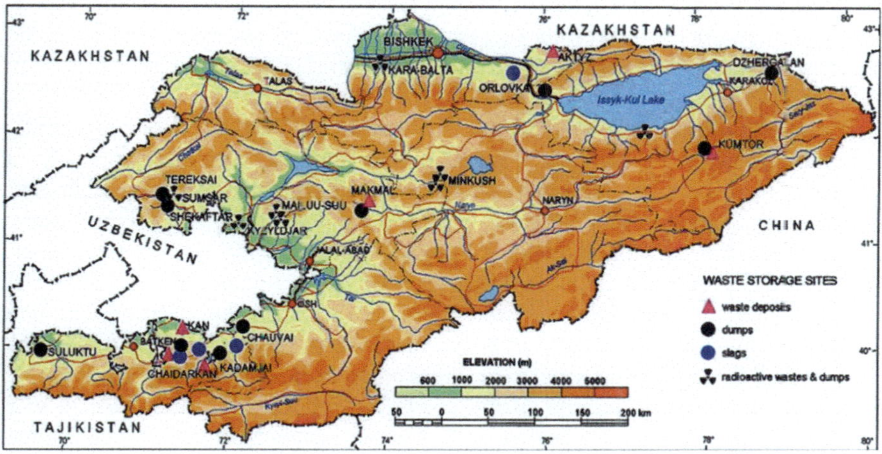

Fig. 14.1 Map of tailings and dumps on the territory of the Kyrgyz Republic

14.3.1 Tailings Ponds and Dumps in the Ak-Tyuz Settlement

Ak-Tyuz tailing dumps are located in the Kemin district of the Chui region. 42° 52′46″ N; 76° 08′03″ E; Height (m) 2205. Since 1942, ore containing lead, zinc, and rare earth elements have been mined and processed here. Ak-Tyuz village is located 150 km from Bishkek city. The processed ore in this area contains radioactive elements from minerals containing thorium (turnerite, thorite, zirconium, and others). Near the Ak-Tyuz settlement, there are four tailing dumps with a total volume of about 3.4 million m^3 and three waste rock dumps, the total volume of which exceeds 50 million m^3.

In December 1964, there was a catastrophic, seismic-synchronous destruction of the tailing dump No. 2 of the Ak-Tyuz mine. By the time of the catastrophe, more than one million m^3 of "tailings" of processing of rare earth elements had been stored in it. In terms of chemical composition, salts of heavy metals with a very high content of lead, zinc, copper, molybdenum, arsenic and beryllium represented Ak-Tyuz tailings. In addition, the tailings contained increased concentrations of cadmium, tungsten, and yttrium. Of the radioactive elements, the tailings contained high concentrations of thorium (800–7000 ppm) and zircon (110–4800 ppm). Because of the destruction of the unstable alluvial dam of the tailing dump, initiated by an earthquake, about 600 thousand m^3 of tailings (or 60% of its volume) were thrown into the Kichi-Kemin River. Tailings streams, containing increased concentrations of thorium and heavy metals, spread in the form of a radioactive mudflow along the river channel and valley. Kichi-Kemin at a distance of up to 40 km, up to its confluence with the river. Chu on the territory of Kazakhstan. Waste stored on a 22-hectare site contaminated agricultural land, settlements with a total area of 3600 hectares.

The consequences of this transboundary catastrophe were eliminated for many years, however, even now, i.e., more than 50 years later; they continue to negatively affect the environment and health of the population living in the Kichi-Kemin valley. This is evidenced by the results of comprehensive environmental studies in the valley of the river. Kichi-Kemin, performed in 2004–2006 by the Czech company "GeoMin".

The results of the analyzes show that the pollution of water, bottom sediments, soil and vegetation in the valley below tailing dump #2 is characterized by a variegated association of various elements (thorium, beryllium, lead, zinc, cadmium, molybdenum, antimony, arsenic, rare metals), half of which are Eco toxicants. Because of environmental and microbiological studies in the valley of the river. Kichi-Kemin, performed 2003–2007 dissertation work (Omurgazieva 2007): the main pollutants of this production were identified, which include Pb, Zn, Cu, and Cd. It was found that the content of heavy metals in the soil on the territory of the plant, around the tailing dump and in the areas adjacent to it (3 km) exceeds the MPC by 100 or more times. Because of the conducted microbiological studies, it was established that the selective pressure of the long-term action of heavy metals on the soil biota manifests itself in a change in the number and species diversity of soil microorganisms.

14.3.2 Tailings Dump with Uranium Waste in the Technogenic Province of Kazhi-Sai (42° 08′24″ N; 77° 10′50″ E). Height (m): 1750

The tailing dump is located 1.5 km from the shore of Lake Issyk-Kul. It was formed in the period from 1952 to 1966 in the process of extracting uranium concentrate from ash burned at a thermal power plant of brown coals with a high uranium content from a nearby deposit. The volume of accumulated tailings of uranium production, as well as other industrial waste, is about 150 thousand m^3. On the surface of soil-covered ash dumps, the exposure dose of gamma radiation (DER) averages 30–60μR/h. However, a significant part of the tailing dump remains uncovered. According to the Ministry of Emergency Situations of the Kyrgyz Republic, there are areas with abnormally high levels of exposure dose rates of 600–1500μR/h (up to 15μSv/h). High EDR levels are observed in areas of damaged protective cover as a result of excavations carried out by local residents or water erosion.

14.3.3 Tailings Dumps in the Area of the Min-Kush Settlement

There are currently 4 tailing dumps and 5 industrial sites in the village of Min-Kush (41° 40′48″ N; 74° 27′36″ E; Height (m): 2260). The ore complex was operated from 1958 to 1969. According to experts, the radiation balance here ranges from 80μR to 740μR/h. The surface of the tailing dump is not reinforced with any of the currently existing stabilization or reclamation methods. Consequently, it is subject to wind and water erosion and leads to soil and atmosphere pollution with dust and solid particles, heavy metal salts.

14.3.4 The Uranium Burial Site in Kara-Balta Is One of the Five Largest and Is One of the Most Potentially Dangerous Objects in the Kyrgyz Republic

This tailing pond is one of the largest tailing ponds in the world. Geographic coordinates of the object: 42° 10′00″ N; 73° 45′00″ E; Height (m): 865. Area (thousand m^2): 550 m. Seismicity of the area: 8 points (although it should not be higher than 6 points). The tailing dump is of a reclaimed type, the total volume of radioactive waste disposed here is about 39,000 tons (thousand m^3), and the design capacity of the tailing dump is almost twice as large—63.5 million cubic meters. Type and types of waste from processing uranium, molybdenum ores. Radioactive material represented by radionuclides of uranium (U 238 + 235 + 234), thorium (Th 232), radium (Ra 226). The radius of the contaminated zone is 10 km, the situation with groundwater in this area is not clear enough. The local population is building new residential buildings right on the border with the territory of the tailing dump, access to which is open for livestock: sheep and cows [http://www.caresd.net/img/docs/5586.pdf].

14.3.5 Objects of Uranium Heritage in the Area of Mailuu-Suu

Geographic coordinates of the object: 41° 160.01 N; 72° 27′ 0 E; Height (m): 1100. Mailuu-Suu uranium deposit was operated from 1946 to 1968. For more than 20 years of operation of the Western Mining and Chemical Combine in the Mailuu-Suu region, 10,000 tons of the final product—uranium oxide-oxide (U3O8)—were obtained. Currently, on the territory of the former enterprise, including directly in the city limits, there are 23 tailing dumps and 13 mining dumps. According to the data of numerous measurements, the indicators of the average exposure dose rate (EDR) of gamma radiation on the surface of the covered tailings are in the range from 60 to 100μR/h (up to 1μSv/h). In the areas of tailings where the coverage is broken, there are high fluxes of radon exhalation, and the exposure dose rate of gamma radiation reaches 1500μR/h (15μSv/h). The analysis shows that only in the south of the country, in the zone of ecological threat, there are constantly 26 thousand people in Kyrgyzstan, more than two million people in Uzbekistan, 900 thousand in Kazakhstan and 700 thousand in Tajikistan, given that the watercourses contaminated with radionuclides Mailuu-Suu flows through the Fergana Valley and connects with the large rivers Karadarya and Syrdarya (Bykovchenko et al. 2005). Living near uranium tailings can affect human health and the characteristics of the course of diseases, as a result of the chronic intake of radionuclides into the body through the food chain, air, etc.

In 2000–2002, in the area of Mailuu-Suu, a group of specialists from Uzbekistan and Kyrgyzstan carried out geoecological studies in the area of Mailuu-Suu, which included geo- and hydrochemical analyzes of samples of soil, soil, water, bottom sediments, vegetation and human biosubstrates (hair, nails). As a result of these studies, it was found that the main pollutants present in almost all natural environments of the Mailuu-Suu area are selenium, uranium, and chromium (Aitmatova and Aparin 2003).

Prospects for improving the ecological situation in these cities and towns are possible only if a set of measures is taken, including: rehabilitation of contaminated areas; reclamation of dumps and tailings using more accessible, proven, and environmentally friendly methods.

In many countries of the world (Europe, USA, Japan, Russia) there is already a special program for cleaning up the territories of contaminated military training grounds, with industrial and household waste, which includes familiarization with the methods of bioremediation of soil, water bodies and air. In 2001, three large US Department of Defense test sites used microbial bioremediation technology to treat soils contaminated with explosives, as this technology showed 99% soil cleaning after the first two phases of treatment. Now the task is to maximize the use of the potential of microorganisms, which have a huge variety of enzymes that carry out the destruction reactions of various xenobiotic. The use of this potential is "bioremediation."

14.4 Characterization and Bioavailability of Heavy Metals

Metals with a density over 5 g/cm^3 are classified as heavy metals (HM). Among the 90 elements found in nature, there are 21 non-metals, 16 light metals and 53 heavy metals (including As). Heavy metals are elements with variable valence and with incompletely filled *d*-orbitals. The ability of heavy metal cations to form complexes that may or may not have a redox activity is provided by *d*-orbitals. All these 53 heavy metals have no positive or negative biological function for a living cell, since they are not available in a normal ecosystem. To be bioavailable, heavy metals must be present in a given ecosystem in at least nanomolar concentrations, since a concentration of 1 nM means that in a 109/mL cell suspension, each cell can receive approximately 600 ions. Metal ions, usually present in lower concentrations, can be used by the microorganism for very specific purposes; however, the lower the average concentration of a metal ion in an ecosystem, the less likely it is that a species is able to use or detoxify that specific heavy metal. Heavy metals are divided into four classes based on their concentration in the medium:

- Frequent trace elements with concentrations between 100 nM and 1μM—Fe, Zn, and Mo;
- Moderately occurring trace elements with concentrations between 10 and 100 nM—Ni, Cu, As, N, Mn, Sn, and U;
- Rare trace elements—Co, Ce, Ag, and Sb;
- Cd, Cr, W, Ga, Zr, Th, Hg, and Pb—below 1 nM.

Other elements, such as 55.8 nM Au in seawater, are unlikely to become trace elements. The relative solubility of heavy metals under physiological conditions determines the differences in the biological significance and toxicity of heavy metals in relation to their affinity for the environment and other interactions with macroelements in living organisms. Due to their low solubility, the trivalent or tetravalent cations Sn, Ce, Ga, Zr and Th have no biological significance. Of the remaining 17 heavy metals: Fe, Mo and Mn are important trace elements with low toxicity; Zn, Ni, Cu, V, Co, W, and Cr are toxic elements of high to medium significance as trace elements and As, Ag, Sb, Cd, Hg, Pb, and U have no useful function, but are considered toxins for cells.

In addition to an important catalytic role, heavy metal cations, forming complexes with biomolecules in complex biochemical reactions, such as nitrogen fixation, water photolysis during photosynthesis, oxygen or nitrate respiration, one-electron catalysis, rearrangement of C-C bonds, assimilation of hydrogen, decomposition of urea, transcription of genes into mRNA, etc., can also form nonspecific complex compounds in cells at increased concentrations, which leads to toxic effects. Heavy metal cations such as Hg^{2+}, Cd^{2+}, and Ag^{+} are highly toxic complexing agents and are often hazardous to any biological function. Even the most important trace elements such as Zn^{2+} or Ni^{2+}, and especially Cu^{2+}, are toxic at high concentrations. Therefore, each life form was forced to develop a certain system of homeostasis in order to ensure tight control over the concentration of heavy metal ions inside cells (Bagaeva et al. 2013).

14.5 Basic Approaches and the Role of Bioremediation in the Restoration of Contaminated Soils

Bioremediation It is one of the most effective methods for cleaning the environment from technogenic pollution (Gradova et al. 2003; Grishchenkov and Shishmakov 2003). The method of bioremediation of environmental purification using effective, environmentally competitive strains of microorganisms is recognized by many scientists around the world as environmentally safe and economically beneficial in comparison with mechanical and physicochemical purification technologies. Existing mechanical, thermal, and physicochemical methods of cleaning soil from oil pollution are expensive and effective only at a certain level of pollution (as a rule, at least 1% of oil in the soil), are often associated with additional contamination and do not ensure the completeness of cleaning. Currently, the most promising method for cleaning technogenic soils, both economically and ecologically, is a biotechnological approach based on the use of various groups of microorganisms, characterized by an increased ability to biodegradate components of oil and oil products (Loginov 2000). The ability to utilize hardly degradable substances of anthropogenic origin (xenobiotics) has been found in many organisms. This property is provided by the presence of specific enzyme systems in microorganisms that catabolize such compounds. Since microorganisms have a relatively high potential for destruction of xenobiotics, exhibit the ability to rapidly metabolic rearrangement and exchange of genetic material, they are of great importance in the development of ways of bioremediation of contaminated objects.

The term "bioremediation" is commonly understood to mean the use of technologies and devices designed for biological treatment of soils, i.e., to remove pollutants already in it from the soil (Gilyarov 1999).

Bioremediation includes two main approaches: (1) Biostimulation—activation of the degrading ability of aboriginal microflora by introducing biogenic elements, oxygen, and various substrates; (2) Bio-replenishment—the introduction of natural and genetically engineered strains-destructors of foreign compounds.

In situ biostimulation (biostimulation at the site of contamination). This approach is based on stimulating the growth of natural microorganisms living in contaminated soil and potentially capable of utilizing the pollutant, but unable to do so effectively due to the lack of basic biogenic elements (nitrogen, phosphorus, potassium compounds, etc.) or unfavorable physicochemical conditions. In this case, in the course of laboratory tests using samples of contaminated soil, it is established which components and in what quantities should be added to the contaminated object in order to stimulate the growth of microorganisms capable of utilizing the pollutant (Loginov 2000).

In vitro biostimulation. The difference between this approach is that biostimulation of samples of natural microflora of contaminated soil is carried out first in laboratory or industrial conditions (in bioreactors or fermenters). This ensures the predominant and selective growth of those microorganisms that are able to most effectively utilize this pollutant. The "activated" microflora is introduced into the contaminated object simultaneously with the necessary additives that increase the efficiency of the pollutant utilization (Loginov 2000). The existing two ways to intensify the biodegradation of xenobiotics in the environment—stimulation of natural microflora and the introduction of active strains, not only do not contradict, but complement each other (Coronelli 1996). The unfavorable physicochemical conditions limiting the degradation of xenobiotics by microorganisms in the environment include low or excessive soil moisture, insufficient oxygen content, unfavorable temperature and pH, low concentration or availability of xenobiotics, the presence of alternative, more preferable substrates, etc. Some of these problems can be solved by creating genetically engineered strains-destructors and their consortia, improving methods of introduction, optimizing the conditions for the existence of natural microbial populations. Thus, the introduction of microorganisms leads to positive results only when appropriate conditions are created for the development of the introduced population, for which it is necessary to know the physiological characteristics of the introduced species, as well as to take into account the emerging microbial interactions (Yankevich 2015).

Phytoremediation It has been proven that plants can be used to clean soil, groundwater, and wastewater from both inorganic (heavy metals, radionuclides) and organic (hydrocarbon pollution, pesticides) toxicants.

There are several main phytoremediation mechanisms (Kuznetsov et al. 2010).

Phytoextraction is the absorption, translocation, and accumulation of a pollutant in plants. The process underlies the cleaning of soils from heavy metals and radionuclides.

During *rhizofiltration*, the pollutant is sorbed on the roots or other parts of plants, which is also important when eliminating inorganic contaminants.

The phytodegradation processes of organic pollutants are under intensive research.

Rhizodegradation is the destruction of impurities in the root zone of a plant under the influence of root exudates and the activity of rhizospheric microflora. A wide range of pollutants, including persistent organic pollutants (POPs), can be destroyed with the participation of introduced microorganisms-destructors of POPs and plants with active rhizospheric microflora, creating new plant-microbial cleansing complexes. The ability of a plant to hyperaccumulate is determined by the bioaccumulation coefficient—the ratio of the concentration of the toxicant in the shoots to the concentration in the soil. Hyperaccumulators are capable of absorbing certain metals in amounts up to a few percent of their dry biomass. The collected plant biomass can be eliminated by incineration and subsequent disposal of ash in landfills. If the toxic elements are of commercial value, they can be recovered using an extraction procedure. Plants that produce a sufficiently large amount of terrestrial biomass that can be mowed several times per season to remove toxic elements are the best candidates for phytoextraction. To be most useful for phytoremediation, these plants must also accumulate toxic elements in plant tissues in an amount of about 2–5% of dry weight (Yankevich 2015).

14.6 Efficiency of Using Aboriginal Soil Microorganisms for Bioremediation of Contaminated Ecosystems in Kyrgyzstan

14.6.1 Selection of Resistant Strains of Microorganisms with the Aim of Using Them for Bioremediation of the Environment from Heavy Metal Pollution

Microorganisms have a variety of mechanisms to deal with elevated concentrations of contaminants and are often accurate for one or more metals. In addition, microbial activity plays a major role in the mobilization or immobilization of metals, although heavy metals cannot be degraded, they can be reduced or biooxidized to less toxic forms (Cortez et al. 2010). Diverse microorganisms cause different responses to heavy metals, giving them a range of metal resistance (Aka and Babalola 2017). This is achieved in a variety of ways through biological, physical, or chemical systems that include precipitation, complexation, adsorption, transport, accumulation, elimination, pigments, polysaccharides, enzymes, and specific metal-binding proteins (Hashim et al. 2011).

However, the ability of microorganisms to utilize heavy metals in different climatic conditions and the mechanisms that ensure these processes at the physiological, biochemical and genetic levels remain poorly studied.

The introduction of microorganisms—destructors into polluted environments is extremely necessary under unfavorable climatic conditions, for example, in the northern regions with a short growing season or when heavy metals enter water areas, where the formation of an enrichment culture based on natural microflora is extremely slow even under optimal conditions (Evdokimova 1995).

Microorganisms resistant to high concentrations of heavy metals were isolated from soil samples taken in the territories of a mining industrial plant and uranium radioactive tailings in northern Kyrgyzstan (Omurgazieva 2007; Omurgazieva et al. 2016).

For the selection of particularly resistant strains of bacteria to high concentrations of heavy metals, experiments were carried out to determine the degree of accumulation of lead salt and mercury by microorganisms in a liquid medium. We focused on two isolated strains H-5-8 *Bacillus megaterium* and H-5-2 *Bacillus cereus* as accumulators of high concentrations and transformers of heavy metal salts. By the intensity of growth and accumulation of biomass, we judged the inclusion of metal ions in certain metabolic processes of bacteria. The ability of cells of a microorganism to maximally absorb metal molecules from the environment and transform them into other harmless compounds, predetermines the use of such bacteria for cleaning from pollution. At the same time, it is advisable to identify endemic forms of microorganisms for each technogenic province. The implementation of such a purification method requires the isolation of specific strains of microorganisms, as well as the determination of their reduction activity in the processes of biodegradation of heavy metals of various hazard classes (Doolotkeldieva and Omurgazieva 2008).

The inoculum—the *Bacillus megaterium* H-5-8 strain, *Bacillus cereus* H-5-2 was cultivated on meat-peptone agar (MPA). The obtained inoculum was introduced under aseptic conditions in the amount of 10^6; 10^7 cells/mL, giving the initial optical density OD = 0.1 (FEC—56 m, 540 nm, cuvette 1 cm) into flasks with a sterile liquid culture medium. The experiments were performed in shaking flasks with a capacity of 250 mL (medium volume 50 mL), into which concentrated solutions (exceeding the MPC 6, 10, 15 times) were added $HgSO_4$: 0.3; 0.5; 0.75 mg/L at a temperature of 28–30 °C, shaken on a rocking chair at 200–220 rpm, the pH of the culture medium was brought to 7.5. As a control in all variants, a culture liquid without adding metals was used (Gerhard 1983; Zvyagintseva 1980).

Every 6, 12, 24, 48 h. the accumulative activity of the strains to the content of mercury salts in the nutrient medium was monitored by the change in the culture biomass by measuring the optical density through FEC-KF-2 (OD 540 nm, cuvette thickness 1 cm), and was also expressed by the character growth and development of colonies and the number of colony-forming units (CFU), by seeding from the last two dilutions (10^{-5}, 10^{-6}) in Petri dishes with nutrient agar and placed in a thermostat at 27–28 °C. After 2–3 days, the number of colonies was counted. The grown colonies were counted in two dilutions. The experiment was carried out in 2 replicates (Methods 1991).

In our research, we selected cultures of microorganisms capable of accumulating and transforming high concentrations of heavy metals in order to use them in the future for bioremediation of the environment from pollution. We used bacterial strains isolated from soils contaminated with heavy metals on the territory of the Ak-Tyuz concentrating plant.

The isolated strains are stored in the laboratory collection under the numbers H-5-8 (*Bacillus megaterium*) and H-5-2 (*Bacillus cereus*). The isolated destructor strains were adapted and selected according to their ability to grow at high concentrations of metals, mercury up to 0.75 mg/L on liquid and $1 * 10^{-1}\%$ per 100 mL of agar media, and lead at a concentration of up to 1.5–2 mg/L.

Analysis of the activity of individual bacterial strains and their associations with respect to the transformation of lead and mercury salts in a liquid medium showed that associations consisting of two bacterial strains accumulated heavy metals most effectively.

The strains are characterized by the following features:

Strain H-5-8 *Bacillus megaterium*

Nomenclature data:

Family *Bacillaceae* Fischer, 1895, 139, genus *Bacillus Cohn* 1872,174, Species *Bacillus megaterium.*

Morphological and Cultural Characteristics

Cells are rod-shaped, mobile 3.5–4 * 1–2.0μm with rounded ends, located singly, in pairs and in short chains. Central spores, 1.5–2.0 * 0.8–1.25μm. Gram stain is positive. Peritrichs. Old rods are irregular in shape and often larger than 2.5–3.0 in diameter.

Colonies on gelatin: grayish-white, raised, shiny, whole.

Gelatin by injection: grayish-white, superficial plaque, liquefaction from crater-like to saccular.

Colonies on agar (MPA): round, thick, off-white to dark cream in color, 8–10 mm in diameter, with a smooth surface and a smooth edge. They do not form fluorescent pigments.

On oblique agar: dark cream color, smooth, slimy plaque, medium turns brown.

Physiological and Biochemical Properties

Chemoorganotroph, optional aerobic. Hydrolyzes starch and gelatin. It uses glucose, lactose, inositol, arabinose as a source of carbon and energy. Does not use maltose, dulcite, mannitol. The source of nitrogen uses inorganic and organic forms of nitrogen (amino acids, polypeptides that make up the BCH).

Indole is not formed.

The optimum temperature for growth is 25–28 °C. Grows in the pH range of 7–9.5.

The culture of strain H-5-8 on agar slabs is stored in a refrigerator at (−2 + 5 °C) with periodic reseeding 3–4 times during the year. If these conditions are met, the stability of the H-5-8 strain is maintained for 4–5 years.

***Bacillus cereus* Strain H-5-2**

Nomenclature data:

Family *Bacillaceae* Fischer, 1895, 139, genus *Bacillus Cohn* 1872, 174. *Bacillus cereus* species.

Morphological and Cultural Characteristics

Rods: 0.8–1.0 * 2.5–4.0μm, located singly and in chains, mobile. Peritrichs. Gram-positive; spores 0.7–1.0 * 1.1–1.5μm, oval, central, or paracentral.

On meat-peptone broth (MPB), a rapid turbidity occurs, then a delicate film, rings, sometimes flakes are formed.

On potatoes, a thick, soft, whitish-creamy layer with a slight pinkish-cream shade.

On meat-peptone agar (MPA) forms off-white to cream colonies, 13–15 mm in diameter, with uneven edges, with a rough, dry surface.

On oblique agar forms a yellowish-green fluorescence.

Physiological and Biochemical Properties

Chemoorganotroph, aerobic, optional, catalase-positive. Ferments glucose, sucrose, glycerin. Starch hydrolyzes. Indole is not formed. Nitrates are reduced. The optimum temperature for growth is 25–28 °C. Grows in the pH range of 6.8–9. Optimal for growth pH 7.0–8.0.

14.6.1.1 Characterization of the Accumulative Activity of the Association of Bacterial Strains (H-5-8 *Bacillus megaterium* + H-5-2 *Bacillus cereus*) of Mercury Salts

Earlier we have shown that bacterial strains of the genus *Bacillus*; *Bacillus megaterium*, *Bacillus cereus* shows marked resistance to high concentrations of mercury and lead (Omurgazieva 2007). In our experiments, the maximum concentration of mercury is 0.75 mg/L (15 times higher than the MPC). This dose is several times higher than the bactericidal effect of this metal. However, even under these conditions, the cells remained viable. We have carried out studies to determine the degree of accumulation and consumption of various concentrations of mercury salt by the above strains for the growth and accumulation of their biomass. By the intensity of growth and accumulation of biomass, we judged the inclusion of metal ions in certain metabolic processes of bacteria. The ability of cells of a microorganism to maximally absorb metal molecules from the environment and transform them into other harmless compounds, predetermines the use of such bacteria for cleaning from pollution. Many microorganisms, including bacteria and fungi, are capable of biochemically modifying mercury compounds (Bogacheva 2011; Ivshina 2012). The bacteria carrying out this process convert mercury salts that are not part of the complexes into methyl—and dimethylmercury, using methylcobalamin (CH_3B_{12}) as a methyl group donor. Mercury ions can be reduced to metallic mercury (Hg^0) with the participation of Pseudomonas, enterobacteria, Staphylococcus aureus, Cryptococcus, and the presence of more than 200 mercury resistance plasmids has been shown (Landerner 1971; Komura et al. 1970).

When studying microbial growth in the presence of a high concentration of mercury, it was shown that strains H-5-8 *Bacillus megaterium*, H-5-2 *Bacillus*

Table 14.1 Effect of pH on the accumulation of high concentrations of mercury in *Bacillus megaterium, Bacillus cereus* (H-5-8 + H-5-2)

	pH indicators of the medium after 6, 12, 24, 48 h of growth			
	Initial pH—7.60			
Mercury concentration (mg/L)	6 h	12 h	24 h	48 h
0 (control)	7.01	6.78	7.28	6.72
0.3	4.21	4.12	4.46	4.51
0.5	6.37	6.34	6.86	6.92
0.75	3.47	3.30	3.38	3.56

cereus showed different degrees of reproduction intensity and yield of their biomass in continuous conditions.

In all variants of the experiment, the percentage of the use and transformation of mercury and lead salts by associations was higher than by individual strains. According to the literature data, it can also be seen that associations of up to 4, even 5–6 strains led to an increase in destructive activity. In modeling the composition of a microbial association, it is necessary to take into account its stability, since the introduction of an unstable community into a polluted environment leads to a sharp drop in the titer of its individual components, and therefore the effectiveness of the use of this association decreases.

The research results showed (Table 14.1) that with a high initial content of mercury (0.75 mg/L) in a neutral medium, the pH of the medium decreased to 3.56, but there was no complete inhibition of biomass growth, growth was inhibited, the first 6 h, the adaptive activity of the strain decreased. The cell wall became loose and deformed. However, after 12 h of incubation, the culture of microorganisms adapted over time, and after 48 h of contact in the entire studied concentration range, there was a uniform, good growth of cells. At a concentration of mercury in the medium of 0.3 mg/L (6 times higher than the MPC), the growth rate of cells in the initial period of contact (6–10 h) was low, but after 48 h, the biomass content, especially at the highest concentration of 0.75 mg/L (15 times the MPC) was 60 and 71.4% of the control. However, at a mercury concentration of 0.5 mg/L (10 times higher than the MPC), the D_{540} values of the experimental and control variants did not differ significantly by 48 h. In environments with higher mercury concentrations, the D_{540} indicator also increased by 2.5 orders of magnitude higher than in the control (Fig. 14.2). This indicates the use by the cell of the above-mentioned strains of metal ions for metabolic reactions and biosynthesis of cell structures, thereby showing the destructive ability of these bacteria in rather high concentrations of metal.

In the dynamics of the number of associations of strains H-5-2 + H-5-8 at a concentration of 0.3 mg/L of mercury from 6, 12, 24 h of growth did not differ much, but already 0.5 mg/L from 6 h and 12 h of growth, as well as high concentration of 0.75 mg/L, the number of cells increases, and are 59×10^5 CFU/mL (12 h). And already by 48 h of growth, the number of cells exceeds 4.5 times than in the control (Table 14.2). This is apparently due, first of all, to the gradual transformation of

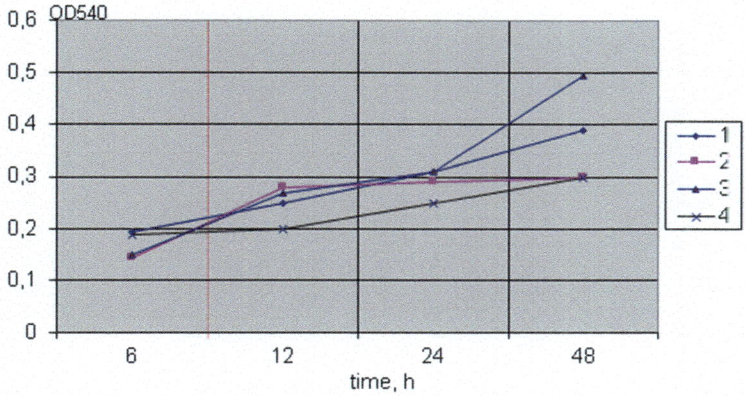

Fig. 14.2 Degree of accumulation capacity of high mercury concentrations in *Bacillus megaterium, Bacillus cereus* (H-5-8 + H-5-2)

metal salt ions and their involvement in different metabolic pathways of bacterial cells, and, second, to a high degree of adaptation of the strain associations to heavy metals.

It should be assumed that the ability of a strain to degrade is determined both by the degree of activity of the corresponding enzymes and by the number of cells of microorganisms carrying out this process. In this regard, in the process of metal transformation, we took into account the ratio of strains of increasing metal concentrations. It should be noted that it is difficult to conduct such studies with unmarked strains. In our case, such an analysis was possible due to the cultural and morphological differences of the studied strains (color, shape, consistency and size of colonies, and cells).

In the experiments, each flask was inoculated simultaneously with the indicated strains, and it is possible that individual strains with an advantage in growth rate could dominate the association.

As follows from the data in Table 14.2, the dynamics of the number of individual strains at different concentrations was determined by the combinations used. At the same time, changes in the number of a particular strain depended on the concentration of the metal and the time of incubation. Thus, at a dose of 0.3 mg/L of mercury, the number of grown colonies of the H-5-8 strain of *Bacillus megaterium* by 6 and 48 h of cultivation exceeds by 2 orders of magnitude than the cells of the H-5-2 *Bacillus cereus* strain, but, already starting from 12 h to 48 h, the growth of the strain H-5-2 equals approximately 40%. During the first 6 h of cultivation at a dose of 0.5 mg/L of mercury, an active growth of H-5-2 strains was observed, and by 12 h of cultivation, only H-5-8 strains of *Bacillus megaterium* were decomposing the metal salt.

At a high concentration of mercury (0.75 mg/L), the ratio of the two strains was not stable, or only one H-5-8 *Bacillus megaterium* actively dominated until the end of the experiment (Fig. 14.3), while the relative number of cells of the H-5-2 *Bacillus*

Table 14.2 Dynamics of the number of strains (H-5-8 + H-5-2) in the process of accumulation of mercury salt ($HgSO_4$) in a liquid medium

Association	Increased concentrations of mercury in the nutrient medium	Amount (cells/mL) and ratio of microorganisms in the association, %			
		Cultivation, time, h			
		6	12	24	48
B. megaterium + *B. cereus* (H-5-8 + H-5-2)	0.3 mg/L	34×10^5 100/0	33×10^5 66.6/33.3	38×10^5 60.5/39	48×10^5 58.3/55
B. megaterium + *B. cereus* (H-5-8 + H-5-2)	0.5 mg/L	44×10^5 68/31.8	43×10^5 97/2.9	41×10^5 56/43.9	36×10^5 72.2/27
B. megaterium + *B. cereus* (H-5-8 + H-5-2)	0.75 mg/L	38×10^5 73.6/26.3	59×10^5 81.3/18.6	49×10^5 82.8/17	72.6×10^5 100/0
B. megaterium + *B. cereus* (H-5-8 + H-5-2)	Control	12.3×10^4 72.9/18.9	38×10^5 61.4/38.5	27×10^5 28.3/71	16.6×10^4 76.9/23

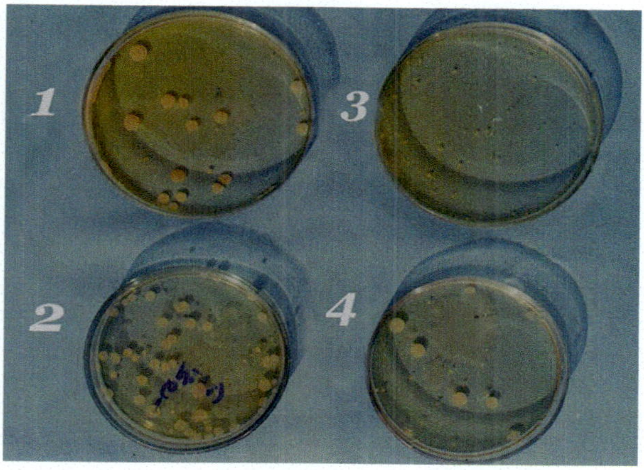

Fig. 14.3 Growth of colonies of strain H-5-8 *Bacillus megaterium* at high mercury concentration (15 MPC) depending on time: *1—growth after 6 h; 2—after 12 h; 3—24 h; 4—48 h*

A B

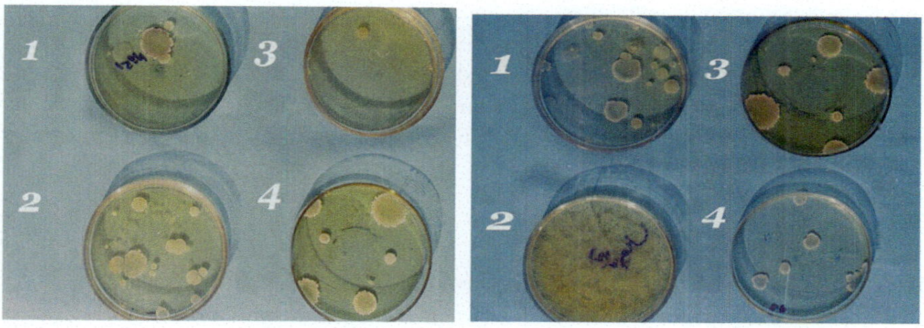

Fig. 14.4 Colony growth of H-5-2 *Bacillus cereus* strains at high concentrations of mercury (**a**) and lead (**b**) depending on the exposure time: *1–6 h; 2–12 h; 3–24 h; 4–48 h*

cereus strain (in 4, 5 times), this was especially noticeable by 12–24 h of cultivation. But this does not mean that this strain (H-5-2) is not capable of biotransformation of metals, on the contrary, this strain continued to actively develop together even at the highest metal concentrations. As can be seen from the figure, these strains have high resistance and accumulation capacity, which decreases with aging of the culture.

It should be noted that in our experiments in the variant 0.3 and 0.5 mg/L of mercury caused active dissociation of the cells of the *Bacillus cereus* strain H-5-2, which formed S-colonies (Fig. 14.4). Against the background of a regular decrease in the number of colonies with an increase in the concentration of introduced mercury 10–15 times the MPC, and also with a high concentration of lead (5 mg/L) at 6 h of growth, an increase in the proportion of cells giving S-colonies of strain H-5-2 was observed. Since these colonies were characterized by slower growth and fewer generations, as evidenced by the size of the colonies, it could be assumed that they have other distinctive features, and the population dissociation of this type, arising under the influence of metals, may be aimed at increasing stability or stress resistance of cells.

This property is realized both through the stabilization of cell membranes and macromolecules in their complexes with heavy metals, and through the cleavage of clones (Kaprelyants and Suleimenov 1987; Kolpakov and Ilinskaya 2000), characterized by a low growth rate, a change in synthetic activities, and an increase in resistance to adverse external influences.

Since the relationship between the colony-forming ability and phenotypic dissociation of the population revealed in our experiments is due to the same factor—the action of heavy metals, it can be assumed that, in addition to the stabilizing effect on subcellular structures, mercury salt complexes directly affect the cellular genome.

The accumulative dependence of mercury on the pH of the medium has a pronounced maximum, lying in the range of pH 6.34–6.92 (Table 14.1). With a decrease in pH, the adaptive capacity of biomass gradually decreases, and the growth of crops practically stops in the pH range of 2.0–3.1. If the pH is maintained at 7.0–7.5 by titrating with a sterile NaOH solution, biomass growth is possible until the substrate is completely consumed.

Thus, the active growth of *Bacillus megaterium* + H-5-2 *Bacillus cereus* strains H-5-8 at high mercury concentrations ends finally at pH 2.9–3. The general regularity for the studied associations of strains was that the accumulative capacity of biomass decreased with decreasing pH. The highest sorption capacity was observed in the pH range 6.9–7.

14.6.1.2 Characterization of the Destructive Abilities of the Association of Bacterial Strains (H-5-8 *Bacillus megaterium* + H-5-2 *Bacillus cereus*) Lead Salt

As you can see in the Table 14.3, at a lead dose 10 times higher than the MPC (1 mg/L), growth stimulation and a high concentration of culture biomasses up to 60–114% were observed to the control variant after 12–48 h of growth. At a concentration of 2.5 mg/L, the number of cells (6 h) is significantly lower than at a concentration of 5 mg/L, but higher than in the control medium (Omurgazieva et al. 2016).

During the growth of culture associations, the *D* of the medium was also measured, however, at a lead concentration of 1 mg/L, the *D* values 540) and at 2.5 mg/L, the concentration did not differ significantly by 12 h, but a higher lead concentration (5 mg/L) increases the OD value, even more than in control (Fig. 14.5). The highest rate of change in cell growth was observed at 12 h of cultivation at concentrations of 1 and 2.5 mg/L, except for 5 mg/L of lead. This testifies to the inequality of the metabolic potential of the cultured cells, which is responsible for the accumulation of metal, at all concentrations during this cultivation period. Therefore, it can be assumed that the number of metabolically active bacterial cells in the culture medium is not the same for all variants of the experiment.

Table 14.3 Dynamics of the number of strains (H-5-8 + H-5-2) during the accumulation of (Pb) lead salt in a liquid medium

Association	Increased concentrations of lead in the environment	Amount (cells/mL) and ratio of microorganisms in the association, %			
		Cultivation, (h)			
		6	12	24	48
B. megaterium + *B. cereus* (H-5-8 + H-5-2)	1 mg/L	89×10^5 65.1/34.8	176×10^5 100/0	173×10^5 48.3/53.6	175×10^5 51.2/47.6
B. megaterium + *B. cereus* (H-5-8 + H-5-2)	2.5 mg/L	59×10^5 47.5/52.5	108×10^5 98.7/1.2	119×10^5 28.5/71.4	116×10^5 22.8/77.1
B. megaterium + *B. cereus* (H-5-8 + H-5-2)	5 mg/L	61×10^5 3.3/74.5	118×10^5 2.5/79.3	198×10^5 41.8/49.07	89×10^5 42/53.1
B. megaterium + *B. cereus* (H-5-8 + H-5-2)	Control	12.3×10^4 72.9/18.9	38×10^5 61.4/38.5	27×10^5 28.3/71	86.6×10^4 76.9/23

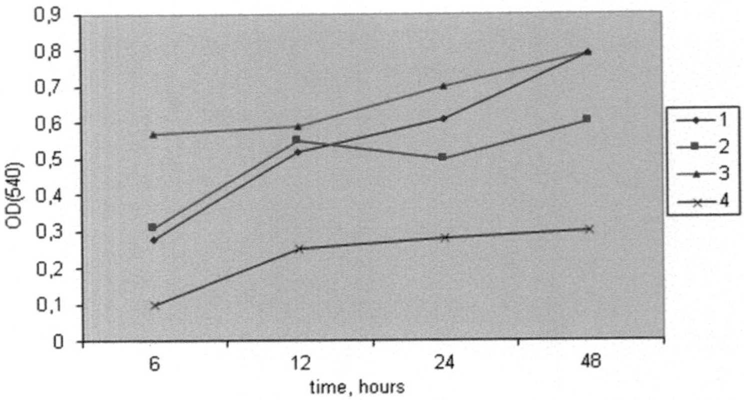

Fig. 14.5 The degree of accumulation capacity of high lead concentrations in cultures of *Bacillus megaterium, Bacillus cereus* (H-5-8 + H-5-2). Lead salt concentration *1—1 mg/L; 2—2.5 mg/L; 3—5 mg/L; 4—control* (without adding metal)

Table 14.4 Effect of pH on the accumulation of lead by associations of cultures *Bacillus megaterium, Bacillus cereus* (H-5-8 + H-5-2)

Lead salt concentration mg/L	pH indicators of the medium after 6, 12, 24, and 48 h of growth			
	6 h	12	24	48
0 (control)	7.01	6.78	7.28	6.72
1	7.21	6.78	7.21	7.33
2.5	7.23	6.87	7.04	7.40
5	6.77	6.82	7.24	6.72

Considering the ratio of bacterial strains in a pair of *Bacillus megaterium* + *Bacillus cereus,* it can be seen that 2 associations at a dose of 1 and 2.5 mg/L of lead, for 6 h to 12 h of cultivation, the relative number of cells of the strain H-5-2 decreased. Starting at 24 h, the number of H-5-2 strains at increasing concentrations of 1; 2.5; 5 mg/L of lead increases, and amounted to 53.6; 71.4; 49.07%. The most dramatic changes in the number of strains were observed at a high concentration of lead (5 mg/L) during the entire period of cultivation. Thus, from 6 h to 12 h of cultivation, the growth of the H-5-2 strain dominates. The number of strains H-5-8 *Bacillus megaterium* in associations was the smallest, precisely in the presence of a high concentration of lead in the nutrient medium during the first 6 and 12 h of contact. This is probably due to the intense absorption of lead salt by the cells of the H-5-2 strain than the cells of the H-5-8 strain (Table 14.3).

Thus, the highest rate of accumulation of the studied metals by the biomass of strain associations are observed at the exponential and early stationary growth phases.

One of the factors affecting the efficiency of lead accumulation is the pH of the cultivation medium. At a dose of 2.5 mg/L (25 times higher than the MPC) of lead, acidification of the medium during contact was not observed (Table 14.4). For the biomass of the association of H-5-8 *Bacillus megaterium* + H-5-2 *Bacillus cereus* strains, the optimal pH for transformation lies in the range of 6.78–7.33, it is 1 mg/L Pb, and with a decrease in pH, a gradual decrease in adaptive capacity is observed, at pH 3–4 and further, there is a noticeable drop in the biomass concentration characteristic of all strains.

Thus, the used strains H-5-8 *Bacillus megaterium* + H-5-2 *Bacillus cereus* can grow at sufficiently high lead concentrations up to 5 mg/L in the medium, however, the optimal concentrations are up to 1 mg/L, where the specific growth rate and biomass of cultures have the maximum values.

14.6.2 *Biological Activity of Strains H-5-2* **of Bacillus cereus** *Resistant to Heavy Metals*

Many researchers note that with deterioration of soil properties as a result of pollution for a long time with heavy metals, phytotoxic forms of micromycetes begin to dominate in the depleted complex of micromycetes. Thus, with constant exposure of the soil to heavy metals that cause toxicosis, the presence of toxin-forming fungi can be an additional factor that negatively affects the development of plants on contaminated soils.

The proposed strain H-5-2 *Bacillus cereus* has not only the destructive properties of heavy metals, but also has a high antagonistic activity against phytopathogenic fungi—causative agents of root rot, as well as strong toxin-forming micromycetes, which expands the range of their application (Table 14.5, Fig. 14.6).

The antagonistic properties of the *Bacillus cereus* strain H-5-2 are important when used for plant protection in a contaminated environment with heavy metals. In addition to antagonistic properties, they are able to continue their vital activity at extremely high concentrations of metals such as mercury and lead. This suggests the possibility of using biopreparations based on *Bacillus cereus* strains in areas experiencing heavy metal contamination in areas heavily contaminated with heavy metals simultaneously in order to bioremediate contamination and to protect plants from phytopathogenic microorganisms.

14.7 Conclusion

There are many specific remediation technologies that are separately or in combination used to solve specific problems in the restoration of contaminated soils. In this work, we considered the method of soil bioremediation based on the metabolic activity of indigenous soil microorganisms. In the conditions of Kyrgyzstan, such studies have not been conducted before. In this regard, we are conducting research to develop an environmentally friendly method of cleaning soil from contamination,

Table 14.5 Antagonistic activity of *Bacillus cereus* strain H-5-2 against phytopathogenic fungi

Bacillus antagonist culture, zone of inhibition in *mm*	Test culture					
	Fusarium oxysporum	*Fusarium graminearum*	*Botrytis cynerea*	*Cladosporium* sp.	*Penicillium funiculosum*	*Nigrospora* sp.
0 (control)	Abundant growth	Abundant growth	Abundant growth	Abundant growth	Abundant growth	Abundant growth
1. *Bacillus cereus* H5–2	+	++	+	++	++	+

Note: "+" test culture is strongly suppressed
"++" the growth of the test culture is almost completely absent

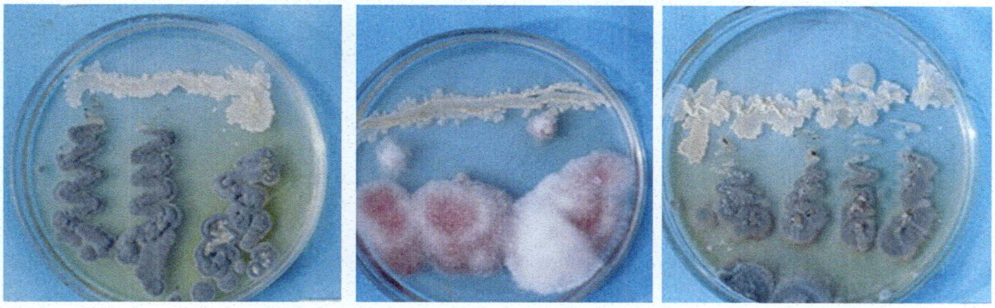

Fig. 14.6 Inhibition of the development of colonies of phytopathogens—*Penicillium* sp. (**a**); *Fusarium oxysporum* (**b**); and colonies—*Trichoderma lignorium* (**c**) culture *Bacillus cereus*

searching for and isolating effective strains of microorganisms resistant to high concentrations of heavy metals and radionuclides in the zones affected by industrial emissions from mining enterprises of Kyrgyzstan. The proposed types of spore-forming strains of bacteria *Bacillus megaterium* and *Bacillus cereus* are concentrates of mercury and lead (Patent KR No. 815). These strains represent a certain perspective in their practical use for eliminating pollution and biodegradation of heavy metals in industrial waters and soils. In addition, knowledge about the nature of the interaction of heavy metals with microorganisms is of great interest in environmental regulation of the maximum permissible concentrations of heavy metals in soils and water bodies, which cause the death of certain types of microorganisms and the destruction of their community, as well as a drop in the productivity of the community and a decrease in the rate of microbial degradation of organic matter. In the future, a valuable product of this research will be highly resistant strains of microorganisms for the creation of environmentally friendly products—biological products, with the aim of using them for cleaning contaminated areas with heavy metals and radionuclides.

References

Aitmatova JI, Aparin VB (2003) Tailings storage facilities for radioactive waste and their impact on environmental components in the territories of uranium mines in Mailuu-Suu and Charke-Sar. J Sci New Technol 3:112

Aka RJ, Babalola OO (2017) Identification and characterization of Cr-, Cd-, and Ni-tolerant bacteria isolated from mine tailings. J Bioremediation 21(1):1–19

Alexander M (1999) Biodegradation and bioremediation. Academic Press, New York

Bagaeva TV, Ionova NE, Nadeeva GV (2013) Microbiological remediation of natural systems from heavy metals: training manual, pp 1–56

Baryshnikova LM, Grishchenkov VG, Arinbasarov MU (2001) Biodegradation of petroleum products by destructor strains and their associations in a liquid medium. Appl Biochem Microbiol 37(5):542–548

Bekasova OD, Orleansky VK, Nikandrov VV (1999) Accumulation of cadmium, titanium and aluminum cyanobacteria Nostoc muscorum J. Microbiology 68:851–859
Bogacheva AS (2011) Sensitivity of cyanobacteria to the toxic effect of salts of heavy metals: abstract of thesis Dr. p 26
Bykovchenko YG, Tukhvatshin RR, Bykova EI, Belekov T, Kadyrova AI, Zhunushov AT, Yushida S (2005) Technogenic uranium pollution of the biosphere of Kyrgyzstan (monograph), p 169
Coronelli TV (1996) Principles and methods of intensification of biological destruction of hydrocarbons in the environment (review). Appl Biochem Microbiol 32(6):579–585
Cortez H, Pingarron J, Munoz JA, Ballester A, Blazquez ML, Gonzalez F, Garcia C, Coto O (2010) Bioremediation of soils contaminated with metalliferous mining wastes. In: Trends Bioremediation Phytoremediation, pp 283–299
Davydova MN, Mukhitova FK, Ibatullin RR (1988) Anaerobic transformation of oil by the action of cell extracts Desulfovibriodesulfuricans. Microbiology 2:202–207
Doolotkeldieva T, Omurgazieva CM (2008) The role of soil microorganisms in bioindication and bioremediation of environmental pollution with heavy metals (monograph), p 112
Egorova AV, Mamontova VN, Nikiforov VA, Yankevich MI (2010) Bacterial degradation of polycyclic aromatic hydrocarbons in urban soils. Bull St Petersburg State Technol Inst:77–84
Evdokimova GA (1995) Ecological and microbiological foundations of soil protection in the far north, Apatity, p 342
Feruke-Bello YM, Babalola G, Olu O (2018) A comparative study of the wild and mutated heavy metal resistant Klebsiella variicola generated for cadmium bioremediation. J Bioremediation 22 (1–2):28–42
Gerhard F (1983) Methods of general microbiology, vol 1. Mir, Moscow, p 536
Gilyarov MS (1999) Biology. In: Big encyclopedic dictionary, p 864
Gradova NB, Gornova IB, Eddaudi R (2003) Use of bacteria of the genus Azotobacter in bioremediation of oil-contaminated soils. Appl Biochem Microbiol 39(3):318–321
Grishchenkov VG, Shishmakov DA (2003) Growth of bacteria-destructors of naphthalene and salicylate at low temperatures. Appl Biochem Microbiol 39(3):322–328
Hashim MA, Mukhopadhyay S, Sahu JN, Sengupta B (2011) Remediation technologies for heavy metal contaminated groundwater. J Environ Manage 92:2355–2388
Ismailov NM (1988) Microbiology and enzymatic activity of oil-contaminated soils. In: Restoration of oil-contaminated soil ecosystems, pp 42–56
Ivshina IB (2012) Bioremediation of soils disturbed by hydrocarbons and heavy metals using Rhodococcus-biosurfactants and immobilized rhodococci. J Agrarian Bull Urals 8:65–68
Kaprelyants AS, Suleimenov MK (1987) Structural and functional changes in bacterial and model membranes under the action of phenolic lipids. Biol Membr 4(3):254–261
Kobzev EN, Petrikevich SB, Shkidchenko AN (2001) Study of the stability of the association of microorganisms-oil destructors in an open system. J Appl Biochem Microbiol 37(4):413–417
Kolpakov AI, Ilinskaya ON (2000) Stabilization of enzymes by autoinducers of anabiosis as one of the mechanisms of resistance of dormant forms of microorganisms. J Microbiol 69(2):180–185
Komura I, Izari K, Tarahashi H (1970) Vaporization of inorganic mercury by cell-free extracts of drug resistant Escherichia coli. J Agric Biol Chem 34:480–482
Korolchenko DA (2007) Modern bioremediation technologies. J Fire Explosion Saf 16(5):75–78
Koronelli TV, Komarova TI, Porshneva OV (2001) Extracellular metabolites of hydrocarbon-oxidizing bacteria as a substrate for the development of sulfate reduction. J Appl Biochem Microbiol 37(5):549–553
Kuznetsov AE, Gradova NB, Lushnikov SV (2010) Applied ecobiotechnology. BINOM, Moscow
Landerner L (1971) Biochemical model for the biological methylation of mercury suggested from methylation studies in vivo with Nigrospora crassa. Nature 230:452–454
Loginov ON (2000) Biotechnological methods of cleaning the environment from technogenic pollution. State. Ed. Scientific and Technical. Literature “reagent”, p 100

Loginov ON, Boyko TF, Kostyuchenko VP (2002) On biological cleaning of technological dumps from oil products. J Pochvovedenie 4:481–486
Methods (1991) Methods of soil microbiology and biochemistry. Zvyagintseva DG (ed), p 304
Muratova AY, Bondarenkova AD, Panchenko LV, Turkovskaya OV (2010) The use of complex phytoremediation for cleaning soil contaminated with oil sludge. Biotechnology 1:77–84
Oborin AA, Kalachnikova IG, Maslivets TA (1988) Restoration of oil-contaminated soil ecosystems, pp 140–159
Omurgazieva CM (2007) The influence of heavy metals on soil microbocenoses and their functioning. Author's abstract. Dis Cand Biol Sci:22
Omurgazieva CM, Beishekeeva GD, Kenenbaeva GM (2016) Accumulative activity of the association of bacterial strains H-5-8 Bacillus megatherium and H-5-2 Bacillus cereus lead salts. J Probl Mod Sci Educ 34(76):6–9
Sharshenalieva ZS, Koleno VA (1990) Environmental pollution and means of protecting the body from their effects. J Izv AN Kirg Rep Ser Chem Tech Biol Sci 2:30–36
Torgoev IA, Aleshin YG (2001) Ecology of the mining complex of Kyrgyzstan. Ilim:141
Trindade PVO, Sobral LG, Rizzo ACL, Leite SGF, Soriano AU (2005) Bioremediation of weathered and recently oil-contaminated soils from Brazil: a comparison study. Chemosphere 58:515–522
Turkovsvaya OV, Muratova AY (2005) Biodegradation of organic pollutants in the root zone of plants. In: Moleecular basis of the relationship of associative microorganisms with plants, pp 180–208
Vel'kov VV (1995) Bioremediation; principles, problems, approaches. J Biotechnol 3(4):20–27
Wang Y et al (2012) Coiled-coil networking shapes cell molecular machinery. Mol Biol Cell 23 (19):3911–3922
White C, Sayer JA, Gadd GM (1997) Microbial solubilization and immobilization of toxic metals: key biogeochemical processes for treatment of contamination. FEMS Microbiol Rev 20:503–516
Yankevich MI (2015) Bioremediation of soils: yesterday, today and tomorrow. Biosphere 2:199–208
Zharikov KA, Marchenko AI, Kraynova OA, Kapranov VV, Zharikov MG (2013) Development and field testing of technologies for bioremediation of territories contaminated with toxic chemicals. J Med Extreme Situat 2(44):42–51
Zvyagintseva DG (1980) Methods of soil microbiology and biochemistry. Mir, Moscow, p 224

The manufacturer's authorised representative in the EU is Springer Nature Customer Service Centre GmbH, Europaplatz 3, 69115 Heidelberg, Germany. If you have any concerns regarding our products, please contact ProductSafety@springernature.com

Printed and bound by CPI Group (UK) Ltd, Croydon, CR0 4YY
30/09/2026
02225241-0008